German and United States Second World War Military Cemeteries in Italy

Transatlantic Aesthetics and Culture

Vol. 8

General Editors

Boris Vejdovsky, Agnieszka Soltysik & Jürg Schwyter

This series hosts works coming from both sides of the Atlantic that offer multicultural and interdisciplinary perspectives on modern literary, aesthetic, and cultural issues. It embraces studies of literature, theatre, cinema, visual arts, or dance. Defamiliarizing subjects by adopting an outsider's view or bringing to bear different aesthetic or theoretical discourses on particular cultural spheres are among the privileged approaches of Transatlantic Aesthetics and Culture. The series aims to foster dialogue and encourage different cultural and critical discourses. It welcomes monographs and collections of essays.

Contributors are invited to submit projects to the Editors.

PETER LANG

Bern · Bruxelles · Berlin · New York · Oxford · Warszawa · Wien

German and United States Second World War Military Cemeteries in Italy

Cultural Perspectives

Birgit Urmson

Foreword by
Richard Ingersoll

PETER LANG
Bern · Bruxelles · Berlin · New York · Oxford · Warszawa · Wien

Bibliographic information published by die Deutsche Nationalbibliothek
Die Deutsche Nationalbibliothek lists this publication in the Deutsche Nationalbibliografie; detailed bibliographic data is available on the Internet at ‹http://dnb.d-nb.de›.

Library of Congress Cataloguing-in-Publication Data: A catalogue record for this book is available.

Cover illustration: "The Angel of Peace" by Sidney Waugh in the Florence American Cemetery

ISSN 1661-805X
ISBN 978-3-0343-3516-4 pb.
ISBN 978-3-0343-3595-9 MOBI
DOI 10.3726/b14489
ISBN 978-3-0343-3593-5 eBook
ISBN 978-3-0343-3594-2 EPUB

Wabernstrasse 40, CH-3007 Bern, Switzerland
info@peterlang.com, www.peterlang.com

Printed in Germany

Contents

Figures

Abbreviations

AARome	American Academy in Rome
ABMC	American Battle Monuments Commission
AGRS	American Graves Registration Service
CFA	United States Commission on Fine Arts
FRG	Federal Republic of Germany (West Germany)
GDR	German Democratic Republic (East Germany)
LL PP	Ministero dei Lavori Publici
NSDAP or NS	National Socialist German Worker's Party (Nazi)
QMG	Quartermaster General of the Army
US	United States of America
VDK	Volksbund Deutsche Kriegsgräberfürsorge
WW I	First World War
WW II	Second World War

Abbreviations for Archives

AA	Auswärtiges Amt, Berlin
AdK	Akademie der Künste, Berlin
NARA	National Archives and Records Administration
NYHS	New York Historical Society

Acknowledgements

This book could not have been written without the help and support of many people. I am very thankful to them all.

My husband, John Urmson, for his unceasing support and indispensable editorial help.

Prof. Dr. Christof Mauch, Professor of American History at Ludwig-Maximilian-Universität, Munich and director of the Rachel Carson Center, Munich for his support and generous help at the right moments. PD Dr. Christian Fuhrmeister, Zentralinstitut für Kunstgeschichte, Munich, for his support and many inspiring conversations. Renata Catambas and Thomas Lemaître at Peter Lang Publishing for their de3dicated stewardship during the production process. The late Garrett Eckbo, landscape architect, for his inspiration and friendship. Dr. Hermann Glaser, for his inspiring conversations and sharing his insights into FRG's post-WW II culture. Professor Dr. Joachim Wolschke-Bulmahn and Professor Dr. rer. hort. habil. (emeritus), Gert Gröning for their enrouragement to pursue this subject. Dr. David H. Wright, Professor (emeritus) of Art History, UC Berkeley for his generous sharing of his knowledge of AARome. Dr. Richard Ingersoll of Florence, Italy, architectural historian, for first suggesting that I visit the Futa Pass cemetery. Prof. Dr. Paul Groth for his inspiration and support.

The late J.B. Jackson for opening my eyes to the American Space. Professor Clare Marcus (emerita) of UC Berkeley for encouraging me to explore unusual subjects. Prof. (emeritus) Marc Treib, Prof. Dr. Margret Lovell. Dr. Gray Brechin for bibliographical suggestions. Special thanks go to Monsignore Timothy Verdon for his interest and support, and to Prof. (emeritus) Peter Selz for his enthusiasm and vision.

Dr. Anne Schmedding for generously sharing her knowledge and work on Dieter Oesterlen. Dr. Kate Lemay for her early support and sharing important sources; Dr. Giacomo Calandra di Roccolino for sharing his views on the Futa Pass cemetery. Professor Kai Kappel, Humboldt Universität, Berlin for his encouragement.

Special thanks go to archivist Peter Paessler at VDK Kassel for his unceasing support and for providing me with vital material. Jürgen Kaulfuß and Tanja Morgenstern, at the Baukunstarchiv of the Akademie der Künste, Berlin. Timothy Nosal and Michael Shipman of the American Battle Monuments Commission for their generous provision of images.David Grinnell, University of Pittsburgh Archives for his generous help. Margaret Smithglass, Avery Architectural Library. Francesca Maroncelli for helping me maneuver through archives in Rome. Jason Miller and David Eifler, librarians at the Environmental Design Library, UC Berkeley; Waverley Lowell, curator of the Berkeley Environmental Design Archives; Miranda Hambro, Archivist. Sim Smiley for her indispensible research help at National Archives and Records Administration. Nadine Granoff for finding and copying material at the Smithsonian Institute. Domenico Tomei, His Honour Nicholas Philpot and Caterina Salvi for usage of their photos. Robyn Adams for editing photos.

Anastasia Meadors, Director of Art and Design at the UC Berkeley Extension, Kathleen Bryant, Osher Lifelong Learning Institute, California State University. Inge Horton and Dr. Johan van der Zande for reading early drafts. Dr. med. Guy Micco, Convener, Berkeley Life and Death Project, for inviting me to lecture on this subject. Prof. Dr. Thomas W. Laqueur, Marilyn Yalom, author of the "American Resting Place". Prof. Dr. R. John Gillis.

I also wish to thank the following: Sara Russell for helping me to translate labyrinthine Italian official documents. The late Dale Johnson for his editing an early version. Prof. Michael Schneider, Senior Adjunct Professor at California College of the Arts, for sharing his knowledge on the symbolism of ratios. Special thanks got to Petra Lander for sharing insights into her fathers' works. Helgard Kühn for sharing information. Dr. Rolf Wernstedt for a crucial conversation on the VDK in September 2014.

Tedice Santini, Georgio Barbarino and Vasco Galliotti for sharing their wartime and partisan experiences in Italy. Dr. Steffi Roettgen, Professor (Emeritus) of Art History, Professor Dr. (med.) Peter Barglow, Dr. Margret Schäfer and Prof. Gayle Greene for their early encouragement; Dr. Larry Marietta, John Walcko, Marco Tovani, Illaria di Giangirolamo, Eliza O'Malley and Dr. Georg Biester, for sustaining me through music. Also thanks for support go to Graziella Cosimini, Isabella Negri, Grazia Santini, Kiki Lambrou, Caroyl Labarge, Susi Gauld, Kitty Hughes, Anne Kroeber, Dr. Anne Maclachlan, Birgitt Claus, Gerd Gauglitz, Gudrun Wolf, Bill and Solace Welch, Professor Dr. Wendy Martin, Irena Chrul.

Special thanks to my family: My sister, Ute Armanski, my brother, Professor Dr. Gerhard Armanski, to my children Carl, Claire and Julian andgrandchildren Helen, John, Carmen, Grace, Bradley, Willa and Elizabeth for sustaining my joy in life.

Finally, I would like to dedicate this work to my mentally disabled son, Carl, and the male members of my German family who were killed or misssed in WW II.

Foreword

Remember to Forget, Not

Cemeteries in Italy, especially in provincial settings, tend to be very lively places. I should know, I've been living next to one for the past two decades. Widows and widowers don their overcoats and amble over here almost daily, and relatives of the recently deceased do their best to visit frequently. Along the pathways conversations develop, flowers of all sorts, real and plastic, abound, variopint candles, lots of paper, and shiny tin wrappers frame the edges. I am forced to notice that the conventional good taste that once accompanied tomb design has been replaced by gaudy inscriptions, prefab statues of the Madonna, and infinitesimal addenda that include multiple photographs of loved ones and representations of things like motorcycles and sewing machines. Occasionally the undertaker arrives with a new corpse and an animated concourse of cars takes over the parking lot almost like going to a football match. The graveyard has assumed an uncanny festivity, and very few visitors come dressed in black.

About military burial grounds, if we desired to make comparisons, the mood appears quite the opposite. These austere landscapes remain almost always empty: phantom necropolises. No flowers, real or plastic, litter their serene order. And nothing about them will ever change. We hear no gossip in the background; we witness no kitsch applied to the tombs. Military cemeteries impose an immutable order that mirrors the rigors of military life. No tombs will be added or subtracted. The rows of regularly placed graves exude a feeling of cosmic certainty. And that's why they qualify as a superb subject for art history. They were built with the highest design criteria, invested with profound political intentions, and in their formal magnitude force us to remember: no matter which war, or what side we preferred, we must contemplate the sad story that countless young men's lives were arbitrarily interrupted in the name of national duty.

The grandest war cemeteries in Europe were built to commemorate the First World War. Edwin Lutyens' Thiepval Memorial in Picardy, a colossal variation of a triumphal arch is perhaps the most famous, but the

Italian site of Fogliano Redipuglia, near Gorizia, surely is the most sublime, an interminable granite stairway. No one was ready for that war, in which 16 million, including 7 million civilians, lost their lives. This tragic, fratricidal conflict made no sense at all. When the dust cleared and the poison gas evaporated, Europeans found their lands riddled with trenches instead of plowed fields, their cities pock-marked with bomb craters. The daunting memorials confirmed that the smug complacency of the Belle Époque had vanished forever.

Italy had been on the winning side, but in fact experienced a pyrrhic victory, since their loss of human life, 1,240,000 fatalities, was much greater than any territorial compensation could supply. Yet the Italian sense of national pride generated profuse monuments, and the battlefields of northern Italy were strewn with grandiose marble statuary. No town in Italy is complete without a prominently positioned war memorial.

The Second World War, as we learn from Birgit Urmson's invaluable study, conditioned a completely different attitude, especially for Italians. They could be blamed for starting it, since the first aggressions occurred with Mussolini's invasion of Ethiopia in 1935, during which the fledgling League of Nations demonstrated its complete incapacity to arrest the attack. Then Italy found itself isolated with the only option of the unholy alliance with Hitler. Despite Germany's aggressions since 1939, Italian troops did not enter on the side of the Axis officially until 1941. Soon after, in late 1942, the Fascist regime was marginalized, followed by the armistice in the summer of 1943. Luigi Commencini in his bittersweet film *Tutti a casa* (1960) captured the overall ambiguity of this moment, sending poor sergeant Alberto Sordi with his men through territories in which no one understands on which side they are fighting. At the conclusion of the war, there was wide-spread equivocation, Italians had neither won nor lost, they had mostly just suffered.

Thus, instead of grand Italian monuments to the war effort after 1945, this time Italy lost 450,000, nearly half of them civilians, the Italians preferred to make additions of names to the existing monuments. The most impressive World War II memorials built in Italy emanated from the erstwhile enemy countries of the United States and Germany. The former triumphant, economically booming, and radiant with new power; the latter battered and ashamed of Nazi atrocities. The surprising aesthetic results of their respective funereal landscapes command our full attention as gestures of political ideology. Here Urmson reveals that a strange inversion has taken place in which the world's largest democracy and most modern country opted for a lapsed classical vision that verged on the style preferred by the fallen dictators, while the shell-shocked Germany rejected any recall of Speer's classical set pieces, seeking a contemporary version of the picturesque. If World War II had been fought to make the world safe for modernism, so aptly displayed in the design of the United Nations Headquarters in New York (1947–52), the lesson did not quite register among the effete cronies of the American Battle Monuments Commission (ABMC) in the design of their cemeteries in Italy. Under the aegis of the military, the commission deferred exclusively to alumni of the American Academy in Rome, whose mindset at this time was lodged in Greco-Roman grandeur. The architecture, sculpture, and axial layout of the cemeteries designed by Eric Gugler and James Kellum Smith were closer to the cemeteries style of Marcello Piacentini than to that of an American modernist, such as Eero Saarinen. The Germans commissioned their projects without the involvement of the military, which by this time no longer existed, hoping that the originality of a more abstract approach, in which design is interwoven with nature, would win the sympathies of those who obviously felt anti-German after the fall of Hitler's regime.

Urmson explores another telling detail in the choice of locations, with the US insisting on valuable valley properties, near major roads, while the Germans preferred lonely hilltops, remote, hard to find, discrete. That a nascent colonial mentality motivated the Americans corresponds to their establishment of military colonies in Livorno and Vicenza and their political maneuvering of the Italian elections during the late 1940s and early 1950s to keep Italy from entering into the communist sphere. Thus the seemingly innocent job of honoring the dead had deeper connotations that the German designers and bureaucrats would never dare to imply at this time.

The ABMC provided competent design service, heavily influenced by the grand old man of Beaux Arts education, Paul Cret, who from the two previous decades had supplied the models for war memorials, including Gettysburg. Yet the projects are more remarkable for their scale than anything else and the statuary, which contains the most explicit iconography, is exceptionally stale. Not so with the Passo della Futa, initiated by the accomplished veteran of German war memorials, Robert Tischler, but designed by the young modernist, Dieter Oesterlen, who with the landscape architect Walter Russow produced a truly memorable landscape.

Tischler had designed cemetery projects of great novelty during the previous two decades, and preferred medievalizing rather than classicizing visions. We might also thank a woman, Christel Eulen, a member of the VDK commission, for pursuing such an extraordinary landscape, and comment that no women were ever decisively involved with the American projects. The crowning sculptural piece recalls Walter Gropius's zigzag Monument to the March Dead in Weimar (1920), as distant from classical precedents as possible. Throughout the book we are invited to judge the considerable talent involved on both sides, but even more become aware of the intricacies of the bureaucratic administrations behind the works, painstakingly hunted in the archives of the respective countries. Honor, respect, or propaganda remain questions that pervade the documentary research, and we witness each side struggling with the classical past and the modernist present.

War cemeteries, due to the uniformity of their tombs about which individuals or families have no control, retain a serene beauty that other graveyards lack. They also are finite, complete, while normal cemeteries witness constant evolution caused by new arrivals and the removal of old tenants. Because of their coherence in iconographic programs, involving monumental architecture, and sculpture in particular, these works beg to be looked at as art. Urmson thus leads us through landscapes of mourning, introducing us to the taste and talents of the designers and bureaucrats behind them, and analyzing the prodigious sculptural and architectural results.

The soldiers who were buried in these hallowed grounds, be they Americans or Germans, had little to say over why they were fighting. Some might have been convinced Nazis, similar to the repentant Luftwaffe pilot turned artist Joseph Beuys, while others may have held sincere anti-fascist convictions, determined to stop Hitler. Yet as a group, they had no control over the reason they went to war, and they were randomly sacrificed to the perpetual madness of the human quest for dominion, and the equally human response of defense, to which we always answer "never again." Alas, hostility still abounds, and the war machine continues to move on with ever more sophisticated devices of destruction. Urmson coaches us on the appreciation of war cemeteries as having a significant role in modern consciousness, because when one perceives the vastness of the destruction of young human lives, this wealth of lost potential so neatly distributed across the landscape sends a shiver of recognition, not of the individual's demise but of the whole. To feel pride here does not make as much sense as feeling regret and hoping that war will remain only as a memory and not as an incumbent necessity.

Richard Ingersoll

Author's Preface

The rugged Italian Apennines north of Florence are quiet and lonely, but in the final days of the Second World War, as Allied forces sought to breach the Axis Gothic Line, these mountains witnessed fierce battles. German defenders, ill nourished, ill-supplied and out-gunned, perished by the tens of thousands for a lost cause. They lie buried in the German soldiers' cemetery at the Futa Pass, an extraordinary experience for anyone who visits it. It is not simply a burial ground, but a powerfully articulate work of art. Endless rows of graves bespeak catastrophe. I, a German child during those dark days, sensed this catastrophe. All had a share in it, that longing pain for those lost and missed. The nearby US military cemetery at Impruneta couldn't be more different. This manicured place of beauty is optimistic in style and message. The young men buried here died as well. Are we understanding the messages? The historian draws from information and contextualization and speaks in terms of confluences, politics, finances, and diplomacy. The art-historian, nourished by style and iconography, speaks of visions and aesthetics, but our emotional response to art allows us to complete the message. Military cemeteries in foreign lands are national showcases. German and US examples are so different that they invite comparison, but they differ so fundamentally in cultural meaning and political message that juxtaposition is the more correct word. This work juxtaposes the two groups of cemeteries to reveal important aspects of the post-war cultures that produced them, of America-the-victorious and Germany-the-vanquished.

Introduction

At war's end, Italian soil bedded about a quarter million dead soldiers from many nations. Most were strewn throughout the land in makeshift graves. Axis and Allied losses in Italy were unequal. US Americans accounted for less than half of Allied losses. About 60% of US remains had been repatriated prior to Italy's two US military cemeteries' being built. These two cemeteries hold the remains of ca.14,000 and German cemeteries ca.107,000. US cemeteries were begun shortly after war's end, whereas the four German cemeteries were begun a decade later, after an accord with Italy's post-war government in 1955.

The two sets differ in style and message, the US celebrating victory and Germany mourning senseless loss and destruction. ("Germany" refers to the Federal Republic of Germany). The theories developed herein are based on archival evidence that these carefully designed artifacts were intended to express political as well as memorial messages. This work points out that in the aftermath of WW II, US American society became more conservative in tastes and values, while Germany's society, in shedding its NS past, produced innovative forms and meanings.

This work's principal method is art historical analysis of styles and iconography, supplemented by more traditional historical sources such as archival research, secondary literature and oral histories. Importantly, this work sees each nation's art as presaging its post-war future, the US remaining a global military power and a conformist, religious society and Germany becoming a largely pacifist society, less concerned with formal religion, grounded on individual rights and oriented toward commerce. The work focuses on the two-decade period beginning shortly after war's end, and includes relevant background. It portrays the agencies that were involved with US and German military cemetery design and construction. It assesses each nation's aesthetic and ideological assumptions of the time as well as its political circumstances.

This work is organized mostly chronologically. During the first post-war decade, when the US planned and began its cemeteries, Germany was not allowed to move forward with its re-burial program in Italy. It did, however, formulate the guidelines that underlie the actual forms its cemeteries would later take. Therefore, a chapter devoted to this preparatory process is inserted between those dealing with site acquisition and the actual construction on the US side. Treatment of the German cemeteries also includes examples built during the Mussolini period, when Germany and Italy were staunch fascist allies, in order to highlight later aesthetic and ideological changes. All translations are my own, intended to be as accurate as possible while expressing in contemporary US English the spirit of a phrase or statement.

The US and Germany erected some of their finest examples of commemorative architecture in these Italian cemeteries. (US refers to its cemeteries as "military cemeteries" while Germany refers to its WW II examples as "soldiers cemeteries"). Much has been published on commemorative culture within the US. James Mayo's important work recognizes military cemeteries as cultural phenomena, noting the importance of their being on actual battlefields (Civil War) and theorizes that beauty was purposely used to make war more palatable to the American people.[1] Beauty and vicinity to battlefields would remain a prime concern for US cemeteries abroad for both World Wars. Researchers of note who have worked on US WW II military cemeteries include art-historian Kate Lemay, who explores cemeteries in France in the context of the confluence of art, ideology and politics.[2] Also, cultural historian Ron Robin presents a critical and provocative analysis of US WW II cemeteries abroad, imputing to them an overriding ideological/militaristic purpose based on the use of soldiers' death to advance US triumphalism and hegemony, concluding that their intended message fails.[3]

A systematic scholarly work on German cemeteries and war memorials abroad has not yet been written. There are, however, important studies on singular sites and specific ideological features, such as Christian Fuhrmeister's essays on the VKD (*Volksbund Deutsche Kriegsgräberfürsorge*) and on Robert Tischler's "*Totenburgen*" (castles of the dead).[4] The subject

1 James M. Mayo, *War Memorials as Political Landscape*. (New York: Praeger, 1988).

2 Kate Lemay, "Forgotten Memorials: The American Cemeteries in France from World War II ," (Ph.D Diss., Indiana University, Bloomington, 2011).

3 Ron Robin, *Enclaves of America: The Rhetoric of American Political Architecture Abroad 1900–1965* (Princeton, NJ: 1992).

4 Christian Fuhrmeister. "Die 'unsterbliche Landschaft', der Raum des Reiches und die Toten der Nation. Die Totenburgen Bitoli (1936) und Quero (1939) als stategische Memorialarchitektur" p. 64 in *Kritische Berichte, Zeitschrift für Kunst-und Kulturwissenschaften,* Heft 2 Jahrgang 29, 2/2001; and "Klatchmohn und Ochsenblut. Zur

and method used in this work approach Fuhrmeister's, where a deeper understanding of ideology is gained through stylistic analysis. It is to be noted that much of the scholarship devoted to German military cemeteries abroad focuses on NS traditions, allegedly promulgated into the postwar period by the VDK and expressed through architect Robert Tischler's designs. Of central importance are the indispensable essays by Meinhold Lurz and Monika Kuberek. Kuberek focuses on VDK's ideology and WWI cemeteries' aesthetics, pointing out their persistence after WW II, taking as an example the *Totenburg* and Tischler's group crosses, to be discussed herein.[5] Lurz makes a strong point that old structures of thought and aesthetics determined Tischler's works after WW II, evident in particular in his "*Totenburgen*."[6] Both authors provide excellent bibliographical material, indispensable for any study of the subject. Contrary to these authors' views, this work focuses on the VDK's and Tischler's distancing themselves from such traditions in search of new beginnings.

Germany's cemetery at the Futa-Pass has attracted the most interest. Anne Schmedding discusses the cemetery and its architect Dieter Oesterlen in her important study on the revival and reinterpretation of expressionism, followed by a chapter in her book on Oesterlen.[7] A sensitive essay on the aesthetics of the Futa-Pass cemetery by an Italian architect is referenced herein.[8] More generally, interest in comparing national styles of commemoration has sprung up in recent years. Of note is a compilation of essays

Ikonographie der Kriegsgräberstätten des Volksbundes Deutsche Kriegsgräberfürsorge." in *Gartenkultur und Nationale Identität.* eds. Gert Gröning and Uwe Schneider. Worms: Grüne Reihe, 2001.

5 "Leiden und Sterben in den Kriegsdenkmälern des Ersten und Zweiten Weltkriegs." In *Unglücklich das Land, das Helden nötig hat:* Michael Hütt ed. et al., (Marburg Jonas, 1990).

6 Meinhold Lurz, "...ein Stück Heimat in Fremder Erde." Die Heldenhaine und Totenburgen des Volksbunds Deutsche Kriegsgräberfürsorge, *archplus_ausgabe_71,* Aachen, Arch+198.

7 Anne Schmedding, "Die Moderne als Tradition. Architektur in der Nachkriegszeit der BRD am Beispiel Dieter Oesterlens 1911–1994" *Architecture in the Age of Empire. Die Architektur der neuen Weltordnung.* Tagungsband 11 (Weimar: Bauhaus-Kolloquium, Bauhaus-Universität, 2011); and in, *Dieter Oesterlen, Tradition und zeitgemäßer Raum,* (Wasmuth: Tübingen-Berlin 2012), pp. 26–36.

8 Giacomo Calandra di Roccolino, "Paessaggio per la memoria. Il cimitero tedesco del Passo della Futa di Dieter Oesterlen", *La Rivista di Engramma*, 95, (December 2011).

edited by Manfred Hettling and Jörg Echternkamp.[9] A ground-breaking symposium of international scope titled *War Graves/die Bauaufgabe Soldatenfriedhof,* 1914–1989, was held in Berlin/Munich in 2014. The symposium's presentations are published in E-Journal form (RIHA, 2017). Jane Loeffler's study on US embassies abroad interrelates aesthetic decisions with diplomatic and political considerations, and is an excellent model for this form of scholarly presentation.[10] Rather than with analysis of the final designs, this work begins with the etiology of the designs. Art-historical analysis is constantly related to diplomacy and politics.

Altogether, data from thirteen archives are included in this work. Material pertinent to the ABMC (American Battle Monuments Commission) is found in the National Archives and Records (NARA) under the Record Group 117, a body of letters, memos, minutes, sketches, photos, and insider discussions revealing decision-making criteria, political attitudes, and assumptions, both aesthetic and ideological. Archives reveal that decision-making was no simple matter as financial, political, diplomatic, and aesthetic considerations were balanced.

The Washington-based ABMC presumed it could and should represent "America" in foreign lands through a beauty reminiscent of Washington. It's elite group of architects and artists assumed that only classicism derived from Greek/Roman antiquity and the Italian Renaissance could represent "America" in Europe (and as far away as Manila, Philippines). WWI cemeteries allowed for a variety of styles, but ABMC's vision for WW II was more restricted, as pointed out in Elizabeth Grossman's essential study.[11] The US's fifteen foreign cemeteries share a normative aesthetic: A wide axis of grass stretches between a sumptuous entrance and a memorial (often in form of a pillared temple or tower) at the distal end. Graves are arranged in arcs or straight lines of identical marble crosses on both sides of this mall. Classicism had come to connote officialdom in

---

9 In Manfred Hettling and Jörg Echternkamp, eds. *Gefallenengedenken im globalen Vergleich, Nationale Tradition, politische Legitimation und Individualisierung der Erinnerung*. (München: Oldenbourg, 2013).

10 Jane Loeffler, *The Architecture of Diplomatic Building: America's Embassies* (New York: Princeton University Press, 2011).

11 Elizabeth G. Grossman, "Architecture for a Public Client: The Monuments and Chapels of the American Battle Monument Commission," *Journal of the Society of Architectural Historical,* vol. 43, May 1984.

American public architecture. This work theorizes that this represented a nationalistic bulwark against the onslaught of "international modernism" introduced to the US by foreigners, and therefore connoting subversion, if not communism. The ABMC insisted that all artists and architects be American-born, and prior military service was preferred.

Cemeteries needed to be built on "hallowed grounds", meaning actual battlefields. Italian archives, in particular the State Archive in Florence, and those of the Department of State and of the Ministry of the Interior in Rome reveal the drama and agony that land acquisition (expropriation) visited upon Italian families and landowners who, to ABMC's surprise, did their best to resist expropriation. While the ABMC was ultimately granted its requested lands thanks to co-operative Italian political partners, it was nevertheless disquieted (but not dissuaded) by the growing strength of Italy's Communist Party, which was not well disposed toward the erection of such US cemeteries. Thus, Italian archives yield insights into Washington-based attitudes of the late 1940's, a presumptive "right of conquest" in Italy.

The US's military cemeteries were intended to present Italy with symbols of American unity, might, values, and a vision of a democratic nation in complete accord with itself. It is noted that all artists had been fellows at the American Academy in Rome, were friends, and moved in the same upper circles of society in New York, Washington and Philadelphia. This work explores ABMC's selected architects, designers, artists and sculptors as individuals as well as artists, and reveals the elitist world-view they shared. Architect Eric Gugler was a patriot, a dedicated monument builder with a grand vision and ambition, and a devout classicist who curried favor among Washington power circles. (His papers at the Smithsonian Institute Archives were unorganized as of this writing, and information presented herein is drawn from other archives, mostly ABMC's (at NARA under Record Group 117). Landscape architect Ralph Griswold could be relied upon to produce the "piece of America" that General Marshall called for in military cemeteries abroad. Attention is given to how this masterful designer created an American Space with a sense of "wide openness", transcending the classicist formulas in which he was well versed. His complete papers are kept at the University of Pittsburgh Archives and allow insights into his careful collaboration with Gugler. Paul Manship, then among the most famous of America's sculptors, likewise could be relied upon to strip away death's unpleasantness, despair, ugliness, and suffering. (The Smithsonian

Institute has an extensive collection of his work). Some of his letters, that give insight into his vision of his work for the Anzio/Nettuno cemetery, can be found in the ABMC papers at NARA. The Anzio/Nettuno's cemetery's chapel ceiling locates America's "manifest destiny" in the stars while in the obligatory museum, prominent and permanent maps with verbiage celebrate its Armed Forces' achievements, in both Italian and English. In this collective triumphal celebration, a statue of Orpheus, a personal gift from Gugler and Manship, plays his lyre in the monument's south garden, a private voice praising civilization and peace. Archives reveal that Gugler's lifelong passion to express "American exceptionalism" in the form of grand-scale monuments led to later attempts to revive nationalistic exuberance through gigantic monuments to American history. None of these was realized, but they are cited in order to reveal how quickly his generation's experience and mindset became outdated.

The cemetery at Impruneta's architect, James Kellum Smith, was a veteran of the air force (Army Air Corps at the time) in World War II. His papers are to be found primarily at ABMC and the New York Historical Society archives under the McKim, Mead & White Collection (PR 042). Smith was an Italophile and a well-connected gentleman. His design expresses gentle impulses for peace through beauty. It will be shown that his work, with that of his collaborators, landscape architect Michael Rapuano, painter Barry Faulkner, and sculptor Sidney Waugh, presents the same message of triumph as that at Anzio/Nettuno in slightly different form. Through art and inscriptions, death's finality is denied and fallen soldiers are transported to a transcendental hero's universe where there is no pain. One wonders whether this elitist world-view the architects and artists shared might have impeded bolder invention.

Post-war Germany's surprising economic recovery (*Wirtschaftswunder*) was preceded by a period of struggle for survival on an individual level and on a societal level, a struggle to accommodate refugees in their millions in its much-diminished territory occupied by foreign armies, plus a host of other challenges, among which was dealing with fallen soldiers, German as well as other nationalities, strewn throughout the land. Germany's non-governmental VDK took charge of this. Founded by private initiative in 1919, it resumed its work in 1946. Through building cemeteries in Germany and collecting information on the fallen in foreign lands, the VDK was immediately faced with the reality of mass death, and the bewildering task of giving ideological and artistic shape to such loss. Records of its meetings, found primarily in its archives in Kassel as well as those in Berlin's Foreign Ministry (AA, *Politisches Archiv des Auswärtigen Amtes*) under Befund 92 reflect the VDK's search for a new orientation in architecture, landscaping, and for whatever meaning might be found in soldiers' mass death. After having erected hundreds of cemeteries in Germany, it was chartered (in the "Adenauer letter" of 1954) to commission and maintain Germany's soldiers' cemeteries in foreign countries. Unlike the US's ABMC that was funded by the government, a substantial part of VDK's funding came from small individual donations. As a consequence, it was constantly involved with outreach efforts such as collections, exhibitions, excursions to foreign cemeteries, and its own magazine publication. Keeping in touch with German citizenry at large, it needed to remain sensitive to public attitudes.[12] It follows that its public presence played an important role in Germany's painful relationship with its collective Nazi and WW II trauma.

Germany became more secular in the immediate post-war era. This work theorizes that the VDK's adoption of Christianity's ethical message reveals Germany's desire to rejoin the western community of nations that share a Christian history. As it is generally held that the German people were unable to confront the causes of their historical trauma before the late 1960's, the VDK has been criticized as being retrograde. This work illuminates its overlooked progressive side, tracing the VDK's search for new and democratic messaging, a complex step-by-step process that required more than a decade and that is revealed through the several revisions of its guidelines (*Richtlinien*). These encapsulate lengthy deliberations on what appropriate forms honoring Germany's fallen might take. This work traces the evolution of acknowledging survivors' right to suffer, right to mourn, and to hope for consolation.

Analyzing German WW II cemeteries in Italy means studying the idiosyncratic work of VDK's chief architect Robert Tischler, a prodigious talent. During the NS period he was VDK's effective plenipotentiary in Italy, and this continued after the War. As a consequence, his designs offer a unique opportunity to compare his post-war cemeteries with his works of the Mussolini period. Tischler's pre-war designs were medieval castle-like

12 VDK *Sonderdruck*, Juni 1959 about an exhibition commemorating the *Volkstrauertag* (National Day of Mourning) in the *Beethovenhalle* in Bonn that displayed the VDK's works with large photos, plans and models. Germany's President Lübke paid an official visit. See "Unüberhörbare Mahnung" in *Kriegsgräberfürsorge*, 1959, Heft 8 (December 2), pp. 12–13.

structures, often denoted as the *Totenburg*. This work explores the continuity of his blend of modernist cubism and heavy medieval stonework as he adjusted it to the VDK's new ideological orientation. Four designs are presented in chronological order to illuminate, through Tischler's aesthetics, VDK's post-war developments. The most grandiose of his designs did not go forward. This would have been a large crypt below the monastery of Monte Cassino. First welcomed, then politely rejected by Italian authorities, its design was abandoned. The project is included because its history casts light on diplomatic and political issues, and because it prompted the VDK to revisit thorny ideological issues such as crypt burial versus earth burial, pompous versus modest architecture, and mastery over nature versus respect for it. This work argues that Tischler did respond with noticeable aesthetic changes. His signature sacredness of interiors became spatial immediacy, his signature mystery became sobriety, and his grandeur gave (reluctantly) way to modesty. He refused, however, to relinquish his view that mass death was best expressed in groups of symbolic crosses. This work examines the controversy surrounding these small groups of nameless crosses randomly distributed over grave fields. Seen by some (e.g. Monika Kuberek) as a distraction from the reality of mass death, Tischler persuasively defended them on aesthetic, but not ideological grounds.

Archives reveal how difficult was the challenge of expressing mass death for a wrong cause. The VDK addressed this by parts, such as the common grave, or "comrades grave," present in all German cemeteries (but absent from US cemeteries). Through art and design, this work traces "comradeship" changing from together serving the nation to the more personal meaning, together serving one's fellow soldiers. The issue of a monument befuddled both the VDK and Tischler, whose style was monumental even in restricted dimensions. Tischler had proved that he was able to elicit powerful effects and strong emotions with modest means. His urges toward the monumental re-emerged with his last commission, the cemetery at Caira near Cassino, where the hilltop cried out for a monumental feature. The VDK disagreed, and refused to budge on the issue, obviously fearing that it might send a wrong message. This attitude casts light on Germany's deference to Italian sensibilities. The tone and content of letters reveal relations becoming more cordial, and this work theorizes such being facilitated by Italians in great numbers finding work in Germany, and Germans traveling in great numbers to Italy as tourists.

This work explores yet other designs that reveal Tischler responding to post-war sensibilities. At Costermano, for example, traces of his dramatic pre-war *mise-en-scene* are recognizable, but mysticism is toned down through architecture and countered through sympathetic, personalized sculpture, in which a delicate youth gazes tenderly upon the comrades' graves. Its message is comfort in quiet mourning. Its style recalls Lehmbruck's work, a sculptor driven to suicide because of WWI's horrors, and whose pacifist message was anathematic to the Nazis' cruel demands.

Unlike the ABMC, the VDK adopted non-classical, abstracting post-war styles in figurative sculpture. The theory is put forth that preference having shifted from the idealized to the abstracted reflects the VDK's changing focus, from the heroic to the individual. Further examples are explored to support this, such as at Pomezia, where a sculpture group of a family in silent grief presents the diametric opposite of a sculpture titled "Germany Awake," a work by the same sculptor during the NS period.

The VDK ended Tischler's reign by awarding Dieter Oesterlen the commission to design its cemetery at the Futa-Pass near Florence, Germany's last and largest in Italy, and costliest anywhere. Oesterlen made a striking leap with his new monumental language and his design's clear pacifist message. It will be shown that he solved many lingering conflicts that had plagued the VDK, such as between designed and untamed nature and appropriate forms for grave markers and monument. Oesterlen's papers are in the Dieter-Oesterlen Archiv in the Akademie der Künste, Archiv für Bauwesen, AdK, Berlin. The Archive also holds those of landscape architect Walter Rossow, who brilliantly designed the Futa Pass landscaping such that mass death dialogues with wild surrounding nature. This work points out the design's great pathos despite Oesterlen's espoused intention to avoid the pathetic.

Oesterlen's bold abstract sculptural architecture is explored as an aesthetic whole, visible from afar, monumental yet free from lingering shadows of the NS heritage that had plagued Tischler's designs. The Germanic tradition of dialoguing with nature is reclaimed, strongly so, but without the notion of sacredness with which it had been burdened. Nature will eventually master the monument, as intended, in contrast to US cemeteries' control over nature, where separate park-like islands inserted into foreign land demand constant maintenance and irrigation.

This work explains the Futa Pass design as a stroke of genius, but standing on the shoulders of others. Oesterlen's expressive dynamism achieved through irregular and abstract forms is linked to Germany's expressionism of the Weimar republic, and to the aesthetics of the "crystalline." It presents designs, both executed and non-executed, as sources of Oesterlen's inspiration, especially Gropius's monument to the *Märzgefallenen*. It links the Futa Pass's design and message to the *Darmstädter Gespräche*, a forum to define new values for man and his environment in which prominent thinkers, even some that had emigrated during the NS period, took part. It postulates that Oesterlen's bold departure responds to the progressive segment of the German population that thirsted for renewal of humanist values in art and life. Soldiers' death was to be quietly mourned as a tragedy, and not as sacrifice. It interprets Oesterlen's design as evidence that the VDK had cast its lot with the progressive side of Germany's late 1950's Zeitgeist. It postulates that the VDK's acceptance of cost overruns evidences Germany's rising economic prowess as well as wider cultural attitudes in motion.

This work concludes that through art and design, these cemeteries politicize both intentionally and unintentionally. The two US cemeteries, intended to showcase democratic virtue and leadership among nations backed up by a benign citizens' army sends a mixed message, one that would become a contradiction and rend US society during Vietnam War of the 1960's. German cemeteries' art witnesses post-war attempts toward renewal and desire to work for peace and reconciliation, finally attaining an interim of restless peace that, in like fashion to the US, would be shattered in the late 1960's by challenge and questioning from its youth.

I. US Organizing for Remembrance

The Players

> "We have urged adherence to beauty of form, to excellence of proportions, and to permanence of materials. We must so continue, not with slavish adherence, but with fundamental basic principles which guided the artistic leaders of the Greek and Roman worlds and of the Renaissance."
>
> Gilmore D. Clarke[1]

This chapter will introduce the "players," those individuals and organizations that planned and executed United States (US) military cemeteries on foreign soil. They are: the Army's American Graves Registration Service (AGRS), the American Battle Monuments Commission (ABMC), and the United States Commission of Fine Arts (CFA).

Burial during the Second World War and the years immediately following was the task of the AGRS.[2] Beginning in 1948, the ABMC transitioned into directorship of these efforts, fulfilling its mandate to create memorializing cemeteries.[3] There was some temporal and ideological overlapping, but the AGRS built modest cemeteries using limited land while the ABMC, with the approval of the CFA, carried out monumental commemorative visions. These were more elaborate, costly, and demanded larger parcels of land.

---

1 Address for a meeting of a Joint Committee of the National Capital, Columbia University Archives, Gilmore D. Clarke papers (February 18,1944).

2 Following the WWI pattern, the AGRS collected, identified, and reburied cadavers in cemeteries. The AGRS also arranged for the repatriation of remains if so desired by next of kin. About 60 percent chose that option. When burial was completed, responsibility was transferred to the ABMC. General Order 125, War Department (December 29, 1945), see <www.qmfound.com/graves_registration.htm>. On the AGRS's reconfirmation of authority overseas see: General Order No. 26, Department of the Army (Washington 25, D.C. (April 16, 1948), General Order No. 26. For the AGRS's reconfirmation of authority overseas see <www.armyupubs.army.mil>.

3 Original authority for the ABMC was established in 1919 and reauthorized in 1947. Law 389 of the 66th Congress, (1919) and Law 368 of the 80th Congress, (1947).

Fig. 1. Memorial at Chateau Thierry, France. Paul Cret's classical design established an aesthetic standard.

The ABMC was established by an act of the US Congress as an independent entity answerable only to the president. The iconic General John J. Pershing had cofounded the ABMC in 1919 and served as its chairman until his death in July 1948.[4] It needed to be revived, re-authorized, and funded following WW II. For this purpose, on May 15, 1946, it's general secretary, General Thomas North, testified before the US House of Representatives Committee on Foreign Affairs.[5] North had served on General Pershing's staff during the first World War (WWI) in France, and had been with the ABMC since 1938. He was appointed general secretary in 1946 by President Truman. North had assisted architect Paul Cret. The French-born Cret had provided Pershing with quick, up-to-the minute maps recording the progress of the US Army's campaigns.[6] This was an aesthetically formative experience for North, who came to greatly admire (and to

4 Edward Shenton, "They Will Never Be Forgotten," *Saturday Evening Post*, vol. 227, (August 14, 1954), pp. 36–7, 59.

5 Committee on Foreign Affairs, House of Representatives, 79th Congress. A copy can be found at Virginia State Library, Y4 F 76/1, AM 3/5, On H.R. 6393, (May 15, 1946).

6 Edward Shenton, "They Will Never Be Forgotten." North's love of maps and insistence on making them as an indispensable part of US cemeteries probably came from his experience with Paul Cret. During WW II Cret prepared the day-by-day military maps for briefings with Franklin D. Roosevelt and General George C. Marshall, his chief of staff.

use as a standard) Cret's rigorous classicism, even in its "stripped-down" form (without ornamentation and using pillars instead of columns). Cret's war memorial near Chateau-Thierry (1930) (near the Thiaucourt cemetery in France), established an aesthetic precedent for commemorative classicist-derived style.[7]

Cret's double colonnade was a brilliant memorial design, a contemporary quotation from an established formal repertory that bespeaks triumph. Wall sections alternating with openings form ribbons and voids, as if pillared.[8] This memorial would become a standard. Cret went on to design many public buildings in the US in this manner, such as the Federal Reserve Building and the Folger Shakespeare Library, both in Washington, D.C., thereby establishing an American aesthetic standard of classicism for civic and war memorial buildings.[9]

Cret had served as the ABMC's consulting architect and North assisted him in the planning stages for memorial projects. As North explained to Congress, the ABMC would have eleven voting members, war veterans as well as a non-military component.[10] The ABMC had erected

---

7 Elizabeth G. Grossman, "Architecture for a Public Client: The Monuments and Chapels of the American Battle Monument Commission." *Journal of the Society of Architectural Historians,* vol. 43, (May 1984). Author reports that not all ABMC WWI memorials were in the classicist style. Neo-Romanesque and Neo-Gothic styles were also used. Grossman's article illuminates the ABMC's post WW II more restrictive aesthetics.

8 This became the standard modern version of classicism, so-called "stripped-down" classicism, a term created by architectural historian A.R. Hitchcock. Cret attempted a modernization of strict Vitruvian classicm, in leaving the columns unadorned. He was able to retain the sense of authority, and his "Ahistorical columns implied a forward-looking concept of national power." Ron Robin, "*Enclaves of America: The Rhetoric of American Political Architecture Abroad, 1900–1965,*" p. 117. (Princeton, NJ, 1992). This would be a model of classical modernity for J.K. Smith in his design in Impruneta, where columns have turned into unadorned wall sections that could be read as posts. See Chapter Chapter VI below.

9 Theo B. White, edit. John F. Harbeson, "Paul Philippe Cret: Author and Teacher," *The Art Alliance Press,* Philadelphia, 1973, p. 21; And see Elizabeth Grossman, "Paul Cret; Rationalism and Imagery in American Architecture" (Ph.D. diss., Brown University, 1980).

10 Ron Robin, "A Foothold in Europe." The Aesthetics and Politics of American War Cemeteries in Western Europe, *Journal of American Studies*, vol. 29, no. 1 (April 1995). Robin sees the military's influence in the ABMC negatively. In his view their cemeteries failed to project American hegemony because of aesthetic flaws and unclear messaging.

several WWI memorials in France, but these were memorial structures without burial grounds. In contrast, WW II memorials would combine and integrate gravesites with memorial structures.[11] Both the Secretary of War and the Secretary of the Navy had asked Congress to reactivate the ABMC and to extend its authority to include WW II cemeteries as well as memorials. Its authority over official foreign memorials was to be exclusive so as to prevent the "popping up" of private memorials. In this hearing, North explained that such had been Pershing's original motivation for founding the ABMC.

Fig. 2. ABMC's Logo. Note stylized Doric temple façade.

11 General John Pershing, "Forever America," *American Legion* vol. 2, no. 5, pp. 14–19, (May 1927). After WWI, eleven memorials were erected on battlefields, some of them rather large and designed by prestigious architects, e.g., Chateau-Thierry, France, by Paul Cret, and a classical high column with a statue by Russell Pope in Montfaucon, France. They were erected to commemorate decisive battles on the grounds where the battle occurred, and be of such construction quality that they would last for centuries so visitors would continue to visit. For more information on all cemeteries and monuments see <www.ABMC.gov> under cemeteries & memorials.

The ABMC's authority would flow from the president (Harry Truman) via executive order and it behooved Congress immediately to supply funding as time was of the essence. A supplemental appropriation was suggested.[12] Congresswoman Francis Bolton of Ohio, a member of the House Foreign Affairs Committee, commented that her son in Berlin had witnessed soldiers' disregard for their own fallen, and articulated the following synopsis of national sentiment: "There is something very definite that we do to give a picture of our belief in eternity in the Deity, in the reverence of those things which are lasting, which are also the qualities of loyalty, patriotism, and so on."[13]

The AGRS had been responsible for the burial of the fallen in WWI and had employed French landscape architects to design the cemeteries. The ABMC was responsible for memorials. French (and one British) architects supervised their construction. North testified that the ABMC had made the decision to commission exclusively American talent to design combined memorials and cemeteries *in toto* (architecture and layout plan) from the outset. Archives reveal that the ABMC began ramping up its post-WW II activities before the end of 1947 under North. The ABMC continued its administration and *modus operandi* inherited from the post-WWI period, but expanded its operational base to include three offices to supervise European affairs: Washington, Paris, and Rome.

Finally receiving a Congressional mandate in 1947, North did not waste any time. A supervising architect was to be chosen, followed by a consulting landscape architect, a painter, and a sculptor, all to be American born. North insisted on "all American" teams for "sentimental reasons."[14] Accordingly, architect John Harbeson, who would design the Coleville-sur-Mer cemetery above Omaha Beach, was appointed as supervising architect. Markley Stevenson became the ABMC's supervising landscape architect and established the scope of specified tasks and fee structures.[15]

---

12 Supplemental funds were immediately approved by the "Committee of Appropriations", page 6 of the document cited in fn. 6.

13 Ibid. Ms. Bolton does not specify the nationality of the soldiers.

14 Letter, North to Harbeson, RG 117, Entry 13, Box 143, Folder 231.24, (March 1, 1948).

15 Letter, Clarke to North and Harbeson affirming that the task of the landscape architect was considered very important. The landscape architect's design would replace the work of the AGRS, which provided the landscaping layout following WWI. AGRS continued to do so in the early years following WW II. RG 117, Entry 13, Box 143, Folder 231.24 (June 26, 1947).

The 1950s was a decade during which landscape architecture rose in prominence in the US, and its practitioners commanded handsome fees.[16] The ABMC adjusted its fees to the standard of New York's department of parks.[17] Gilmore D. Clarke, the CFA's chairman and landscape architect used a negotiating meeting in New York (on June 5, 1950) as a forum to define respective responsibilities of architects and landscape architects. The architect's vision would determine a site's overall plan, its chapel, and its memorial.[18] Harbeson was pleased with this approach: "The present program of having work considered as a whole from the very start should lead to even finer results."[19] Clarke, also a believer in collaborative creative design, held that the character of the total design must include sculpture as a bearer of meaning.[20] Such collaborative design work favored the selection of architects and artists who were already well known to one another, and resulted in a nepotistic "old boys' network."[21]

The US had become a global power as a result of WW II, a war that produced many more casualties than had WWI. Nonetheless, commemorative buildings were to be more modest than their WWI equivalents.[22] A letter from North to architect Eric Gugler, who would design the Sicily-Rome military cemetery (herin referred to as Anzio/Nettuno) declared that, in

16 Ibid. Four percent of the landscaping costs.

17 The issue of the tasks of landscape architects had been discussed before in detail via correspondence. See letter from Stevenson to North. The emphasis was on collaboration. RG 117, Entry 13, Box 143, Folder 231.24, (March 16, 1948).

18 QMG's LC Donelson undated report of a meeting held in January 5, 1950, saying that architects would still control the sign, hire the landscape archiects, and al other talent. RG 117, Entry 13, Box 143, Folder 231.42.

19 Letter, Stevenson to North. in which Stevenson quoted Harbeson. RG 117, Entry 139, Box 143, Folder 231.24, (March 16, 1948).

20 Letter, Clarke to North expressing that sculpture was the bearer of meaning. Howeverr, Clarke gives architects dominance of decision making. RG 117, Entry 13, Box 143, 231.24 (June 26, 1947).

21 In this context, one might note that architect H.R. Shepley did not wish to consider anyone other than Lee Lawrie or Paul Manship as his sculptor, even though each had a commission for another site. The rule was only one sculptor per site. This can be deduced from a letter from Harbeson to Clarke. RG 117, Entry 13, Box 143, Folder 231.24, (March 1, 1948); RG 117, Entry 13, Box 143, Folder 231.24, (March 3, 1948). (This rule was not broken.)

22 As to modesty, see Pershing, "Forever America," *American Legion,* (May 1927), vol. 2, No. 5, p. 15. "We are fittingly discharging the sacred duty entrusted to us by the American people, and the veterans of the World War would have done it, with accuracy, *modesty*, and with reverence." (Emphasis added by author).

contrast to after WWI, where rather elaborate and imposing memorials stood on battle sites, there would be no such separate memorials.[23] Rather, cemeteries themselves would both memorialize and constitute the historical record. Emphasis on the historical record was of great importance. There would be museums on site. The term "historical record" came to connote the "achievements of the Armed Forces."

Congress re-enacted the requirement that the CFA approve all such memorial designs.[24] The agency made it clear that it would remain involved in all design steps and undertook its own study of WWI war memorials.[25] It wielded considerable influence in all aesthetic decisions.[26] All designs were submitted to the CFA for "review, advice, criticism, and unifying spirit."[27] Its chairman, the aforementioned Gilmore D. Clarke, was an Army major decorated with a Purple Heart and who shared experiences and tastes with North.[28] In 1947 he received the Army of Occupation of Germany medal (WWI) for "his honorable services rendered

23 RG 117, Entry 9.

24 S. Kohler, *History of the Commission: The Commission of Fine Arts, A Brief History 1910–1976*, (Washington DC, 1977). The Commission was established on May 17, 1910 and consisted of a chairman, a secretary, three architects, one landscape architect, one sculptor, one painter, and one layman (eight total). President Hoover appointed Gilmore Clarke chairman in 1932. He served until 1950. (Clarke followed Charles Moore, who set the standard of the white gravestones Box 122, Folder 687, (8–22–47 to 13–31–48).

25 Report On War Memorials. United States Commission of Fine Arts, Washington D.C. US Gov. Print. Oct. 1947. Series United States, 79th Cong. 2nd Sess. Enat 3, Doc. 234 (adopted on May 14, 1946). Chapter VI titled Design begins; "The ultimate requirement for a war memorial is that it should be a work of art." Since the CFA and the ABMC favored classicism, there is no indication that aesthetic conflicts arose.

26 Speech by David E. Finely, Director, National Gallery of Art (and member of CFA's board headed by Clarke), to the Municipal Art Society – New York, on May 25, 1949 praising Clarke's work. He reminded listeners that the CFA had been established by an Act of Congress in 1910 and specified that its members must be "well qualified judges of the fine arts" and that "before any design or material for memorials is accepted by the Commission [ABMC], the same shall be approved by the National Commission of Fine Arts." Columbia University Archives, G. Clarke Papers.

27 Aline B. Louchheim, "Memorials to Our War Dead Abroad," *New York Times*, (January 15, 1950).

28 Letter from Adjutant General, War Department (March 6, 1947) mentions that Clarke received a Purple Heart as well as the Medal of Occupation of Germany (post WWI). Columbia University Archives, G. Clarke Papers.

in Germany during the occupation period between November 12, 1918 to July 1923." In addition, he enjoyed much prestige in the world of east coast design.[29] As an engineer and landscape architect, he and his partner, Michael Rapuano, had designed major parkways in and around New York City. Both had graduated from prestigious Cornell University. Clarke had been New York City's consulting landscape architect since 1934, responsible, together with Rapuano, for the concept of "gardens for the machine."[30] Under New York City's parks commissioner Robert Moses, he and Rapuano designed bold thoroughfares, such as Riverside Drive, that changed New York City into a city for cars connected by motorways to the city's suburbs. They applied this experience with cityscapes to the master plan for New York's World Fair of 1939, in which many of America's major designers had a hand. Many of these would be receiving cemetery commissions from the ABMC. (Rapuano would later design the landscape for the Florence Cemetery.)

Appointed by President Herbert Hoover in 1932 to chair the CFA, Clarke had been involved in major public aesthetics controversies. For example, in 1944 he argued passionately before Congress that Washington must preserve its Beaux-Arts heritage and thus take the lead among the world's beautiful cities.[31] He argued that destroyed cities abroad would be rebuilt after the war in lesser styles, and thus Washington could and should

29 Clarke received the Gold Medal of Honor by the Architectural League of New York in 1931 for his work in Westchester County. This and numerous other awards are listed in Wikipedia entry for Gilmore David Clarke.

30 Thomas J. Campanella, "American Curves: The Public Works Legacy of Gilmore D. Clarke and Michael Rapuano." Abstract of lecture given at the Botanical Garden, New York and lecture series of *the Cultural Landscape Foundation* on October 9, 2009, an indispensable article for understanding Clarke's accomplishments. "…Clarke and Rapuano would help bring to life Moses's dream of a modern, integrated landscape: a vision articulalted in the 1920's of "weav[ing] together the loose and frayed edges of New York's arterial and metropolitan tapestry." <https://tclf.org/sites/default/files/microsites/pioneers/nybg/Campanella%20Abstract.pdf>. Accessed March 14, 2018.

31 Clarke addressed the Joint Committee of the National Capital on February 18, 1944 in a speech titled; "Aesthetic Standards for the National Capital." "While recently we have not invited the incorporation of those details on buildings, which particularly distinguish Greek and Roman monuments, we have urged adherence to beauty of form, to excellence of proportions and to performance of materials, all attributes of design exemplary of the art of the architecture of the past." Columbia University Archives, G. Clarke Papers.

shine. He later lamented that his vision for Washington would have to give way to more demanding financial and political realities.[32]

The works and teachings of Paul Cret figured prominently in Clarke's "classical" aesthetic tenets.[33] New York's parks, such as Bryant Park (with M. Rapuano) reveal his adherence to the Beaux-Arts tradition, where "beauty" is art's most important element. "Beauty may not be a queen, but she is not an outcast beyond the pale of protection and respect. She may at least shelter herself under the wing of safety, morality, or decency."[34] Clark viewed sculpture and murals on public buildings as a form of patriotic duty: "Art immortalizes the cause symbolized, the thought embodied, the individual portrayed ... linked to our artistic development are questions of patriotism and culture."[35]

Although Clarke's basic credo paid homage to ancient Greek civilization, it would be incorrect to classify him as a "staunch" classicist.[36] He did not advocate slavish adherence details presented in classic form.[37] He was aware of the need for a modern adaptation of classicism, as exemplified in

32 Clarke's January 10, 1949 speech before the American Society of Landscape Architects, "The Future of Our Profession." By 1949 Clarke realized that Europe was going to be rebuilt and he lamented the "... financial drain to rebuilding Europe, and the Berlin airlift." Columbia University Archives, G. Clarke Papers.

33 Clarke, in a February 6, 1951 speech, quotes Paul Cret. "The abandonment of classical disciplines is neither new nor without its price. Regardless of their use made later on of the forms they proposed as examples, these disciplines had an unquestionable educational value. What is to be substituted for their experience, proved efficacy in training the eye to proportion, to rhythm, to composition, is not as yet divulged, and those who explain it as stifling to originality forget that an originality so easily stifled must not be very robust. Of the men doing original work in this country at the present time, by far the greater number have been classically trained by our schools."

34 Ibid.

35 In an April 20, 1940 speech to members of House and Senate, Clarke stressed the importance of spending money for sculpture and mural painting on public buildings. Columbia University Archives, G. Clarke Papers.

36 Clarke was Dean of the College of Architecture and professor emeritus of landscape architecture at Cornell. In a June 10, 1950 address at Cornell, he evoked the splendor of ancient Greece's civilization, citing as examples the grandeur of the Acropolis, the defeat of Persians, philosophy, and drama. To him, ancient Greece was an exemplary civilization. Columbia University Archives, Gilmore Clarke Papers.

37 Ibid. "In some of the finest buildings of the Capitol the architect departed from a slavish adherence to classic form, but nevertheless, followed the fundamental basic principles of design which guided the artistic leaders of the Greek and Roman worlds and of the Renaissance."

Paul Cret's aforementioned Folger Shakespeare Library in Washington.[38] For him, the library exemplified that beauty might also reside in modernized classicism, that came to be referred to as "stripped-down classicism." He further thought that patriotism would benefit from a simplified architecture because it would "make room for rich embellishment by the sculptor and by the painter, and thus provide a greater distinction in our buildings, a distinction which will tend to make them wholly American in flavor."[39] Thus Clarke, confirming Cret's authority, linked classical art to the American public identity's most cherished ideals.

Indeed, it appears that Clarke and the ABMC rejected all other options for a different style. There is no evidence that they took notice of the symposium titled "A New Monumentality" held by the prestigious *Architectural Review* in the fall of 1948. Famous architects, Walter Gropius among them, and theoreticians and architectural historians, e.g. Henry-Russell Hitchcock, searched for a contemporary form of "monumentality."[40]

As previously mentioned, North had witnessed Cret's effectiveness in his role as ABMC's supervising architect during the period between the two world wars. Following WW II, with more remains to be interred in more countries, he deemed the role of supervising architect to be indispensable.[41] North began his strategy of endorsing Harbeson by reminding Clarke of Cret's excellence. Cret's WWI memorial in France (Chateau-Thierry) was iconic for ABMC's vision of a monument.

Harbeson had been Cret's student and had become heir to his office. Clarke agreed that Harbeson would be a splendid choice, extolling his "most attractive" personality.[42] However there was Henry Richardson Shepley, in high esteem with Clarke, who had already served for four

38 Ibid. "The strict and rigid compliance with the tenets of the classical school in architecture, which have obtained altogether too long in Washington, must be abandoned in favor of a more fresh approach to the problems which will confront the designers of new buildings in the future."

39 Ibid.

40 Lewis Mumford, "In Search of a New Monumentality" in *Architectural Review*, vol., 104 (September 1948), pp. 117–28; For Mumford classicm was the language of despots, p. 127; and see Lewis Mumford, "Monumentalism, Symbolism, and Style," *Architectural Review* vol. 105, (April 1949), pp. 173–8.

41 Shenton, Edward "They Will Never" pp. 36–7, 59. North had worked under Cret during WWI in the latter's production of military cartography.

42 Letter, Clarke to North. RG 117, Entry 13, Box 143, Folder 231.24, doc., (April 29, 1946).

years on the ABMC and had great buildings to his name.[43] And what about James Kellum Smith, heir to the firm of McKim, Mead & White, royalty among American architects? He had been involved with notable projects, had served as a lieutenant colonel in WW II and also had a winning personality.[44] (General John Markey, ABMC member of influence, favoring the military and maintaining close ties with the American Legion, preferred both Smith and Shepley to Harbeson because they were veterans).[45] Shepley had served in the Army Air Corps in France as Captain in WWI. (Markey remained adamant about the issue of veterans' participation and two years later, a somewhat irritated General North wrote to Markey that there would be impressive contingents of veterans among those working on designs of the cemetery projects).[46]

The matter of choosing a consulting architect was raised and brought to a quorum vote. With Charles Shaw and Senator David Reed joining North, the matter was settled in favor of Harbeson.[47] The latter would be North's unerring comrade-in-arms for years to come, mediating skillfully between the ABMC's ideological guidelines and his fellow artists' aesthetic visions. He supervised the erection of fifteen WW II military cemeteries on foreign soil (at the cost of 1950's thirty million dollars). Clarke's characterization of Harbeson as an "attractive personality, tactful in his dealings with others" would prove to be more than correct.

---

43 As Clarke said in his eulogy; "Harry Shepley demonstrated in this monument (Margraten, Holland) that he understood in full measure the great sacrifices that were made by the men in uniform who followed those of his generation to France a quarter century later." Note the emphasis on the soldiers. In its WW II sites ABMC gave equal, if not more, space to the achievements of the Armed Forces. Columbia University Archives, G. Clarke Collection.

44 Smith's reputation rested mainly on having designed buildings for universities and colleges. RG 117. Entry 13. Box 143. Folder 231.24.

45 Brigadier General and WWI veteran (awarded the Distinguished Service Medal) John Markey was a member of the ABMC from 1924 until 1963. He was responsible under Pershing for the building of chapels and memorials for WWI cemeteries. He was also chairman of the military affairs committee of the American Legion. RG 117, Entry 13, Box 143, Folder 231.24.

46 Letter, North to General Markey. RG 117, Entry 13, Box 143, Folder 231.24, (December 8, 1948) in response to the latter's stated preference for veterans.

47 Decisive vote from Senator David Reed for Harbeson. See letter from Reed to Shaw. RG 117, Entry 13, Box 143, Folder 231.24, (December 8, 1948).

Next, suitable architects were needed. North, in a letter classified as "confidential," asked Clarke to assist the ABMC in finding them.[48] Four lists of architects considered suitable were prepared. (It is not clear who prepared the lists, but it was likely North assisted by Harbeson.) The first list named architects who had already worked on WWI memorials and were still active. Among those was the eminent George Howe, who would play such an important modernizing role at the American Academy in Rome (AARome) after WW II.[49] The second were architects known for public works, such as Eric Gugler, who was listed as having designed Chicago's WWI war memorial (which was never built) and the White House restoration, most strikingly the Oval Office. J.K. Smith (again, of the firm, McKim, Mead & White) was *de rigeur*, as was Saarinen, designer of the Cranbrook School. And of course, this list included Skidmore, Owings and Merrill, a Chicago firm rising in prominence. The third list bearing the title "List for Small Monuments" suggested architects experienced with memorial design, among them Henry Tombs, a favorite of Franklin D. Roosevelt.[50] Most of these architects were active on the east coast, principally in and around New York and Philadelphia. The fourth list included "Some able architects not on the other list(s)," such as Frank Lloyd Wright (Taliesin, Johnson Wax Works), Walter Gropius of Harvard, and Mies van der Rohe of the Illinois Institute of Technology.[51] This list also included the west coast architects William Wurster, Richard Neutra, and Gardner Dailey. The latter two pursued more modernist aesthetics. With the exception of Dailey, none of these "able architects" was offered a commission. Dailey would design the cemetery at Manila adhering to the ABMC's classicist style.

By May 6, 1947 in a letter marked "confidential," Clarke submitted another list.[52] From his correspondence, it is clear that Clarke considered

---

48 Letter, North to Clarke. RG 117, Entry 13, Box 143, Folder 231.24, (April 28, 1947).

49 Denise Rae Costanzo. "The Lessons of Rome: Architects at the American Academy 1947–1966," PhD diss. Pennsylvania State University, (August 2000). George Howe designed the US consultate in Naples. For research on AARome archives: Catherine S. Gaines, American Academy in Rome, Smithsonian Archives of American Art, 2001.

50 Henry Toombs, on List III, was based in Atlanta. He designed Warm Springs for Franklin D. Roosevelt and his Hyde Park Library in Hyde Park, New York. RG 117, Entry 13, Box 143, Folder 231.24, (MG_5294).

51 On List IV, Gropius's qualification was "World War I Memorials in Germany." Mies van der Rohe's qualification was the German Pavilion, Barcelona Exhibition, 1929. RG 117, Entry 13, Box 143, Folder 231.24, (MG_5294).

52 Ibid.

it his patriotic duty to urge that all participating architects be both American born and trained in classicism. Indeed, only American born architects would be chosen, and classicist styles would prevail in WW II memorials' aesthetics.[53]

Harbeson was active in the selection of architects. For Italy, he suggested Eric Gugler and J.K. Smith. Both were connoisseurs of Italy, having been fellows at the American Academy in Rome (AARome).[54] His own firm was the first to be commissioned and awarded the prize site, the first to be built, at Colleville-sur-Mer, Normandy (dedicated in 1956) overlooking the wildness of Omaha Beach. Gugler and Smith immediately accepted their invitations to submit preliminary designs for the two military cemeteries to be built in Italy, at Anzio/Nettuno and at Impruneta (referred to as "Florence" by the ABMC) respectively.

Clarke wrote again to the ABMC: "I would also like to add, for your consideration, the names of the three architect members of this Commission (CFA) whom you may or may not wish to consider." Clarke reminded North of the CFA's authority and North, in turn, did not fail to commission all three of Clarke's candidates.[55] Very soon thereafter, architects for the remaining twelve foreign cemeteries were in place.[56]

Attempts of modernist architects to define a new form of monument were completely ignored.[57] Harbeson would be a guarantee that classicism would rule supreme. He pursued his role as supervising architect with great attention from the very beginning. He advised North to adhere to Cret's model, having a single architect design each cemetery. Each architect would provide initial sketches, and if not asked to continue, would

53 All eleven listed names and their works were mentioned. It also included William Wurster, later a proponent of architectural "brutalism" exemplified at the College of Environmental Design at the University of California at Berkeley. RG 117, Entry 13, Box 143, Folder 321.24.

54 Letter, Harbeson to North. RG 117, Entry 13, Box 143, Folder 231.24, (May 21, 1947).

55 Letter, Clarke to North. RG 117, Entry 13, Box 143, Folder 231.24, doc., (May 6, 1947). Clarke's choices were: L. Andrew Reinhard (Reinhard, Hofmeister & Walquist, New York), Neufille en Condroz, Belgium, the firm of W.R. Aldrich, Boston, St. James, France, F. V. Murphy (Murphy & Locraft,) Washington, St. Avold, France.

56 Letter, Harbeson to North. He noted that all architects were in place, including Harbeson for Colleville-sur-Mer. RG 117, Entry 13, Box 143, Folder 321.24 (May 29, 1947).

57 In a letter to North, Harbeson agrees with the choice of a representative architect for a project in France because he was not the "Corbusier type." RG 117, Entry 13, Box 143, Folder 231.24, (March 10, 1948).

be paid $600. Accordingly, Harbeson drafted a standard letter proposing contracts to ABMC's selected architects, which North would use.

The "players" all shared similar tastes and values making for a seamless progression within a harmonious infrastructure. This, in turn, promised smooth realization of the ABMC's unambiguous vision. However, conflicts would soon arise between the ABMC and the AGRS.

The Situation in Italy

> "The responsibility for permanent designs and etc. is a responsibility of the American Battle Monuments Commission. However, Paragraph 20 states that the War Department will make all necessary burials and reburials of World War II dead in permanent cemeteries overseas."
>
> General G.A. Horkan[1]

Historical Timeline:

- May 1943: Military targets in and around Rome bombed by the Allies.
- September 1943: Armistice of Cassibile (Sicily) moved the major battlefields to the Italian mainland. German and other Axis forces were present throughout Italy except Apulia and Sardinia.
- January 1944: US (and Allied forces – mainly British and Commonwealth) land at Anzio.[2] Fighting in the Anzio/Nettuno area continued until the following June.
- June 1944: Liberation of Rome.
- Summer and fall 1944 and winter 1945: The Allies advance slowly northward toward Axis's "Gothic Line."
- May 2, 1945: Hostilities cease by order of Hitler's successor, Admiral Dönitz.
- June 2, 1946: Plebiscite affirms new Italian Republic (effective January 1, 1947).
- September 24, 1946: US-Italian Bilateral Agreement on burying US soldiers in Italy (about 40,000).
- February 10, 1947: US and Italy sign formal peace treaty in Paris.

---

1 Letter, General G. A. Horkan (chief of the QMG's memorial division) to North. RG 117, Entry 38, Boxes 2 and 156 (5962), (April 7, 1946).

2 Operation Shingle, January to June 1944, led by General John P. Lucas. Two landings on Italian soil preceded Operation Shingle: The first, Gela in Sicily, on July 10 1943, under General Patton. The second, Salerno September 3, 1943 under General Mark Clark (US 5th Army) and Field Marshal Bernard Montgomery (British 8th Army).

During the early post-war years, the US enjoyed its moment as an ascendant global power militarily, financially, politically, and morally. At home, sentiments of "wellbeing" and "mission accomplished" prevailed. Erection of military cemeteries in Italy juxtaposed a giant against a small nation with a fledgling democratic government. Inequality lent complexity to an already murky relationship. Audio recordings reveal Italians cheering enthusiastically when Mussolini announced their country had declared war on the US in late 1942. As Italy's fortunes of war changed for the worse, its new government in Rome joined the Allies and declared war on the Axis.[3] Though having suffered destruction through Allied bombings, Italians in general welcomed the US (and its allies) as liberators, and although the US viewed Italy as friend and ally, subsequent actions indicate ambivalence on both parts.[4]

The young Italian Republic struggled to find its footing in the face of divisions along party, regional, and social divides. Its government included Conservatives, Christian Democrats, Socialists, and Communists. The latter formed a strong party and claimed credit for their anti-fascist deeds. The new government needed to govern, but building consensus and making decisions proved to be complicated matters.

The histories of land acquisition for the two US cemeteries reveal that those acting on behalf of the US were self-confident, unyielding, determined, and less than nuanced. Deference to Italy's longstanding social structures, specifically, to Italy's system of almost feudal land ownership and the consequent dependency thereon for working/farming class people was not of high priority.[5] Said histories suggest that the US underestimated the slow and complex decision-making travails of the fledgling Italian government, torn between its desires and needs to oblige the US and its duty to defend the interests of its citizens.

The Italian government needed to expropriate lands from heretofore-rightful owners in order for the ABMC to build its military cemeteries. This led to consternation that tested governmental resolve and authority, particularly in the case of the Florence site. Grievances such as loss of livelihood,

3 Declaration of war against Germany, October 13, 1943.

4 The ABMC insisted on the lands of their choice regardless of negative consequences for the owners. See chapter I section on acquisition.

5 *"Mezzadria"* was a complex form of share-cropping, a left-over from feudal times. Families worked on land they did not own and could keep a percentage of the harvest.

for the lands' owners and occupants alike, were legitimate. Clashing styles of behavior and modes of communications exacerbated the situation, which became further complicated by shifts in global power that challenged the US position as the welcome victor. While non-communists looked to the US for leadership, Italy's Communist Party supported the termination of the US military presence, as demanded by Stalin.[6] Although US influence became sharply curtailed as a result of Soviet pressures, revitalization of Italy's economy relied to a great extent on the US "Marshall Plan."[7]

The American Graves Registration Service (AGRS) had been active in Italy since the beginning of the Italian Campaign, in charge of burying soldiers in temporary graves in areas near where they had fallen. It had consolidated the remains of approximately 32,000 in fourteen temporary cemeteries.[8] Its charter to do so stemmed from the 1929 Geneva Convention.[9] During the immediate post-war years, the AGRS's task was to consolidate and rebury remains in permanent cemeteries.[10] Working from the Army's Mediterranean Theater of Operations (MOUSA) headquarters

6 The peace treaty (Part 5, Article 72), with Italy signed in Paris, February 10, 1947 (effective on September 15, 1947) stipulated that the US Army would leave Italy.

7 Carlo Sforza, "Italy, the Marshall Plan and the 'Third Force,'" *Foreign Affairs,* vol. 26, no. 3, (April 1948). Author discusses the attempts to support economic recovery of Europe, including Italy, through the Marshall Plan. Marshall announced this plan (officially "The European Recovery Program" or "ERP") in a speech at the Harvard University Commencement on June 5, 1947. The plan was funded and operated between April 1948 and to the end of 1952. It also benefitted West Germany.

8 Note, Foreign Secretary Zambrini to plenipotentiary Zoppi, Italian Foreign Ministry, dated September 26, 1947. Archive of the Ministry of Foreign Affairs, Rome, Box *Genio Civile*, busta 123. Burial on the field of battle was a tradition that, for America, went back to the Battle of Gettysburg in the US Civil War.

9 "The dead must be disposed of in a respectful manner and their graves respected and properly maintained." First Geneva Convention, Art. 17 (§ 330); Second Geneva Convention, Art. 20 (§ 330); Third Geneva Convention, Art. 120 (§ 330); Fourth Geneva Convention, Art. 130 (§ 330), 1929 Rule 115.

10 Re-established on September 11, 1943 by the War Department, The AGRS outside the continental US (following WW II) was reorganized into area and zone commands. The Quartermaster General (QMG) was in charge of each theater of operations. Field activities were coordinated at the QMG, Memorial Branch (92.8) in Washington D.C. General Horkan was appointed in 1946 as chief of the memorial division (the follow-on to the memorial branch). General Order 125, War Department dated December 29, 1945.

in Rome, the AGRS dutifully and correctly fulfilled its mandate,[11] which necessitated grading and contour map-making.[12] During the war, the US Army had taken possession of sites for temporary burial by authority of the "right of conqueror" doctrine.[13] In the process, buildings were razed, agriculture damaged, and inhabitants displaced.[14] Bitterness lingered as a result of lack of compensation for losses.[15]

A first cemetery was opened at Anzio/Nettuno (near Rome) on January 24, 1944, during hostilities.[16] The AGRS's work was both ongoing and indispensable. It was also in charge of the repatriation of remains (an option that families could request from the US Government).[17] This brings up the question of why so many remains were not repatriated.[18] Why build US cemeteries abroad in Italy and elsewhere? The Italian government evidently did not choose to embrace the fact that the ABMC was motivated

11 G. C. Marshall and J. A. Ulio, Graves Registration, War Department, pp. 10–63 (January 15, 1945) and General Order 125, War Department dated December 29, 1945. Written and published under Marshall, Army Chief of Staff and Major General Ulio, the AGRS's adjutant general. The AGRS followed the detailed guidelines in this field manual.

12 Contour maps and grading plans were sent from Rome. For source studies of AGRS work, see maps and charts of temporary battlefield cemeteries (1941–45) compiled 1946–51. RG 117, Entry 38, Box 2 (1947–1962).

13 United Nations resolution 3314 formally invalidated this "right of the conqueror" in 1974.

14 The US did not compensate individuals for their land. Article 3, Legislative Decree, No. 88, official Gazette of the Republic of Italy, No. 533, (February 1948). State Archive, Florence, box *Genio Civile*, Cimitero Val di Pera.

15 Letter of protest by land owners. Archive of the Ministry of Foreign Affairs, *Affari Esteri 1950–1957,* busta 123, F.C. P. 2, (November 8, 1948).

16 An AGRS brochure for the dedication of its cemetery at Anzio/Nettuno gives historical data, (March 1948). The brochure informes that: "6,791 United States military personnel and 2,723 German military personnel remains were buried within the wall enclosure." Archive of the Ministry of Foreign Affairs, *Afffari Esteri 1950–1957,* busta 123, F.C P. 2, (1950–57).

17 The Act of May 16, 1946 authorized AGRS to repatriate WW II dead. This provision was terminated on December 31, 1951 (in accordance with an amendment dated August 5, 1947).

18 Ron Robin, *Enclaves of America.* Chapter 5, pp. 109–135. The author theorizes that the ABMC's primary objective was to celebrate the accomplishments of the Armed Forces, and secondarily to commemorate the remains. Hence, it discouraged their repatriation, which was desired by ninty-nine percent of parents. (Through making forms difficult to fill out, persuasion and moral pressure etc.), p. 110.

by ideology, namely to create celebratory monuments to the accomplishments of the Armed Forces (to be further illuminated herein). They assumed that the ABMC's intent was to create simple burial grounds and imagined that only those families of Italian origin might want the remains of their sons to be buried in Italy.[19]

In February of 1946 the Army's Quartermaster General (QMG) issued a document directing the AGRS to collaborate with the non-military ABMC in the construction of permanent memorial cemeteries abroad.[20] In July, their respective responsibilities were defined, and accepted by the ABMC by formal resolution on November 19th of that year.[21] The AGRS, in turn, acknowledged a working relationship with the ABMC: "The responsibility for permanent designs and etc. is a responsibility of the ABMC. However, paragraph 20 states that the war department will make all necessary burials and reburials of WW II in permanent cemeteries overseas."[22] The AGRS would follow ABMC's directives and plans, to be provided by ABMC's supervising architect, John Harbeson. AGRS personnel would work out the details and carry them through with their own resources. In reality, the burden to develop plans for grave fields remained on AGRS's shoulders. "Since your [ABMC] funds have not been allotted, the plans, as you know had to be developed to date by our own forces for the European cemeteries."[23]

General George Horkan wrote this letter because a full year after the original directive, the AGRS still lacked detailed ABMC plans. Horkan reminded North of their shared assignment: "It is requested that you have your architect work up a final grave-by-grave layout plan on this cemetery to be followed by grave registration personnel (AGRS) making final burials."[24] At that point in time, Harbeson had submitted preliminary artistic sketches, but no plans.

Harbeson prepared a revised architectural sketch on June 27, 1947, which shows an additional area to be "purchased" for the Anzio/Nettuno site. It would accommodate the 8,700 plus an additional 1,600 bodies. The

19 Archive of the Ministry of Foreign Affairs, *Affari Esteri* 1950–1957, busta 123, F. A.

20 QMG's General Horkan in a letter to North reiterated this delineation of responsibilities. RG 117, Entry 38, Boxes 2 and 156, (April 7, 1946).

21 Ibid.

22 Ibid. Horkan makes it clear to North that having burial plans ready for September 1947 reburial was of paramount importance.

23 Ibid.

24 Ibid.

addition included a pond and a lake in the low areas, along the small stream "following the suggestion of the AGRS," but it did not provide grave-by-grave detail. He defended himself: "If the Mediterranean Theatre had a group comparable to those working with Major Fry in the European Theatre (Paris office), this sketch would be sufficient to allow them to make a usable layout of graves, area spaces sufficient to permit the commencement of the re-interment program."[25]

Of course, North knew that architectural plans were urgently needed but argued that the fluctuating numbers of remains to be buried delayed the making of such plans. He suggested to Harbeson that they go together to Anzio/Nettuno.[26] In another letter to North (answering North's letter of April 10, 1947), Horkan's frustration seems palpable. He was in charge of a situation that was out of control, disorganized, and needing to be set aright. Reading between the lines, Horkan was asking North whether his AGRS should not step in, work out the details, and produce plans based on Harbeson's sketches using their own architectural resources. With ABMC approval, plans could be ready by September 1947, the target date to begin reburials.[27]

It is obvious that the ABMC did not wish to relinquish control over the cemetery design, and not yet having been authorized funding, it demurred (played for time). Diplomatically, North reminded Horkan of the AGRS's important role in the upcoming selection of cemetery sites: "The ABMC works within future jurisdiction upon the selection of which sites the Quartermaster General and Commission are in agreement."[28]

In terse military language Horkan brought up the reality of taking more land: "The AGRS requires the necessary information relative to the amount of land to acquire (a euphemism for taking possession), which information will be used at the proper time as the basis for the acquisition of necessary land. It is proposed to take sufficient land for the cemeteries themselves and to provide a reasonable amount of zoning protections in the peripheral area to assure that no undesirable activities will be conducted adjacent to these permanent cemeteries."[29]

25 Letter, Harbeson to North. RG 117, Entry 13, Box 156, (June 27, 1947).
26 Letter, North to Harbeson. RG 117, Entry 38, Box 2, (July 3, 1947).
27 Letter, Horkan to North. RG 117, Entry 13, Box 156, (May 22, 1947).
28 Letter, North to Horkan. RG 117, Entry 13, Box 156, (May 22, 1947).
29 Letter, Horkan to North. RG 117, Entry 38, Box 2, (April 7, 1947).

The race against time had begun. September was fast approaching. In May, the AGRS in Rome prepared layout plans, but upon viewing them, North took his time to react. He commented to Harbeson that they "… differ materially from the proposal in your report."[30] Harbeson answered North that he had reviewed the plans and made recommendations, but had not acted further because he expected an architect would be appointed shortly. He was busy preparing his list of recommended architects for North, but for budgetary reasons, individual projects could not be assigned to individual architects until the next fiscal year.[31] Horkan let it be known that engineers and architects at the ABMC's Paris office had developed preliminary plans, and their staff should be used in the future.[32]

The September target date seemed ever more unreasonable. The QMC postponed commencing its reburial until February 1948,[33] and Horkan continued to "push for" the grave layout plan—or at least to be assured there would be one.[34] The AGRS had prepared its own contour map in Paris but lacked sufficient manpower in Rome to prepare a grave-by-grave layout plan. Remains were due to arrive from temporary cemeteries, including Sicily, Latio, and Rome, and Horkan had reminded North that unless there would be plans for February 1948, they would have to be stored above ground.[35] North responded on the same day he received Horkan's communication (May 22) and assured him that they could provide a plan for Anzio/Nettuno before the February 1948 target date, as well as plans for all other sites.[36]

The ABMC had selected its architects by the end of 1947, but detailed grave-by-grave plans were lacking. The AGRS proceeded to take into possession additional lands and carried out its own re-burial based on a grid pattern. Reburials were virtually completed and the enlarged cemetery was dedicated in March 1948. In a telegram of congratulation, Italy's minister, the

30 Letter, North to Harbeson. RG 117, Entry 38, Box 2, (May 1, 1947).

31 Ibid.

32 Letter, Horkan to North: "You would be very happy to consider absorbing this organization for the future development of buildings and permanent structures. All of these people have been carefully selected, have done an excellent job, and are very much interested in continuing on the job until it is finally completed." RG 117, Entry 38, Box 2, (5952/53) (May 22, 1947).

33 Letter, Horkan to North. RG 117, Entry 38, Box 156, (May 13, 1947).

34 Letter, Horkan to North. RG 117, Entry 38, Box 2, (May 22, 1947).

35 Letter, Horkan to North. RG 117, Entry 38, Box 2, (May 13, 1947).

36 Letter, North to Horkan. RG 117, Entry 38, Box 2, (May 22, 1947).

eminent politician Alcide De Gasperi, to US Ambassador James C. Dunn, praised the sacrifice of the US for peace and freedom. Referring to the land as "*terra italiana gloriosa*" (glorious Italian lands) he reminded Dunn that the US victory was achieved with the assistance of Italian soldiers.[37]

Not before 1949 were overseas cemeteries officially transferred to the ABMC for "construction and maintenance" by presidential executive order. On December 16, 1949, there was a ceremony of transfer at the Anzio/Nettuno cemetery. Standing before rows of wooden crosses, the AGRS's Colonel Watson and the ABMC's Colonel Walker delivered speeches.[38] The transfer was "sanctified" with blessings from Roman Catholic, Protestant, and Jewish clergy. Shortly thereafter, the ABMC issued its directive to the AGRS to raze the cemetery, including its surrounding wall, and to commence planning for the construction of a completely new one, a larger, more imposing cemetery. Additional land that was needed was not in yet in US possession, but the ABMC went ahead with planning as if it were. To the ABMC's surprise, a dramatic struggle lay ahead.

Fig. 3. AGRS cemetery. Title page of brochure with photos and historical data published by the AGRS, Rome.

---

37 Telegram, Alcide De Gasperi to Ambassador Dunn. Archive of the Ministry of Foreign Affairs, *Affari Esteri 1950–57,* busta 123, F.C. P. 2. (March 13, 1948).

38 For photos of the ceremony see RG 117, Entry 9, Folder 687, (1949).

The Siting Process

"Our permanent cemeteries should reflect by their size and location, at least in a general way, the scope and progress of the operations of our Armed Forces."
General Thomas North, ABMC[1]

Locating cemetery sites (herein called the "siting process") and laying out graves was a collaborative, albeit somewhat disjointed, effort of the Army's Office of the Quartermaster General (QMG) of which the AGRS was a part, as well as the ABMC. This well-archived process offers insights into the visions that underlay their decision making. On the whole, the AGRS was considerably more modest where land use was concerned, requiring only enough for cemetery plots and some peripheral area for zoning protection.[2]

In the summer of 1946, Harbeson, the ABMC's supervising architect, traveled to Europe. He examined existing WWI cemeteries and monuments in France in order to establish new design guidelines for WW II memorial cemeteries to be built.[3] He and General North, the ABMC's secretary, met in Versailles,[4] and then again in Italy after Harbeson had inspected temporary cemeteries there and found only the cemetery at Anzio/ Nettuno suitable to retain.[5]

From the perspective of the newly formed Italian Republic, the presence of US cemeteries on Italian soil was a murky issue, definitely not yet to be considered a "*fait accompli*." On September 9, 1946, the Italian embassy in Washington wrote to Italy's plenipotentiary, Vittorio Zoppi,

1 Letter, North to QMG's General Larkin. RG 117, Entry 13, Box 156, (February 21, 1947).

2 AGRS claimed only the land needed for burial proper. (See Figure 5). Photo shows that the cemetery did not use additional lands. RG 117, Entry 9, Box 72.

3 Ron Robin, "A Foothold in Europe." See also ABMC report of 1946, Survey in Europe, from August through September. (Cannot be located in the Archives at the time of this writing. Robin, gives NARA RG 66 without further reference).

4 Letter, Harbeson to North. RG 117, Entry 13, Boxes 2, and 156, (February 8, 1947).

5 Harbeson visited Italy in the summer of 1946, (mentioned in his report from Sept. 11, 1947). This visit can be corroborated through a letter from North to General Thomas Larkin, QMG, February 21, 1947. RG 117, Entry 13 Box 156.

(later foreign secretary from 1948 to 1954) reminding him that France had accorded the US use of military cemetery sites in perpetuity and he would recommend that Italy follow suit. He reminded Zoppi that a permanent cemetery at Anzio/Nettuno might result in a wellspring of tourists due to Americans' sentimental leanings (*late sentimentale del carattere Americano*).[6]

On January 21, 1947, the ABMC met in Washington, DC, to discuss cemeteries in Italy.[7] The temporary cemetery at Castelfiorentino, in the Florence region where bodies from fighting along the Gothic Line were interred, had flooding problems.[8] Sicily was not easily accessible. Paestum and Mirandola (Tuscany) would remain under consideration. It was recommended that Anzio/Nettuno be retained because it sat upon an actual battlefield, an important "sentimental" consideration,[9] and that a second cemetery in Italy might be needed.[10] A search for a site in the vicinity of Florence would be undertaken,[11] with local AGRS assistance.

North, in a letter to the QMG's office, revealed a vision that should guide site selection: "Our permanent overseas cemeteries should reflect by their size and location, at least in a general way, the scope and progress of the operations of our Armed Forces." North's recommendations also reveal the importance that it ascribed to the psychological effect upon "alien"

6 Letter, Italian Embassy in Washington to Vittorio Zoppi, *Plenipotenziario*. The diplomat obviously felt a bit upstaged by the generosity of their "Latin cousins" and suggested that Italy follow suit and do as France did: donate land near Anzio. Archive of the Ministry of Foreign Affairs, *Affari Esteri 1950–57,* busta 122, F.B., (September 9, 1946).

7 The ABMC met under the leadership of General Markey and decided which cemeteries to eliminate in order to make recommendations to the QMC in Italy. RG 117, Entry 38, Box 156, (January 21, 1947). See letter from North to General Larkin, QMG, (February 13, 1947) about this meeting in which he urges co-operation between the agencies in the selection process. RG 117, Entry 38, Box 156, (February 13, 1947).

8 Harbeson, in his report, dismisses the existing burial ground at Castelfiorentino because it was prone to flooding. He made sure to mention that AGRS experts had accompanied him. RG 117, Entry 9, Box 32, (September 11, 1947).

9 For that reason, Harbeson agreed to the Anzio/Nettuno site and by that time already had made a preliminary map sketch mentioned in letter from Harbeson to Gugler. RG 117 Entry, 9 Box 122_0082.pdf, (June 25, 1948).

10 Letter, North to QMG General Larkin evidences that members of the Commission (ABMC), its consulting architect, and its secretary visited temporary cemeteries. RG 117, Entry 38, Box 156, (February 21, 1947).

11 Ibid.

peoples of the "evidence of the sacrifices made by Americans that the permanent cemeteries will constitute."[12] This remark is notable for two reasons. It evinces the ABMC's emphasis on celebrating the achievements of the Armed Forces, and it refers to the Italian citizenry living in their own country, which the ABMC wanted to impress, as "aliens." This suggests psychological distance.

The AGRS's preference was for a single cemetery at Anzio/Nettuno. Nevertheless, Colonel Wood, the Mediterranean Theater of Operation's (MTO) chief engineer, had already begun searching for a new site, including the areas near Florence. On April 10, 1947, he dispatched a detailed report to Washington evaluating seven possible sites, designated A through G.[13] His report included an analysis of soil samples, topographical maps, photos, size, descriptions, and opinions about suitability in aesthetic and ideological terms. About site E, later to be chosen, he wrote:

> "A beautiful gently sloping open field fully visible from the highway is backed by the hill mass S. Andrea which is almost theatrical; terrain falls towards river Greve and the highway. Drainage is effective; soil is loam without hard rock; there is an abundance of water. It is in a land-locked valley without distant panorama but withal is lovely and appropriate."[14]

However, in spite of such ideal conditions, Colonel Wood's report recommended against sites E and F because they had not been taken by US troops, noting "…while admirably suited in physical characteristics, they do not conform to sentimental requirements and should be rejected."[15] Apparently, commemoration by building on land captured by Americans was of paramount importance for both the AGRS and AMBC.

Wood's report claims that on site E there were no buildings to be purchased and destroyed, although there were several farms on the site. Apparently, the AGRS imagined a more modest cemetery requiring a smaller

12 Ibid. Parallel impulses to NS and VDK thought, as noted verbally by Dr. Christian Fuhrmeister.

13 Report by chief eng. of MTO, Colonel J.E. Wood, also signed by John C. H. Lee, General, US Army. The selection criteria were: 1) Ownership (the fewer owners the better); 2) Accessibility by highway or road; 3) Views (on the site and from the site); 4) Water availability and little danger of flooding; 5) Topography suitable for burials; 6) Soil (clay to be avoided); 7) Existing buildings (to be purchased and razed); 8) Taken by US troops. RG 117, Entry 9. Box 72, Folder 687, 8–26–47 to 5–31–48 (April 10, 1947).

14 Ibid.

15 Ibid.

piece of land. Indeed, ABMC's vision was to realize a more grandiose, monumental cemetery.

General Lee, MTO's commander, had also visited the sites and expressed his judgment in an annotation at the end of Wood's report, writing that site A was suitable and that its 73 acres would provide for 22,000 graves: "The size of the plot is determined principally, however, by the area required to guarantee proper integrity of the site."[16] He noted that sites in the Florence area should "… not be considered unless the Anzio/Nettuno site is inadequate to needs."[17] This is yet another indication that the AGRS was not in favor of a second cemetery in Italy. The report reached North on May 1, 1947.

At about the same time, ABMC's General Mott traveled from Washington to visit General Lee in Rome and reminded him "… of the functions of the Commission (ABMC) and its relations with AGRS, and that he would send copies of the latest executive orders on the subject."[18] (A gentle slap on the AGRS's wrist.) Lee informed him that Wood, who had commanded a large task force in the Italian Campaign and knew the land, had already undertaken a search, and that a report had been duly submitted. Lee provided Mott with a detailed map of the Florence area sites and ordered Wood to guide Mott to the four sites under consideration, which lay to the north of Florence, near the Futa Pass. Of these, only site A was visible from a highway, but there, water was a problem. Mott wrote, "I think a cemetery should lie where people see it as they go about their usual business and it should not be necessary for a visitor to be obliged to look for it."[19] Hence, interest in the Impruneta site became revived.

A visit to Anzio/Nettuno was also part of this tour.[20] This site fulfilled all requirements: water, soil, accessibility, ease of protection against objectionable encroachments, and the possibility to create a monumental approach from the picturesque village nearby, but not close by. More land could be acquired at reasonable cost "… since all the surroundings are merely farm lands with no houses to buy, and all of it is capable of happy

16 Ibid.

17 Ibid.

18 Letter, Mott to North. Mott represented the ABMC's interests. RG 117, Entry 13, Box 156, (April 15, 1947).

19 Ibid.

20 Ibid. Mott also consulted with AGRS's Colonel King, the Commander of the Registration Commission in Rome (CRC).

treatment by a landscape architect."[21] And, the road from Rome crossed "… the whole length of the flat plain over which our men fought for weeks under withering fire. The beach where they landed is close by. Sentimentally, no place could be more appropriate for commemorating their sacrifice and their gallantry."[22] Also, it would offer the desired "sentimental" reminder to Italians "… as they go about their usual business." Its name should be Anzio rather than Nettuno, because it is better known.[23] General Lee concurred, noting that General Clark agreed as well, for both sentimental and practical reasons.[24]

In Washington, North was pursuing authority and funding, but the outcome of his efforts was as yet uncertain. If he were to be successful, the monumentalizing of Anzio/Nettuno would require undoing much of the AGRS's work and beginning anew with redefining the site's borders, and then new grading and drainage work in order to realize the not-yet-existing plans of a yet-to-be-selected architect. As all this was transpiring, Italy's new government, with its newly aggressive Communist Party, was becoming increasingly defiant, viewing the US less as liberator or as ally, and its military more as an army of occupation. Russia's Stalin insisted on adherence to the Treaty of Paris and the US military was duly ordered to leave Italy. Rome's AGRS personnel were excluded, but individual soldiers were compelled to obtain formal permission to reside and to work in Italy.[25]

The search for a new site near Florence continued, but the requirement that the US Army had fought for the land on which the site was situated presented a problem. Washington-based Colonel Evans (AGRS Memorial Division) suggested that the most important selection factors should be physical characteristics and accessibility, and suggested relaxing said requirement. British South African (SA) forces had captured site E, but US troops had arrived in Florence at approximately the same time. Harbeson agreed with Evans,[26] and with this caveat, sites A and E could

---

21 Ibid. This attitude reveals indifference toward the inhabitants and owners of these lands.

22 Ibid.

23 The ABMC asked Eisenhower to select a proper name. He recommended that they call it "Sicily-Rome American Cemetery" which became the official name. Letter from North to the chief of the memorial division, Colonel Bush. RG 117, Entry 9, Box 112, Folder 687 (8–22–47 to 12–31–48) (February 20, 1948).

24 Ibid.

25 *Permesso di Sojourno.*

26 Letter, Harbeson to North. RG 117, Entry 13, Box 156, (May 19, 1947).

remain under consideration.[27] North made a pencil note on this letter saying the 442nd Infantry had relieved British South African troops (at site E) on August 20 (1944). However, agreement over which of these sites should be chosen was not unanimous.[28]

At Anzio/Nettuno, ABMC's development of a definite grave layout plan needed to accommodate ever-shifting numbers of remains. North wrote to Harbeson that the number of bodies was still uncertain.[29] He estimated that not more than 40 per cent of the bodies would remain in Italy (at least 60 per cent would be shipped home on the request of the family).[30] Harbeson's plan for Anzio/Nettuno reckoned with 5,000 and if "… a little more land to the west and at the N.W. and S.W. corners were taken, a maximum of 6,000 could be accommodated, but the central design feature would need to be narrowed a bit."[31] Harbeson and North would be better able to envision details when visiting the site, but it is obvious that they considered the AGRS's more modest land acquisition to be inadequate: "I think we should plan to acquire whatever is reasonably desirable, but I would not pay particular attention to the suggestion shown on the contoured map for an outer boundary as this was made by someone who has no experience with WWI cemeteries."[32] A joint visit must include representatives of both the ABMC and the AGRS, with someone with authority to make decisions and who "… will endeavor to make our plans coincide with yours."[33]

Thus, another trip was planned, this time with people highly enough placed to make real decisions about the future of the Anzio/Nettuno cemetery and to coordinate its development with that of the future cemetery in the Florence area. In addition to North and Harbeson, ABMC's powerful General Markey decided to go in person, accompanied by fellow ABMC member, retired Congressman Baldwin, and Markley Stevenson, ABMC's consultant on landscape architecture. From the QMG's side, Lieutenant Colonel Donelson was assigned to accompany the group. (The ABMC had

27 Letter, Colonel Evans to North. RG 117, Entry 9, Box 72, (April 29, 1947).

28 The Washington office preferred site E at San Martino to site A at Coli Bruce, which was preferred by the AGRS and the ABMC in Rome.

29 Letter, North to Harbeson. RG 117, Entry 12, Box 156, (July 2, 1947).

30 Ibid.

31 Letter, Harbeson to North. RG 117, Entry 13, Box 156, (July 3, 1947).

32 Ibid.

33 Ibid.

already proposed Smith to the US Commission on Fine Arts [CFA] as the architect for the Florence area's cemetery).[34]

The party arrived in Frankfurt where General Huebner put a large aircraft at their disposal.[35] They flew to Paris on August 9th, and then flew to Rome where General Lee assigned them motor transportation as well as technical support from his QMG operation. They visited potential sites near Florence first, beginning the day after their arrival in Rome, and then returned south to the Anzio/Nettuno burial ground to confirm it should be expanded into a large, permanent memorial cemetery.

The small convoy of vehicles visited five sites north of Florence, and two to the south.[36] Harbeson wrote a report on this scouting trip, diplomatically beginning it by giving credit to General Lee and Colonel Wood for the preliminary surveys.[37] While the area north of Florence, along the fortified "Gothic Line," had witnessed hostilities and destruction, the area south of Florence had been largely spared. Harbeson suggested a cemetery at Fiesole, on the highlands dominating Florence with a view on Brunelleschi's *duomo*, but noted that American and English villa owners might not be pleased by the presence of such a cemetery. Such deference to the wealthy Anglo-American presence is striking, as none such was accorded to the local country folk.

Bitter feuds between Fascist, Republican, and Communist partisans persisted into the post-war era.[38] Italy's fledgling government coalition sought to establish authority as the Communist movement was gaining more support. The US found this disturbing.[39] Immediate post-WW II Italian peasantry was desperately poor. One can only imagine the impact of Americans arriving unannounced, photographing, making measurements,

34 Letter, Harbeson to North. Harbeson informs North that Smith is in place. RG 117, Entry 13, Box 143, Folder 231. 24, (May 21, 1947).

35 Memo no. 115 (unsigned describes the scouting travel). RG 117, Entry 119, Box 72, Folder 687, (October 10, 1947).

36 Report by Harbeson. RG 117, Entry 9, Box 72, (September 11, 1947).

37 Ibid.

38 Tedici Santini, lawyer, was an adolescent during this period. Personal interview, summer 2013, Casteglione (LU) Italy. Georgio Barbaro, partisan in Italy. Interview, in Berkeley, California, 2012.

39 Milton Lehman, "We Visit the Ghosts of Anzio," *Saturday Evening Post* vol. 222, p. 142, (September 24, 1949). Lehman reports that communist politicians in Anzio and Nettuno had spread rumors about the US intentions to build an airport at Anzio and assemble atomic weapons "to fight the imperialist war." ("Fake News" of the day).

walking about, and imagining boundaries while Harbeson sketched. Their evident material abundance and health must have shocked those facing hunger, or living with destroyed property, or needing to provide for barefoot children without running water or electricity. Their mission to decide whose land to confiscate must have caused both fear and envy.

Accessibility favored site E, a cultivated site with plenty of water conveniently located on the State Road linking Florence with Sienna, a road well-traveled and close to Florence. A villa on the crest of the hill bespoke comfortable social privilege. Amid Tuscany's famous rolling hills, vineyards, olive and fruit trees, and eternal stone buildings on the far bank of a gentle river, site E bursts into view as the traveler approached from either direction.

Harbeson appreciated the site's advantages. The burial area would have a slope of about seven percent after some cut and fill and erection of a single retaining wall, and therefore, would not need terracing. It had a slight natural concavity, a pleasant "nestling" character that would create a natural unit, and appropriately, invite a visual main axis. The elevation at the western end of the axis would be a most suitable location for the chapel/memorial, with a forest behind (looking west). The view eastward was framed by hills and in the foreground, a valley with a grassy planar expanse embraced the River Greve on both sides. This water feature bade development into a park-like entry experience, the design of which, as we will see, occupied Smith intensively.

Again in Paris, where Harbeson had written the aforementioned report in which he credited General Lee and Colonel Wood for their preliminary surveys, he made another sketch. The picturesque farm groups shown on this sketch dated August 11 are absent, but the superintendent's home is indicated.[40] In Harbeson's mind, the cemetery was becoming a reality.

James Kellum Smith of the New York architect firm, McKim, Mead & White, was staying at AARome's Villa Aurelia at that time. On August 27, North wrote to him from the ABMC Paris office formally offering the commission to design the Florence memorial cemetery.[41] On the previous day, Donelson had advised General Lee that "Action should be taken by MTOUSA to acquire this land for U.S. military cemetery purposes."[42]

40 RG 117, Entry 9, Box 72, Folder 688, (August 25, 1947).

41 Letter, North to Smith. RG 117, Entry 9, Box 72, (August 26, 1947).

42 Map with red outlines was enclosed, which was agreed upon by the War Department and the ABMC. Survey map of scale 1:500 and data requested should be sent to

Smith then visited site E and concurred that it was what he wanted. Back in Washington on October 20, 1947, in a meeting that reportedly lasted only fifty seconds, the ABMC officially decided to take possession of site E.[43]

There is no archival indication that either North or the AGRS endeavored to grasp the implications of land acquisition (although the AGRS in Rome seemed to have gradually become more sympathetic to local Italian concerns). Expropriation likely meant that farmers would be forced to join the urban poor, and that the Italian government would need to confront landowners as a baronial class.[44] After lengthy and scrupulous consensus building, US representatives single-mindedly pursued their objective, seemingly unaware of the landowning class's power in Italian society.

Commanding General, American Graves Registration Command, European Area, APO 58, US Army with a copy to ABMC's European office, 2 Avenue Gabriel, Paris, 8th district. RG 117, Entry 9, Box 72, (August 26, 1947).

43 Abstract, ABMC meeting. RG 117, Entry 9, Box 72, Folder 687, (8–26–47 to 5–31–48), (October 20, 1947).

44 A study tracing the fates of these displaced people would be interesting. Italy soon experienced an industrial boom that changed the economic situation for many. The new government also forbade the tenant farmer system known as "*medrazzia*."

Site Acquisition at Anzio/Nettuno

> "The pity for the dead must not suffocate the pity for the living which is our duty; the memory of the dead for their country cannot be honored if the very needs of modest farmers, small landowners, share-croppers, and farm-laborers' livelihoods are sacrificed."
>
> Professore Dottore G. Scarpitti[1]

Following the precedent set at the Battle of Gettysburg, US soldiers killed in the Anzio/Nettuno battles were buried near where they had fallen. As Rome was captured and the battlefront moved north to the fortified Apennine Mountains (Gothic Line), about 17 hectares (42 acres, 0.17 km^2) were taken at the Anzio/Nettuno site. This resulted in the destruction of buildings and lands under cultivation. Properties destroyed included an agricultural colony for children and its school building that was under construction with one story finished. Archives contain letters of protest. One such from a landowner describes "unspeakable damage,"[2] "... prospering vineyards, stone quarries, all walls and fences, many forest trees, the railroad's water main, all destroyed,"[3] and "... materials to level ground just simply taken from nearby quarries, beyond the limits of the occupied zone."[4] The AGRS retained this land after the war's end and used it for consolidating and re-burying remains gathered from all of southern Italy.

France had gifted land to the US for funerary use in perpetuity by 1946. (Other countries made separate arrangements for the US foreign

---

1 Letter from Professore Dr. Guiseppe Scarpitti, principal landowner, to the Ministry of Public Works. Archive of the Ministry of Foreign Affairs, *Affari Esteri 1950–1957,* busta 123 F. C. P. 2, (April 7, 1948). His estate was almost completely occupied. See fig. 4.

2 Letter from Scarpitti forwarded on November 8, 1948 to the LL PP, to other Italian Ministries, to US Ambassador and to AGRS. Scarpitti owned an experimental farm, whose land was to be confiscated for the ABMC's enlarged cemetery. He described past military requisition of land and its dire consequences. Archive of the Ministry of Foreign Affairs, *Affari Esteri 1950–1957,* busta 123, F.C. P. 2, (Undated).

3 Ibid.

4 Ibid.

cemeteries.) There were no US cemeteries in Germany proper or to Germany's east. In Italy, it was understood that the US would pay for further land acquisition,[5] but during the immediate post-war period, the Italian Government decrees give an unclear picture as to exactly what was intended. The US was granted the right to erect a monumental cemetery at Anzio/Nettuno at the expense of "communities" and of the Ministry of Public Works (*Ministero dei Lavori Publici*) (LL PP),[6] but there was also mention of reimbursement for ancillary costs and/or expenses associated with such, as well as for other activities.[7]

As the US military's occupation mission in Italy would end in October 1947, the Foreign Ministry informed various other ministries that "military requisition" authority would end on December 16th of that year. Henceforth, AGRS personnel would have civilian status and all subsequent issues regarding expropriation by prefectorial decree (*esproprio, decreto, prefettizio*) would be under the purview of the LL PP.[8] The AGRS was given assurance that its Anzio/Nettuno cemetery would be permanent. An agreement on the matter was joined in Washington, and a decree was issued affirming that

5 Letter from AGRS to Signore Zamboni at the Foreign Ministry. Archive of the Ministry of Foreign Affairs, *Affari Esteri 1950–1957,* busta 123, F.A., (October 9, 1947).

6 Decree saying that the establishment of cemeteries would be the responsibility of the communities and the LL PP. The document specifies that the latter will also bear costs (excluding maintenance). But confusingly, the following US documents reveal that the US was planning to reimburse the Italian government for acquisition costs. Italian State Archive, Rome, busta *Cimiteri Americani* (July 5, 1945).

7 Letter from Colonel George King, Quartermaster General (QMC), Rome, to Minister Zamboni, Ministero degli Affari Esteri. "Adequate and just compensation will be paid for all on a consolidated basis in dollar terms directly to the Italian Government in a method to be agreed upon in all its particulars. Such an arrangement would be of advantage to the Italian Government as it would receive directly any dollar credits involved and could pay any private owners in lire in accordance with its own laws and at the same time would be of advantage to AGRS-WZ." This decree was probably not executed. See fn. 9. Archive of the Ministry of Foreign Affairs, *Affari Esteri 1950–1957,* busta 123, F.A., (October 9, 1947).

8 Note from Foreign Ministry signed by Minister Zoppi to various ministries that determined the civil status of the AGRS. Its members would need identity cards and meet certain qualifications to receive a work permit. They would be allotted their quarters from the Foreign Ministry. Archive of the Ministry of Foreign Affairs, *Affari Esteri 1950–1957,* busta 123, F.A., (October 20, 1947).

Italy would give lands without cost as a token of friendship.[9] (The political background for this decision would make for an interesting study.)

In 1947, in accordance with the original Foreign Ministry decree of July 1945, the Italian government reaffirmed its intention to compensate landowners for lands taken by the AGRS.[10] The AGRS's General Horkan, inspired by the good news, wrote from Rome to North that Anzio (adjacent to Anzio/Nettuno) was appropriate historically and geographically, and that the "… people of the community want it (the US memorial cemetery) there."[11]

In January 1948, the AGRS was completing its first phase of construction (dedicated on March 14), and the US requested official confirmation of the cemetery's permanent status.[12] At that time, the US Embassy and Italian government had neither agreement as to specific terms and conditions of permanent occupancy, nor details of additional land that would be needed for the reburial of thousands of bodies. US Ambassador Dunn delivered the US request for clarification of status to the Foreign Ministry. Said request also included a map indicating an additional 47 hectares. The AGRS was planning for grading to begin on June 1, and permanent interments to commence on or about July 1.[13]

Expropriation procedures went forward and the happiness that Horkan had so auspiciously foreseen gave way to stunned consternation. Eight landowners, a Prince Borghese among them, contacted Italian Ministries as well

---

9 Burial Decree for free usage of land by the US worked out in Washington from December 18, 1947 to April 19, 1948 and published by the Italian Foreign Ministry in 1950. It stated that the US from then on would be given lands free "as a token of friendship." Archive of the Ministry of Foreign Affairs, *Affari Esteri 1950–1957,* busta 122 FB, (December 19, 1948).

10 Decree from Minister of Foreign Affairs Zamboni to the LL PP. Archive of the Ministry of Foreign Affairs, *Affari Esteri 1950–1957,* busta 122, F.B., (December 18, 1947—April 19, 1948).

11 Letter, Horkan to North. RG 117, Entry 9, Box 112, Folder 687 (8–22–47 to 31–12–48), (December 29, 1947).

12 The official request for the Anzio land, from Watson, QMG, Rome via the US Ambassador Dunn to the Foreign Ministry per the Bilateral Agreement of November 24, 1946. Attached was a blueprint delineating the boundaries desired by the ABMC. The Embassy forwarded this request on January 19, 1948, together with the blueprint and proposed plan. It showed additional land pieces to be surveyed by the Italian Government and the AGRS. Legal description was to follow based on that survey. RG 117, Entry 9, Box 112, Folder 687, (January 6, 1948).

13 Ibid.

as the US Embassy, outlining their grievances.[14] They had not been informed of the enlargement plans. Without permission, technicians of the Ministry of Public Works (*Genio Civile*) together with American technicians, had trespassed upon their private lands, driven stakes into the ground, and made measurements.

Unlike Impruneta, some landowners at Anzio/Nettuno did reside on their land.[15] The group formed the "Committee Against Expropriation" (*Comitate Difesa Espropriazione*). With reverence, the Committee acknowledged the US need to bury the dead, but lamented that this duty would result in damages falling on their shoulders, and on those of peasants and their families "… that such a binding necessity needs to be reconciled with the consideration of the damages that are implied for the owners and the numerous families of laborers" (… *che tale inderogabile necessita si concili con la considerazione del danno che essa implica per i proprietarie e numerose famiglie di laboratori*). However, the Committee's letter did not prevent the LL PP from invoking an 1885 law. The provincial LL PP informed the petitioners that permission had already been granted (on April 9, 1948).[16] The Committee contended that this procedure was illegal, and the matter was appealed to the LL PP (Ministry of Public Works) in Rome. Thus far, no compensation for the damage suffered during the war had been received, either from the Americans or from the Italian government.

At this point, it began to become clear that the US's ABMC and AGRS were at cross-purposes. AGRS's architect, Lakart Roth, had written that lands that would not be useful should not be appropriated.[17] Nonetheless,

---

14 Letter from Scarpitti, to the Ministry of Public Works. Later this letter, signed by eight owners, would be sent in English to the AGRS. Archive of the Ministry of Foreign Affairs, *Affari Esteri 1950–1957,* busta 123, F.C. P. 2, (April 7, 1948).

15 In this letter, the landowners point out that; ".... Equal payment for just recovery would satisfy the conscience appropriately in recompense for sacrifice for high purpose." (*Ugual appagamento per al guista rialsa che soddisfi la coscienze del sacrificio per il fino altissimo che l'impone*).

16 This may be deduced from contents of letter from LL PP to Ministero degli Affari Esteri. Archive of the Ministry of Foreign Affairs, *Affari Esteri 1948–1958,* busta 123, F.C. P. 2, (May 10, 1948). At about this time the Marshall Plan was established for Italy. This might have influenced the Italian government to act against the interests of the landowning class.

17 Internal AGRS letter from architect Roth. "Some parts not useful should not be appropriated." RG 117, Entry 9, Box 112, Folder 687 (8–22–47 to 12–31–48), (April 4, 1948).

activities began almost immediately. Newly completed cemetery walls were torn down, bodies were exhumed and grading for the expanded memorial project begun. Design issues were already being discussed. Trees were to be planted as soon as possible.[18]

That November, the Italian Foreign Ministry forwarded a letter (in English) to the AGRS. It was written on behalf of the Committee (*Comitate Difesa Espropriazione*) and penned by Professore Scarpitti, owner of the experimental farm allegedly destroyed.[19] He represented himself as an advocate of the families that depended on agriculture and were being displaced. The US sacrifice notwithstanding, his words express bitterness about land having been confiscated in the first place, his experimental farm destroyed, agriculture suspended during the occupation, and now, all precepts of basic fairness breached by the more recent expropriation. "People entering houses without invitation, pulling down farmers' cottages, leveling fields … an all-destroying, leveling roller passing over everything… sudden removal of farmers without shelter, income, or resources … showing indifference and … cynicism." Although the author blamed Italian authorities for this, he did not spare complaints about the way "US personnel entered land, dug and razed farmers' cottages without warning."[20] Scarpitti could not understand the US need for a larger piece of intensely cultivated land since it already had a "monumental" cemetery in place. The negative repercussions "from an economic, social, and political point of view" should outweigh the US urge for a much larger cemetery. "Pity for the dead must not suffocate pity for the living, which is our duty. The memory of those who died for their country cannot be honored if the needs of modest farmers, small landowners, sharecroppers, and farm-laborers' livelihood are sacrificed."[21]

In Scarpitti's judgment, the US had violated social morality, but could redeem itself if it were to resurrect the agricultural colony, complete with school and experimental station, and renounce its intention to erect a larger cemetery. In an articulate expression of Italian sentiments, he wrote: "This would show, in some way, that life comes from death, and

18 Unsigned report from AGRS in Rome. RG 117, Entry 9, Box 122, (May 17, 1948).

19 Scarpitti was the head of the Committee. See also fn.1. Archive of the Ministry of Foreign Affairs, *Affari Esteri 1950–1957,* busta 123, F.C. P. 2, forwarded to AGRS, (November 8, 1948).

20 Ibid.

21 Ibid.

that the brethren's generous blood was not in vain, as from their sacrifice may now rise, following the return of peace, a great activity of love, and work."[22] Following this letter, Scarpitti and other owners secured a letter from the Agricultural Ministry stating that their cultivation of protein-rich peanuts (*arachide*) was of national importance.[23] Further entreaties to the Italian government followed.[24]

The controversy about property rights and boundary lines resulted in a conference in Rome on November 26, 1948.[25] Issues included the railway's existing water main, electric power lines, Scarpitti's experimental farm, maintaining the existing access road (*Via di Tinozzi*) on the north border, and an aesthetic priority of the ABMC: the straightening of the highway (*Strada Nationale*) at the cemetery's entrance (in order to properly set the cemetery). The Committee expressed its shock at the amount of acreage sought by the US, which included land for a road that would encircle the cemetery "like a holy picture."[26] Neither the owners nor the government could understand why the US needed such a huge piece of land (600,000 M^2), equivalent to about two hundred football or soccer fields. They had envisioned 200,000 M^2.[27]

The Committee stated that such excessive expropriation would result in financial ruin for many peasant families. It passionately defended the agricultural farm and their hopes for its survival, as a symbol of brotherhood.[28] The Committee even went so far as to offer a design and suggested that *less* land than intended to be expropriated.

22 Ibid.

23 Letter from the Ministry of Agriculture and Forestry (*Ministero delle Agricolture e delle Foreste*) to Prof. Scarpitti. RG 117, Entry 9, Box 122, (November 10, 1948).

24 Letter from the President of the "*Comitato Difesa Espropriatone Sezione Cimitero Americano di Anzio/Nettuno*" (on its own letterhead) to the Minister of Public Works, LL PP. RG 117, Entry 9, Box 122, (December 1, 1948).

25 Report of ABMC in Rome to ABMC, Washington, DC. The owners, including Prince Borghese, participated in this conference. RG 117, Entry 9, Box 122, Folder 687, (December 2, 1948).

26 Letter from the *Comitato* to LL PP penned by Scarpitti. Addressee not given, but from contents, likely went to the ABMC. RG 117, Entry 9, Box 122, Folder 687, Anzio (12–2–48), (date given as November 1948).

27 Ibid.

28 Ibid.

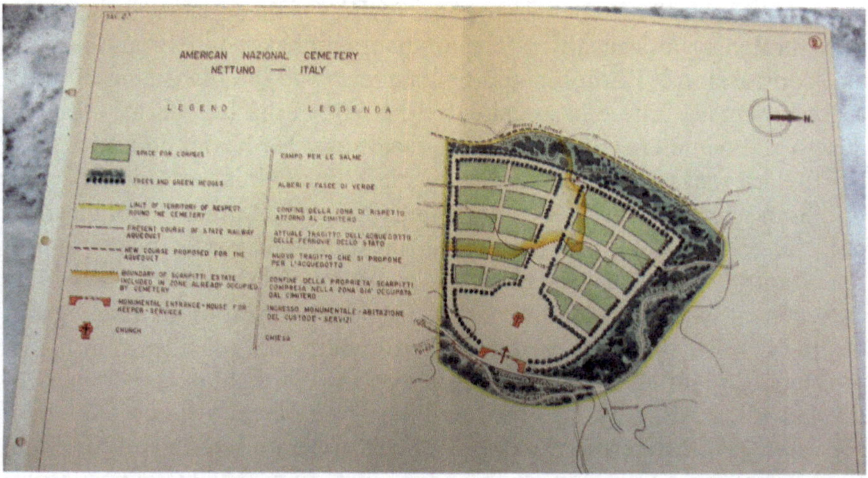

Fig. 4. Design Proposal by Italian land owners superimposed over Gugler's plan of April, 1948.

They sent four plans based on Eric Gugler's GR1 plan[29] and was endorsed by the LL PP in a letter to the AMBC.[30] It would provide space for 15,000 graves and include a church, a semi-circular plaza, an ample entrance portal with two wings, a great central alley, four side alleys and a number of transverse alleys. Around a great alley with trees, beyond the road, a free area would protect the grave areas with silence and shade, as a wood covering a beautiful garden. It was offered with reverence: "This could be a construction really worthy of the piety of the great nation that builds it."[31] The design took the form of a cross and had a prominent chapel. A huge sculpture in marble showing the "Descent for the Cross" by prominent sculpture Ettore Ximenes (who had many public works in Rome to

29 Gugler's Plan 1-A dates from April 27, 1948. RG 117, Entry 9, Box 112, Folder 687 (8–22–47 to 12–31–48). The committee sent four plans. RG 117, Entry 9, Box 112, Folder 687 (11–1–55 to 4–5–56). One shows existing occupied land, the second the additional requested land. Two offer a design, one of which shows a small strip of land to the north, less than the ABMC requisitioned. The second plan (above) is without said extra land to the west. The two first plans show that Prof. Scarpitti had lost most of his estate from the very beginning in 1944 and the demand for more land in Gugler's plan would take even more of his estate.

30 Letter from LL PP to ABMC. RG 117, Entry 9, Box 122 Folder 687 (8–22–47 to 12–31–48), (December 1, 1948).

31 Ibid.

his name) was offered for sale to the ABMC. Of fine quality, it displayed in naturalist style a theatrical Catholicism emphasizing extreme passion and compassion.[32] Harbeson and Gugler reviewed the plans and with irony commented, "There is a desire, on the part of the Italians this time, to design the cemetery." With a note of warning, Harbeson commented to North, "Eternal vigilance is necessary until the work is carried out."[33]

Some gesture was needed to assuage Italian sentiments. Compromising, the ABMC gave in where it did not hurt.[34] It would bear the cost for aligning of the country road "... to fit architect's plans." The Italian Government was obliged to bear the costs for removal of an existing water main: "The Italian Government had undertaken consideration of the prospect of granting use of the site in perpetuity, free of encumbrances."[35] Basically, the ABMC got its wishes for a large parcel of additional land. Nevertheless, Professore Scarpitti's efforts were not completely in vain. Boundaries were retracted on the north side, such that his experimental farm could continue functioning.

By February 1949 the Committee had resigned itself to the inevitable, and thanked the ABMC for their willingness to make adjustments. It acknowledged the "noble manner of US burial." Thus, the Committee saved face, displaying traditional Italian graciousness, "always to appear gracious in public" (*sempre la bella figura*). "I wish to forward the appreciation of the committee for your kindness and to renew the assurances of our great admiration for the noble manner in which you commemorate the heroic American soldiers who found their death in Italy."[36]

Archives reveal much discussion about a proper name for the enlarged cemetery. The ABMC even consulted with General Eisenhower, then Army chief of staff. His decision was "Sicily-Rome American Cemetery" because this title reflected the fighting areas as well as the cemetery's

32 Letter, AGRS to North saying it would forward photos of the group titled "Pieta" by Ettore Ximenes. North replied that the subject did not fit and only American talent would be used. RG 118. Entry 9. Box 122. Folder 687, Anzio, (April 6, 1948); RG 117, Entry 9, Box 122, Folder 687 Anzio, (April 16, 1948).

33 Letter from Harbeson to Gugler. RG 117, Entry 9, Box 112, Folder 687 (8–22–47 to 12–31–48), (December 2, 1948).

34 Letter from Watson to LL PP General Tinti. RG 117, Entry 9, Box 122, Folder 687, (January 3, 1949).

35 Ibid.

36 Letter from Scarpitti in the name of the *Comitato Difesa Espropriazione* to ABMC's Walker. RG 117, Entry 9, Box 122, Folder 687, (February 16, 1949).

location.[37] Italy's Foreign Ministry formalized the agreement on boundaries, justifying the loss of such a large piece of land with the optimistic rationale that such a monumental cemetery would undoubtedly become a place of pilgrimage for Americans.[38] This became a bittersweet irony. Americans never arrived *en masse*. The cemetery's designers lamented that they might not live long enough to see its landscaping fully evolved, and many intuited that it might quickly pass from memory.[39]

This expropriation saga had a final costly postlude for the Italian government. Before the war, Prof. Scarpitti's wife, Carmelia d'Amelio, had donated the majority of the cemetery's land to a Roman Catholic order for the construction of a farm/colony for orphans. WW II had interrupted these plans. After WW II, this land was expropriated. The Signora revoked her gift and sued the government to contest expropriation. She was granted a substantial settlement.[40]

Fig. 5. Aerial View of AGRS cemetery in cultivated landscape.

37 Letter from North to Colonel Bush, acting chief of the memorial division, QMG, Department of the Army. RG 117, Entry 9, Box 122, Folder 687, (February 20, 1948).

38 *..., in ultima analis, ... destinata a diventire una importante meta di pelligrinaggio di cittandini americani.* Archive of the Ministry of Foreign Affairs, *Affari Esteri 1950–1957,* busta 123, F. C. 1, (November 20, 1949).

39 The designers of Anzio/Nettuno as well as Harbeson expressed that they would not live to see the final prospect of their work.

40 Note on settlement was published by *Stars and Stripes,* Fall issue of 1968.

Site Acquisition at Impruneta

> "We have pointed out to the Italian government the importance attached to an early and satisfactory implementation of the agreement concerning this matter conveyed in the note of December 18, 1947 from the Italian Embassy in Washington to the Department of State."
>
> General George Marshall[1]

By late August 1947 the long search had settled on the Impruneta site. The decision was made to acquire it and the US Army's Mediterranean Theater of Operations (MTO) was ordered to request the Italian government to procure it.[2] As at Anzio/Nettuno, brusque military style met with shock and resistance, richly archived in many exchanges.

Before acquisition became a *fait accompli*, it passed through Italy's Ministry of Foreign Affairs (*Ministero degli Affari Esteri*), the Cabinet (*Gabinetto Consiglio*), the Ministry of the Interior (*Ministero dell'Interno*), the Ministry of Defense, Section Honoring the War Dead (*Ministero della Difesa d'Oneranze dei Caduti della Guerra*), the Ministry of Public Works (*Minstero dei Lavori Pubblici* [LL PP]), General Directorate of Urban Construction and Hygienic Works (*Direzione Generale Edilizia Urbanistica e Opere Igieniche*), the Ministry of Agriculture and Forestry (*Ministero delle Agricoltura e delle Foreste*), the Ministry of Public Education (*Minstero della Pubblica Instruzione*), the Public Works Directorate in Florence (*Genio Civile, Corpo Reale* under the old regime), and finally, Italy's General Directorate of Antiquity and the Fine Arts (*Direzione Generale della Antichità e Belle Arti*). Each added and circulated its own memoranda and arguments based on Italy's long history of precedence and laws that left room for contradictions. Some documents are covered with innumerable stamps. The process on the Italian side was slow, purposely so, and also reveals a hint

---

1 George C. Marshall, "Our War Memorials Abroad: A Faith Kept," *National Geographic*, June 1947, pp. 733–755. "I have always felt that establishment of American War Cemeteries abroad was of great international importance." p. 755. Also, letter from Marshall to North. RG 117, Entry 9, Box 72, (6–1–49 to 3–31–49), (June 25, 1948).

2 Abstract of ABMC meeting. RG 117, Entry 9, Box 72, Folder 687 (8–26–47 to 5–31–48), (October 20, 1947).

of despair, as post-war times promised both social progress and dislocation. Nothing would remain the same.

The letter requisitioning the site included a map that outlined its boundaries, evidence of prior surveying that had bode ill for peasants and landowners (*contadini* and *propretarie*) alike. Powerless and war-weary farmers were traumatized when, without warning or permit of any sort, aircraft appeared overhead and Army surveyors arrived and set to work.[3] This violated the ancient custom of asking for permission (*permesso*) where, among all classes, one must ask permission before entering another's land or dwelling, no matter how humble.

Soon after, the project's architect, James Kellum Smith, fresh from meeting General North in Rome and enthusiastic to begin, appeared at the site to inspect and photograph.[4] His photos show an idyllic, highly cultivated landscape nestling in the amphitheater-like site, with lines of fruit trees.[5] A letter of protest from the Provincial Association of Agriculture to the Government in Rome quickly followed, decrying interruption of work.[6] This was the first salvo in what was to become a veritable "Battle of Expropriation."

From the very beginning, the site's principal landowner, *Avvocato Dottore* Giorgio Bandini, was adamant in resisting Smith's even visiting the site, but Smith, admiring the vineyards and fruit trees, ignored this intention. Instead, he and the ABMC regarded the site as if already secured, and Smith began planning and designing. Back in New York, Smith soon would produce a great number of water-color sketches.[7] Smith asked for a more detailed survey, and this was completed on October 5, 1947,[8] but unbeknownst to Smith et al., the US government's request for this land had yet to be submitted to the Italian Foreign Ministry. An unexpected turn in diplomatic affairs had put the US Army's role in Italy in flux. Under

---

3 Letter, Italian Confederation of Agriculturists to the Ministry of Agriculture and Forestry. Italian State Archives, Rome, Folder *Cimiteri Americani*, (October 23, 1947).

4 Letter, Smith to North. RG 117, Entry 9, Box 72, Folders 0014/, (September 1, 1947).

5 Photos of the site. The New York Historical Society, hereafter NYHS, Architectural Records McKim, Mead & White Collection, circa 1875–1950, PR 42, Box 453, (hereafter referred to as McKim, Mead & White collection).

6 See fn. 3.

7 The NYHS, McKim, Mead & White Collection, roll container NSI N137.

8 Letter from Colonel Ralph Tilley to the AGRS in Rome and to Colonel Donelson, AGRS in Washington DC. RG 117, Entry 9, Box 72, (8–22–47 to 5–31–48), (October 15, 1947).

pressure from the Soviet Union's Stalin, the official peace treaty between the US and Italy of February 10, 1947, requiring the US Army to leave Italy, would be strictly enforced. As a result, US diplomatic officers did not deem the moment auspicious to submit the request.[9]

This diminished the AGRS's resources as its personnel were largely military and needed to be reduced in number.[10] Further their status as plenipotentiary became unclear. Colonel George King wrote to the Foreign Ministry requesting a meeting to discuss acquisition of the land, for which Italy had not yet agreed to pay its owners.[11] A personal visit from King and Colonel Scott (also from the AGRS) to Minister Count Vittorio Zoppi at the Foreign Ministry followed. Zoppi subsequently promulgated, via a memorandum to all respective ministries (October 1947), the Government's position, i.e. to cooperate with the US in the spirit of the bilateral accord of September 26, 1946, but that US military personnel would be reduced to civilian status. In addition, they would have to prove identity and qualification in order to get permission to stay and work in Italy. All issues relating to "expropriation by prefectural decree" (*esproprio, decreto prefettizio*) were to be directed to the Ministry of Public Works (LL PP). Further, the AGRS would be required to reimburse costs and related expenses. And still further, Italian State Police (*carabinieri*) would be assigned to supervise American facilities and installations (*depositi gia custoditi da soldati Americani*).[12] Italy was asserting its independence. Zoppi obviously did not consider Italy to be a defeated and "occupied" nation. In the realm of diplomacy, Italy scored a victory and the US would need to adapt.

9 Report from Donelson to the AGRS. "No action had been taken by the headquarters toward acquisition of this site for the United States. In view of the imminent close-out of this theater and termination of the right of requisition of Italian property for military use, it was not considered politically advisable to place this site under requisition." RG 117, Entry 9, Box 72, (8–22–47 to 5–31–48), (October 15, 1947).

10 Letter, Harbeson to Smith saying that General Lee will be leaving Italy along with Colonel Troland's group of engineers. RG 117, Entry 9, Box 72, (8–22–47 to 5–31–48), (September 9, 1947).

11 Letter, Colonel A. King, AGRS Rome, to Minister Zamboni of the Foreign Ministry. Archive of the Ministry of Foreign Affairs, *Affari Esteri 1950–1957,* busta 123, F.A., (October 9, 1947).

12 Letter from Minister Zoppi from the *Gabinetto Consiglio* (hereafter referred to as Cabinet) to various ministries. Archive of the Ministry of Foreign Affairs, *Affari Esteri 1950–1957,* busta 123, F.A., (October 20, 1947).

Archived communications between Rome and Florence reveal the depth of fear and foreboding on-site surveying had caused. For the peasants (*contadini*), it was sowing season and they wondered whether they should proceed.[13] Whether the land should remain under cultivation was also to become an AGRS concern.[14] Bandini, supported by local and state police, first confronted the surveyors on September 27, 1947. A few days later, Italy's war office dispatched a representative to the site to inform them that the land was to become a cemetery. The farmers apparently did not "get the message," and in late October they forcefully ejected US soldiers from the land. Then, landowners and farmers joined forces (about 30 people) calling themselves the "Confederation of the Farmers" (*Confederazione degli Agricoltori*) and sent a letter of protest to the Ministry of Agriculture.[15] This protest made its way to the War Ministry, office of military graves, and then to the Foreign Ministry.

In September 1947 the AGRS's General Horkan had been blithely optimistic: "If there are no objections to the acquisition of this site, it is our intention to push its acquisition as rapidly as possible."[16] In November he learned that not only had his soldiers been forcefully ejected (thrown off the land),[17] but that aerial photography was no longer permitted, and that any further surveying would be by permission only and this would require at least one month to obtain.[18] Horkan wrote again to North, suggesting that there might be need for only one permanent cemetery in Italy.[19] The AGRS persisted, however, and the required permits were apparently obtained, as

13 Letter, Confederation of Owners to Public Works. Archive of the Ministry of Foreign Affairs, *Affari Esteri 1950–1957,* busta 123 F.C., (Jan. 14, 1948).

14 Impact on agriculture concerned also the AGRS. In the summer of 1948 wheat was being harvested and fruit trees were blooming. Sowing another crop should be avoided. Letter from Whitfield Watson (AGRS Rome) to AGRS Washington. RF 117, Entry 687, Folder Florence 1948, (June 14, 1948).

15 See fn. 3.

16 Note from Horkan (AGRS Rome) to Watson, (AGRS Washington). RG 117, Entry 9, Box 72, Folder 0053, (November 6, 1947).

17 Memorandum from a visitor (signed "Metz" presumably from AGRS Washington) to a Colonel Freeman, civilian liaison of the AGRS-MZ saying that he had heard from a friend in the Italian Foreign Ministry about this humiliating incident. RG 117, Entry 9, Box 72, (9–26–47 to 4–31–48), (October 31, 1947).

18 Telegram, Watson AGRS Rome to AGRS Washington. RG 117, Entry 9, Box 72, (9–26–47 to 4–31–48), (November 9, 1947).

19 Memorandum (undated) from Horkan to North. RG 117, Entry 9, Box 72, (9–26–47 to 4–31–48).

the requested additional surveys arrived in Washington and future aerial photos were promised.

Wrangling continued, at various levels. In an August 1947 letter, the Ministry of Defense had made it clear to the Presidency of the Government's Cabinet (*Presidenza del Consiglio de Ministero*) that the Italian-American accord does, in principle, authorize expropriation and the right to use land in perpetuity. The author of the letter, signed as the *Commissario Generale* (the signature is almost illegible, but it looks like Leonid Nacini), emphasized that they had not failed to remind the AGRS that the Impruneta site was under intensive cultivation, that nothing has been definitively decided and that inspection of other sites was an option.[20] The AGRS's own source also reported that the Impruneta site was prime agricultural land, which was scarce in the region.[21] An owners' (Bandini) hand-written inventory painstakingly listed each tree, its size and yield in figs, cherries, peaches and nuts, the acreage of fields, and the income yielded.[22]

The rumor was "leaked" from the US State Department that the Impruneta site would be obtained.[23] Rome had requested exact boundaries.[24] North was confident.[25] On February 11, 1948, Ambassador Dunn delivered the AGRS's official letter requesting the Impruneta site to Italy's Foreign Ministry[26] with a map enclosed indicating boundary lines

20 State Archives, Rome, Folder *Cimiteri Americani,* (November 20, 1947).

21 Letter, Watson to QMG, Washington forwarded to North on December 29, 1947. He acknowledges the fertility of the land. "Half of it corresponds to what we'd call River Bottom Land back in Georgia." RG 117, Box, Entry 9, Box 72, Folder 687 (8–26–47 to 5–31–48, 8–26–47), (December 19, 1947).

22 State Archives, Florence, *Ufficio del Genio Civile di Firenze* 1948–1965, Box no. 536, (October 1948).

23 Memo that Donelson had word from the state department that the Italian government agreed to acquire the Florence site. Donelson requested written confirmation from the Italian government. RG 117, Entry 9, Box 72, Folder 687 (8–26–47 to 5–31–48), (January 13, 1948).

24 Letter, AGRS MZ, US Army to QMG office, Washington Memorial Division. RG 117, Entry 9, Box 72, Folder 687 (8–26–47 to 5–31–48), (January 7, 1948).

25 Letter, North to Smith asking him to provide the design of boundaries of burial plots in order to avoid delay of acquisition. RG 117, Entry 9, Box 72, Folder 687 (8–26–47 to 5–31–48), (January 20, 1948).

26 Colonel Talbot to Colonel Watson, QMG APO 794, Rome, to QMG Memorial Division, Washington. RG 117, Entry 9, Box 72, Folder 687 (8–26–47 to 5–31–48), (February 11, 1948).

in red.[27] Cabinet undersecretary Giulio Andreotti (who later would become an influential political figure) stamped receipt of said request. But an impediment to progress had occurred when Italy's Cabinet informed the Ministry of Defense that surveying had again alarmed the farmers (here upgraded to "*agricoltori*").[28] Bandini had complained again about the unannounced trespass. Andreotti, a master tactician, responded that together with the Foreign Ministry, he would argue to the AGRS that such rich land used as a cemetery would materially harm the nation's agricultural production.[29]

Bandini's complaints began with the statement that he liked Americans very much.[30] That said, other land holdings he owned had been under US military requisition for a number of years and he had received no reparations for damages. He had invested personally to restore these properties.[31] In addition, he resented how the Americans had entered his land without formal permission. He defended the *mezzadria* (sharecropping) system, (soon to be forbidden by the Italian government), and vowed not to sell his property (called the Villa S. Andrea located at S. Casciano, Val di Pesa) at any price. The AGRS thought he might consider selling were it not for his peasants, who knew his diminished estate could not support them and who according to the report, had begun to suspect that Bandini might be prepared to bargain.

At this point, the Government in Rome became more engaged. The official US request having been tendered, it behooved Rome to respond. Rome's Ministry of Labor took up the peasants' cause and sought support

27 Letter, AGRS in Rome to Ambassador Dunn. The map sent could well be an undated survey map. RG 117, Entry 9, Box 72, Folder 687 (8–26–47 to 5–31–48); (February 11, 1948); Archive of the Ministry of Foreign Affairs, *Affari Esteri 1950–1957,* busta 123, F.B., Folder B.

28 Letter, *Gabinetto* of the *Presidio* of *Consiglio* to the Ministry of Defense, to the *Commisario Generale Onoranze ai Caduti,* (January 30, 1948). State Archives, Rome, *Cimiteri Americani, 1957.*

29 Letter, Andreotti to the Ministry of Agriculture and Forestry. Archive of the Ministry of Foreign Affairs, *Affari Esteri 1950–1957,* busta 122 F.A., (March 2, 1948).

30 Letter, AGRS's commanding officer to AGRS's architect Roth. RG 117, Entry 0105, Box 72, Folder 687 (8–26–47 to 5–31–48), (March 8, 1948).

31 Mentioned in Report by AGRS's Carl Hunter, 1st Lieutenant, Inf. Rl. Est. Off. RG 117, Entry 0105, Box 72, Folder 687 (8–26–47 to 5–31–48), (February 2, 1948).

from the Ministry of Defense.[32] The Foreign Ministry needed to honor the Bilateral Accord of 1946, but the Ministries of Agriculture and Forestry begged the Cabinet to intervene on behalf of Bandini and the peasants.[33] The Cabinet sought to do so, but found no ally at the prefectural level in Florence, where the provincial *Genio Civile*, ostensibly answerable to the LL PP in Rome, echoed their position on the matter, i.e., in favor of expropriation. Its supervisor (capo) was unwilling to overlook the law that served as the legal basis of expropriation (*Expropriazione per Causa di Pubblica Utilita. Legge 25 Guigno 1865 N. 2359 e 18 December 1879 N. 45188 Serie 2*).[34]

Florence rejected the arguments put forth by the various ministries in support of Bandini and the peasants, specifically, that a cemetery would disturb the panoramas of the area,[35] or that the loss of woodlands would result in damage. Florence's capo directed that the values of the land and the structures thereon be assessed.[36] The result was a valuation of about half of its usual tax base, a strong indication that they were preparing the way to pay Bandini the lesser value.[37]

Meanwhile, Bandini had personally undertaken his own search for another site, and claimed that he had located one. The *Genio Civile* compiled names of the site's landowners and sizes of their parcels, and determined the aggregate to be inadequate.[38] Bandini then suggested making a new road, so that at least one of his farms might remain intact.[39] Working through such a request would delay receiving acquisition papers for months.

32 Ministry of War Archives in Rome were not available for research at the time of this writing.

33 Letter, the Ministero degli Affari Esteri to the Presidency of the Cabinet (*Presidenza del Consiglio dei Ministeri-Gabinetto*), (February 10, 1948). State Archives, Rome, *Cimiteri Americani*. busta 1 (27.2.48).

34 State Archives, Florence, *Ufficio del Genio Civile di Firenze* 1948, busta 9, Box 536, (February 27, 1948).

35 Ibid. Law pertaining to "*vincolo panoramico acceso fin dal 1925*," would be violated.

36 Ibid.

37 Letter from *Capo di Genio Civile*, Florence, to the LL PP office in Rome. It says that tax payment had been based on 55,000,000 lire. That sum corresponded to value estimate commissioned by Bandini. State Archives, Florence, *Ufficio del Genio Civile di Firenze* 1948–1965, busta 9, Box 536, (July 23, 1948).

38 Archive of the Ministry of Foreign Affairs, *Affari Esteri 1950–1957*, busta 123, FB, (March 11, 1948).

39 Archive of the Ministry of Foreign Affairs, *Affari Esteri 1950–1957*, busta 123, FB, (March 14, 1948).

Additional impediments were put forth, one after another, likely enabled by the wealthy Bandini family's substantial influence. It was alleged that the presence of the cemetery would be deleterious to Italy's artistic patrimony. The office of superintendent (*Superintendenza*) of monuments in Florence requested the involvement of the Ministry of Public Education's General Directorate for Antiquities and Fine Arts in Rome. A Professor Armando Vené was dispatched to visit the site and rendered his judgment that a cemetery would indeed disturb the beauty of the landscape, which his report described in most poetic and laudatory terms.[40] Florence's *capo* once again became involved, commissioning another such report, which contravened that of Professor Vené, and again supported expropriation, the position taken by his superiors in Rome.[41]

As Bandini seemed determined to use all legal recourses, an AGRS staff member alerted Washington to expect serious delays.[42] Ministry after ministry was drawn in. Perhaps another site should be chosen, Harbeson suggested to North.[43] Smith's designs had been approved by the US's CFA, but Smith, having been a fellow at the American Academy in Rome (AARome), and cognizant of Italian ways and the Bandini family's influence, informed North that Bandini might yet find a way to avoid loss of his land.[44] At yet another visit to the region, ABMC and AGRS personnel stopped to view the site, lauded its beauty, but refrained from entering the property.[45] North, however, remained firmly determined, almost defiant. All US agencies had approved the choice of the site and the Quartermaster General had "empowered" him.[46] Not only would he go to Italy himself,

40 Letter, Prof. Armando Vené to the Ministry of Public Education and General Directorate of Antiquity and Fine Arts. State Archives, Florence, *Ufficio del Genio Civile di Firenze 1948–1965*, Box 536, (April 3, 1948).

41 Archive of the Ministry of Foreign Affairs, *Affari Esteri 1950–1957,* busta 123, FB, (April 23, 48).

42 Extract, ABMC meeting. RG 117, Entry 9, Box 72, (8–26–47 to 5–31–48), (May 24, 1948).

43 Letter, Harbeson to North. RG 117, Entry 9, Box 72, (8–26–47 to 5–31–48), (May 25, 1948).

44 Letter, Smith to North. Smith was planning to spend six months in Rome beginning in July of that year and was counting on the land being secured for him to be able to work on the site. RG 117, Entry 9, Box 72, (8–26–47 to 5–31–48), (May 27, 1948).

45 Report from Rome by the ABMC's Colonel Walker. RG 117, Entry 9, Box 72, (8–26–47 to 5–31–48), (July 1, 1948).

46 Letter marked "confidential', North to Watson. RG 117, Entry 9, Box 72, (8–26–47 to 5–31–48), (July 8, 1948).

but he would also enlist the assistance of Marshall, at that time the US Secretary of State, (and author of the plan bearing his name that was so important to the economic recovery of WW II-ravaged Europe).[47] On June 3rd North sent a memorandum to Marshall exhorting him to "bring to the attention of the Italian authorities the necessity that this site be acquired, and also that permission to possession be granted."[48] Italy, he argued, would receive back at least as much land as it would lose as the result of the US returning its seventeen temporary cemeteries. (The Impruneta site is 70 acres). In addition, marble for the crosses would be ordered from Italian quarries.[49] Marshall sent an official request for acquisition to Italy.[50]

Indeed, Marshall's intervention moved matters forward. A promissory note was given.[51] Watson immediately telegraphed North, who at that time was in Paris cribbing for "the Italian campaign" about the good news.[52] The US Embassy hosted a luncheon, gathering all US personnel. Representatives from the Italian Ministry of Agriculture and Foreign Ministry promised to write a definitive letter acceding to the gifting of the *Val de Pesa* site. The ABMC agreed to all desired changes. Niceties were exchanged, one being the promise to hire all the peasants (*contadini*) displaced by the expropriation, and adding that the American people would be delighted with the news.[53] The Foreign Ministry demanded boundary changes that would maintain productivity at one farm (*podero)*, and that no change be made to the state road.[54] The AGRS drew new maps incorporating said

47 Letter, Harbeson to Smith tells of North seeking assistance from the Secretary of State. RG 117, Entry 9, Box 72 (8–26–47 to 5–31–48), (May 26, 1948).

48 Memorandum, North to General Marshall. RG 117, Entry 9, Box 72, Folder 687 (6–1–48 to 3–31–49), (June 3, 1948).

49 Ibid.

50 Letter on official Department of State letterhead from Marshall to North informing him he wrote to Italy. RG 117, Entry 9, Box 72, Folder 687 (6–1–48 to 3–31–49), (June 25, 1948). Marshall was adamant that the world owed the American nation, through their sacrifice of life, "a little something." *New York Times*, (May 30, 1945) p. 20.

51 A *nota verbale* (promissory *note*), Foreign Ministry to the Ambassador Dunn with English translation and two attached plans. RG 117, Entry 9, Box 72, Folder 687 (6–1–48 to 3–31–49), (July 12, 1948).

52 Telegram, Watson to North. RG 117, Entry 9, Box 72, Folder 687 (6–1–48 to 3–31–49), (July 12, 1948).

53 Minister Zamboni wrote a report on this luncheon meeting. Archive of the Ministry of Foreign Affairs, *Affari Esteri 1950–1957,* busta 123, F.D., (July 19, 1948).

54 A "*Nota Verbale*", the Foreign Ministry that sets down conditions for expropriation. State Archives, Rome, *Cimiteri Americani* (July 10, 1948).

conditions and prepared itself for a presentation to work out the details.[55] The AGRS's Watson informed the US Embassy in Rome about the agreement to Italian conditions and about boundary changes.[56]

Yet happiness was premature. On July 28, 1948, some days after the luncheon, the Ministry of Agriculture again solicited the Cabinet and the Foreign Ministry to appeal to the War Ministry, the main decision-maker on matters concerning war cemeteries.[57] The appeal was to cease and desist the expropriation process on the basis that the land formed a complete organic farming community that would be lost. Three hundred ninety persons lived on this land, and there was also livestock to be considered. WW II-related damage to the land had been restored, and since 1925 the land had been designated to be of "noteworthy public interest" (*notevole interesse publica*). In addition, it was argued that choosing the site had been a unilateral process because Italy's post-WW II government had not been involved. The argument was put forth that there should be a re-choosing in collaboration with Italian authority. Another piece of land should be chosen nearby, by the US and Italy together, possibly outside the monumental area around Florence, unused land near the main road, perhaps the provincial road to Cassia.

Decision-makers and designers were, at this point, gathering in Italy to promulgate the long-awaited cemetery planning in person. Soon General North, ABMC members, and Smith would visit the site, acting as though the expropriation was a done matter. Harbeson, ever watchful, admonished that despite changes in boundaries, to which the US agreed,[58] the axis remain as planned and not be moved to the north.[59] Setting eyes on the site once again, Smith noted that the entrance area should be redesigned, and additional land for it be made available.[60] The request amounted to

55 Letter, Horkan to Charles Shaw (memorial division officer of the QMG) in Washington, DC. RG 117, Entry 9, Box 72, (6–1–48 to 3–31–49), (July 22, 1948).

56 Letter, Watson to the American Embassy in Rome. RG 117, Entry 9, Box 72 (6–1–48 to 3–31–49), (August 16, 1948).

57 Letter, the Ministry of Agriculture and Forestry to the Cabinet and the *Ministery of Foreign Affairs,* State Archives, Rome, *Cimeteri Americani,* (July 28, 1948).

58 Telegram, Shaw to North mentioning boundary changes. RG 117, Entry 9, Box 72 (6–1–48 to 3–31–49), (July 23, 1948).

59 Letter, Harbeson to Shaw. RG 117, Entry 9, Box 72 (6–1–48 to 3–31–49), (July 22, 1948).

60 Report, AGRS Rome office to Harbeson. Walker recommends accepting Italian conditions. RG 117, Entry 9, Box 72 (6–1–48 to 3–31–49), (July 26, 1948).

having additional land expropriated. For this, he gleaned no support at the time, as nobody wanted to endanger an agreement so tenuous. All recommended accepting Italy's conditions and allowing momentum toward expropriation to build until it could no longer be stopped.[61]

In August of 1948, Italy's Foreign Ministry officially accepted the ABMC as an agent of the American government.[62] August 13 found Smith and North in Rome's public works office to express their agreement with the compromise negotiated by AGRS's Watson. They agreed to small changes, and not to expropriate the additional land that Smith wanted for designing an entrance zone between the state road and the river Greve. (The US did eventually secure this land).

At long last, the official expropriation decree was issued on September 18, 1948, and Minister Tupini sent the *Genio Civile* office in Florence an execution with an admonishment to urgency.[63] By September 27, 1948 the Foreign Ministry had confirmed the decree and so informed the Cabinet, the Defense Ministry, the Agricultural Ministry, and the *Prefettura* in Florence, but added that resistance in Florence persisted.[64]

Professor Vené entered the picture once again on behalf of authorities at the provincial level, in Florence. On September 28, he again objected to expropriation of the Impruneta site, this time arguing that it had been chosen unilaterally during the Armistice, without consulting the Italian Administration. Now, with the Armistice period in the past and the Italian government in place, there should be collaboration with Italian authority and another piece of land should be chosen.[65] This same argument had been put forth previously in Rome and had been dismissed. It is noteworthy that the governments in Rome and Florence did not strictly respect one another's purview.

---

61 Letter, Walker to Harbeson and letter, Shaw to Horkan. RG 117, Entry 9, Box 72 (6–1–48 to 3–31–49), (July 26, 1948); RG 117, Entry 9, Box 72, (July 26, 1948).

62 Promissory note, the Ministero degli Affari Esteri to the American Embassy in Rome. Archive of the Ministry of Foreign Affairs, *Affari Esteri 1950–1957*, busta 122, F.B., (August 20, 1948).

63 Decree, LLPP's Minister Tupini. RG 117, Entry 9, Box 72 (6–1–48 to 3–31–49), (September 18, 1948).

64 Letter, the Foreign Ministry to the ministries involved and to the Prefecture in Florence. Florence State Archive, *Ufficio del Genio Civile,* busta 536 Florence, (September 27, 1948).

65 State Archive, Florence, *Ufficio del Genio Civile*, busta 536, (September 28, 1948).

It is also noteworthy that the Italian Civil Code can be contradictory. Although handed a copy of the decree of September 18, 1948, Bandini dispatched yet another appeal on the basis that he had played no part in the site selection process.[66] He argued that his rights and his wellbeing as an Italian citizen had been violated, as had been his trust in the law. Ironically, a law dating from 1928, that held "declaration of urgency" to be illegitimate grounds for expropriation, contravened the 1885 law. In effect, Italy's contradictory body of law, in which he claimed to have no trust, formed the legal basis for his appeal.

Alas for Bandini, his appeal addressed to *Genio Civile* in Florence had little clout in Rome, and the Florentine Prefect issued his the order for expropriation "*Occupati d'Urgenza*" on October 15, 1948.[67] Bandini's demands, appeals, hindrances, and obfuscations seeking to annul expropriation had been officially and finally rejected.[68] On October 18, 1948 the binding decree was issued from Florence (by the *Prefetto di Firenze*) for the construction of an American war cemetery (*Cimitero di Guerra Americano*) by authority of the afore-mentioned 1865 law (*legge 25/6/1865 n. 2359 "Espropriazione per Cause di Pubblica Utilita Considerati Urgenti e Indifferibili"*).[69]

North could now report the success to Smith.[70] Expropriation had happened, and they had the land between the street and the river as well. The right to enter was officially granted to the AGRS on October 20, 1948.[71] The delay that had begun in the spring of 1947 ended in the autumn of 1948 as AGRS surveyors recommenced their work. Yet, Harbeson did not

66 He cites violation of two laws: *Violazione dell art. 3 del. D.L.* Del. 23.6.1918 no. 896 and *Eccesso di potere saviamento di poter in relazione anche all art. 4 del. D.L.* Del. 23.6.1918 no. 896. State Archive, Florence, *Ufficio del Genio Civile,* busta 9, Box 538, (October 13, 1948).

67 State Archive, Florence, *Ufficio del Genio Civile,* Box 536, (1948–1965).

68 Letter, capo, *Genio Civile* office to the LLPP and the General Directorate of Urban Construction and Hygienic Works. State Archive, Florence, *Ufficio del Genio Civile,* busta 9, Box 536, (October 18, 1948).

69 Official Decree. State Archive, Florence, *Ufficio del Genio Civile,* busta 9, Box 356, Folder 8, (October 18, 1948).

70 Letter, North to Smith. RG 117, Entry 9, Box 72 (6–1–48 to 3–31–49), (October 1, 1948).

71 Telegram, Walker to North. RG 117, Entry 9, Box 72 (6–1–48 to 3–31–49), (October 20, 1948).

completely trust the peace: "It will be well to wait until interments are made to be certain that the owners cannot make real trouble."[72]

The LL PP assessed the value of Impruneta's land at 12,216,216 lire in 1950.[73] Other landowners accepted recompense totaling approximately 10,000,000 lire[74] but Bandini, the principal landowner and rebel, refused an offer based on the same per hectare apportionment and eventually accepted an offer of 17,094,000 lire.[75] (Archives in Florence do not reveal accounting procedures). Having "fought the good fight," Bandini adjusted to the inevitable and made the best of the situation. He also cleared vines and olives at his own expense, and recovered a good salvage value.[76]

72 Letter, Harbeson to Smith. RG 117, Entry 9, Box 72 (6–1–48 to 3–31–49), (November 19, 1948).

73 *Genio Civile* estimated this value, (a "far cry" from the 55,000,000 lire on which the tax payment had been based). Letter from *Capo Genio Civile,* Florence to the LLPP office in Rome. State Archive, Florence, *Ufficio del Genio Civile,* busta 9, Box 536, (November 22, 1950); busta 9, Box 536, (July 23, 1948).

74 Letter, the LL PP to *Genio Civile*, Florence. State Archive, Florence, *Ufficio del Genio Civile,* busta 9, Box 536, (December 11, 1959).

75 Compensation to Bandini was published in the *Foglio Annunzi Legali della Provinzia di Firenze Genio Civile 1952. Ufficio del Genio Civile,* State Archive, Florence, busta 8, Box 536, (December 29, 1952).

76 Report by Mr. Townley, AGRS, to the AGRS in Washington, DC. RG 117, Entry 9, Box 72, (6–1–48 to 3–31–49), (December 5, 1948).

II. Germany Organizing for Remembrance: Post-WW II VDK's Mission

"The VDK knows its task. After the plight that two heavy wars have visited upon the German people, it sees its mission as working, in the spirit of anti-fascism, to avoid the misery of future wars."
VDK, Bundesführer Manfred Zimmermann (1945)[1]

The first commemorative soldiers' cemetery in the western world was established at Gettysburg, Pennsylvania in the United States, which was dedicated in 1863.[2] Subsequently, the genre evolved to include works of art and monuments that express national identities. After the "Break of Civilization" (*Zivilisationsbruch*)[3] that the First World War brought about, that cost millions of lives, many nations erected major cemeteries on battlefields in foreign countries.[4] In the victors' art and architecture, one reads nationalist ideology of triumph while German cemeteries were initially mere burial fields.[5] This chapter will introduce the "People's Association

---

1 Quotation from VDK's president Manfred Zimmermann (July 1945), (having been a NSDAP party member, he had to give up his position in 1946 to Wilhelm Ahlhorn).

2 Ron Robin, *Enclaves of America,* p. 31. Author discusses Gettysburg as the first soldiers' cemetery in the western world where "graves of the fallen of all ranks were laid out in in formal patterns to achieve national designs."

3 The term *Zivilisationsbruch*, coined by Dan Diner, refers to the Holocaust, but may also be applied to WWI per a conversation with cultural historian Professor Dr. Gerhard Armanski in August 2014.

4 Erection of soldiers' cemeteries (German in France) began after the war of 1870/71. See Manfred Hettling and Jörg Echternkamp, eds., *Gefallenengedenken im globalen Vergleich, Nationale Tradition, politische Legitimation und Individualisierung der Erinnerung,* (München: Oldenburg, 2013). In 1871 European powers formalized an agreement guaranteeing the "right to eternal rest," meaning that dead soldiers could be buried in the countries where they had fallen. This concept was again iterated in the Versailles Armistice that ended WWI.

5 Gunnar Brands, "From World War I Cemeteries to the Nazi 'Fortresses of the Dead: Architecture, Heroic Landscape and the Quest for National Identity in Germany," pp. 215–265 in *Places of Commemoration: Search for Identity and Landscape Design,* Wolschke-Bulmahn ed., (Dumbarton Oaks, 2001). The article discusses

for the Care of German War Graves" (*Volksbund Deutsche Kriegsgräberfürsorge*, or VDK), exclusively charged with erecting and maintaining military cemeteries, both in Germany and abroad. Founded by Dr. Emmo Eulen *et al.* in 1919 and endorsed by Albert Einstein, Walter Rathenau, Konrad Adenauer, and other notables from different political camps, the VDK transformed these into places of commemoration, bearing an appropriate form and message.[6] Following the Second World War, the so-called "Adenauer Letter" (1954) renewed VDK's mandate to build and care for soldiers' cemeteries abroad.[7] Its president, Wilhelm Ahlhorn, accepted this mandate, calling the organization "custodians of the German people" and committing German soldiers' cemeteries in foreign countries to the cause of peace: "May their death be an admonition for peace" (*Mahnendes Vermächtnis der Toten*).[8]

Research on the VDK's post-WW II activities has tended to focus on the continuity of Nazi thought and personalities. As Manfred Wittig's important critical study proposes, the VDK's post-WW II reluctance to expound on its past may be viewed as intention to salvage (Nazi) or NS-friendly views, and this in turn may have had a negative influence on Germany's emergent post-WW II democratic identity.[9] Per Rolf Wernstedt,

German foreign soldiers' cemeteries after WWI. It traces the relationship between nature, tamed and untamed, and architectural elements, in particular the "Totenburg" and puts them into relation with political shifts during the Weimar Republic and the early Nazi era.

6 Treaty of Versailles, Article 225. Under the War Graves Commission, the VDK was recognized as a Commission, but in reality, it was given little authority after WWI.

7 Adenauer's letter of November 23, 1954, referred to by VDK's president, Ahlhorn. The Politisches Archiv des Auswärtigen Amtes, Berlin, (hereafter AA Archive), Befund 92, 602, (All material relating to War Cemeteries Archive can be found under Befund 92).

8 Ahlhorn responded to Adenauer on December 16, 1954, reminding him that he, Adenauer, had belonged for many years to the administrate council of the VDK (*Verwaltungsrat*) after its founding in 1919 and therefore would know that the VDK's duty was to carry out the legacy of the dead, namely the admonition to peace. ("*Wie Sie aus Ihrer früheren langjährigen Zugehörigkeit im Verwaltungsrat des Volksbundes aus eigener Anschauung wissen, geht es dem Volksbund bei seiner Arbeit nicht nur um die Erfüllung einer selbstverständlichen Pflicht der Pietät, sondern auch um das mahnende Vermächtnis der Toten.*") AA Archive, Befund 92, 602.

9 Manfred Wittig, "Der Tod hat alle Unterschiede ausgelöscht." In *Unglücklich das Land, das Helden nötig hat: Leiden und Sterben in den Kriegsdenkmälern des Ersten und Zweiten Weltkrieges.* Michael Hütt et al., (Marburg: Jonas, 1990), pp. 91–98. Wittig sees the VDK as a lingering rightist organization that in essence did not change its

politician and author (SPD, German Social Democratic Party), because Adenauer endowed the VDK with the mandate to erect and curate cemeteries in foreign countries, his party and its coalition partner (CDU/CSU, Christian Democratic Union/Christian Social Union) saw the VDK as if chartered to communicate the coalition's conservative views.[10] This study traces an evolution toward more democratic tenets, evident in the protocols of the VDK's sessions and publications about aesthetic and iconographic guidelines.

The immediate post WW II period brought the burden of historical trauma overlaid by a sense of new beginnings. The VDK played a curatorial, palliative role during this time of repressed mourning by facilitating private processing of individual catastrophe, giving recognition to each individual soldier's death.[11] Anyone who came of age in those years will likely remember the VDK's annual fundraising campaigns, where schoolchildren were asked to contribute.[12] Every coin was important. Additional funding came from memberships, subscriptions, and other donations (exclusively private), but the amounts raised from schoolchildren were astonishing.[13] A child would ask parents or relatives to provide them with coins, and would sacrifice their own meager pocket money. A child learned that the VDK would find and provide a decent burial for fallen fathers, uncles, and often

---

fundamentally nationalist, speak NS, orientation.In his interpretation, the VDK did not undergo a successful "*Vergangenheitsbewältigung*," (coming to terms with the NS past).

10 Author's interview with Rolf Wernstedt (in Munich, September 2014). See also his "Der Volksbund Deutsche Kriegsgräberfürsorge im Auftrag der Politik" in *Loccumer Protokolle*, ed. Corinna Hauswedelle, 25/08, Evangelische Akademie Loccum, Rehburg Loccum.

11 Werner Fuchs-Heinritz, *Todesbilder der Modernen Gesellschaft*, (Frankfurt: Suhrkamp, 1969), p. 174. The author quotes Leopold Ziegler, "Death, the most powerful founder of societal ties." (*Tod, mächtigester Stifter gesellschaftlicher Bindungen)*. On p. 201 author uses this quote by Leopold Ziegler, *Vom Tod. Essays.* Leipzig, 1937, p. 48.

12 Author's personal recollection. Collections began after the currency reform of 1948.

13 Many small coins made large sums. The cemeteries at Lommel (Belgium) and Cassino/Caira (Italy) became so-called "*Betreuungsfriedhöfe*" (cemeteries under the care) of the German school youth so that they might feel as co-curators. Building costs were shared between the VDK, the German Foreign Ministry, and collections. The latter amounted to DM 1,890,370. See the AA Archive for cost-sharing negotiations with the *Ausswärtiges Amt* on October 1, 1955. Also, session on April 4, 1959, VDK Archive, A.100–857.

brothers whose presence was sorely missed in almost every family. These fundraising campaigns gave the dead and the missing a ritualistic presence. They helped forge a collective identity from the pain, the loss and most poignantly, from the emptiness in which most had a share.[14] In this way, the VDK had a considerable civic influence in post-WW II German society. The yearly collection ritual reminded people that mourning was their moral right. It contributed to the evolution of the new moral national identity that arose from total defeat. New categories and values were formulated.[15] Germany's population learned to live free from fear, a new experience for many.[16]

The VDK's management style differed from that of the US's ABMC. The VDK made decisions internally, largely independent of governmental oversight, and had virtually no military component. It was formed of several committees. For the scope of this study, of primary importance was its Committee of Experts (*Sachverständigenbeirat*). This included architects, landscape architects, designers, and construction professionals who would review designs and make recommendations. Project approval lay with its Directorate (*Vorstand*), and project "go-ahead" with its Chairmanship (*Präsidium*), whose president was VDK's official figurehead. Although headquartered in Kassel, the VDK's artistic center (*Bauleitung*) remained located in Munich under the direction of Robert Tischler until his death in 1959.[17] All the tasks of the US's AGRS, ABMC, and CFA were fulfilled by Germany's VDK.

Italy hosts VDK cemeteries dating back to the post-WWI period. The last to be completed in Italy was dedicated in 1969. Directly following WW II, the VDK was not allowed to build in foreign countries. It concentrated its efforts on soldiers' cemeteries in Germany (in the hundreds) while formulating guidelines for aesthetics and messaging of their future

14 Author's personal recollection.

15 Alex King, *Memorials of the Great War in Britain: The Symbolism and Politics of Remembrance,* (New York: Oxford, 1998), p. 9.

16 Even former Nazi's realized that they had lived with pervasive fear. Experiences shared with author.

17 Consensus building was a new procedure and was welcomed even by old-timers, such as Werner Linder in 1950. Sonderdruck, Der Friedhof 41. Jahrgang, Nummer. 1, (1950) AA Berlin, B 92, 259, vol. 271, p. 1.

cemeteries abroad. By the time the VDK was allowed to re-commence its foreign building activities (in Italy, 1955), the US was completing its two monumental military cemeteries and finalizing their messaging. Visual analysis of VDK cemeteries in Italy reveals underlying attitudes reflecting German societal developments. Beginning with fascist Germany's alliance with Mussolini's Italy in the early 1930s and its emergence as one of Europe's leading democracies in the 1960s, we can observe Germany's ideological and aesthetic development. Post-war cemeteries form a temporal sequence, from Pomezia to Costermano to Caira, to Futa Pass, that reveals progressive distancing from the NS past.

Post-WW II re-democratization was not a seamless process for Germany. The VDK's archives reveal shifts in semantics. For example, there persisted a reference to soldiers as "seed." In 1933, the VDK's Dr. Eulen wrote, "On the day of the uprising, the bloody seed will germinate,"[18] meaning that German soldiers would take arms to avenge the calamity of their WWI defeat. A VDK brochure printed in May of 1960 contains the invocation: "These crosses witness that we have entrusted our precious seed to the earth."[19] Thus, Nazi terms "slip in," but their meanings change.[20]

18 Siegfried Emmo Eulen: ("... *ein Tag der Erhebung werden, ein Tag des Aufgehens der blutigen Saat.*") On Memorial Day 1933, the *Volkstrauertag* (People's Day of Mourning) officially became the Heroes' Day, quoted in; Meinhold Lurz: "...Ein Sück Heimat in Fremder Erde. Die Heldenhaine und Totenburgen des Volkbunds Deutsche Kriegsgräberfürsorge." *Archplus_ausgabe 71.* Aachen, Arch+, 1983, p. 67. The article's title refers to soldiers' cemeteries in foreigns lands being considered equivalent to pieces of homeland, referring to WWI cemeteries as well as to Tischler's post WW II cemeteries. Nature played a principal role. It stood for homeland and heimat in the German mindset. See Mauch, *Nature*, p. 3.

19 ("... *die verteilten Kreuzgrupen verweisen sinnbildich auf die wertvolle Saat symbolisch hin, die hier der Erde anvertraut wurde.*") From a VDK publication on Pomezia. The AA Archive.

20 Letter, Oesterlen to Rossow, uses the terms "*Einsatz*" and "*Umbettungskommando*" in connection with coordinating building progress with interments. AdK Archive, the Dieter-Oesterlen Archive, Ordner –AR-Nr. 74, (February 16, 1961).

The VDK Resumes its Work

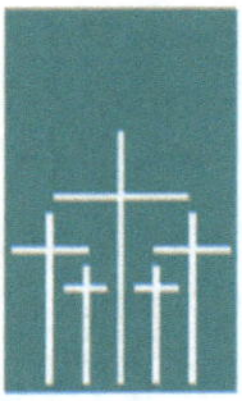

Fig. 6. VDK's Logo. Note design of crosses emphasizes death and suffering without ornamentation.

After WW II, the VDK's staff quickly organized for action. Its personnel had expertise and contacts in foreign countries, and began planning foreign projects before 1950, the year it was allowed to be active in many former enemy lands.[21] There was much to be done and time was of the essence. Inconvenient questions about its past were relegated to the background.[22] The VDK's task at hand was to gather, identify, and rebury fallen soldiers,[23] and to do so in foreign countries, as was its right.[24] This essential

21 In 1946 the VDK published its first "Guidelines" for soldiers' cemeteries in Germany (*Grabstättenordnung des Volksbundes Deutsche Kriegsgräberfürsorge e.V. für das Inland, August 23, 1946*). The VDK Archive, A 10–124. This document gives insight into the immediate post war activities of the VDK. In September 27,1950 "Tasks for the Construction of World War II War Cemeteries in Foreign Countries" followed. ("*Aufgaben der deutschen Kriegsgräberfürsorge im Ausland*"), The VDK Archive, A 100–885.

22 Meinhold Lurz in his indispensable essay "…ein Stück Heimat" interprets the VDK's self-representation as resistors as a distortion of history. He quotes the deputy leader (*stellvertretender Bundesführer*) Zimmermann (July 1945): "… always energetically and successfully resisted" (*… immer energisch und mit Erfolg entgegen getreten*), p. 68.

23 The AA Archive, Group B 92 602–85.20/271 contains information about war graves in foreign countries up to 1972. A list of equipment from the early years needed for the burial of remains in Italy painstakingly gives the kind and numbers, e.g., 20 shovels, 20 spades, plastic sacks, etc.

24 Document GA/Dr./D/V *Bundesgeschäftsstelle* of the VDK mentions the revival of the law (§5) from 1922 titled *Gesetz über die Erhaltung der Kriegsgräber aus dem Weltkrieg*. On September 3, 1952, *Gesetz über die Sorge für die Kriegsgräber*. See Bundesgesetzblatt Teil 1 1952 Nr. 23 from May 28,1952. (The VDK's activities continue, and as of 2010 it had reburied about 800,000 remains in Russia since the fall of the Berlin Wall. (Author's interview with VDK professional in Munich, September 1, 2014).

and unglamorous work represented enormous logistical accomplishments that have been paid very little homage. Their work began by partnering to make lists with the International Red Cross, and corresponding with German soldiers that remained war prisoners in foreign lands. At times antipathy and hostility toward Germany exacerbated this.[25] Particularly, Communist Party adherents harbored animosity against the German (as well as the American) nation. However, Italy's population at large was sympathetic and respectful.

When about 2,000 German bodies buried in the US cemetery at Anzio/Nettuno needed to be removed (per request of the US) the Italian Government provided a new burial ground in nearby Pomezia.[26] The VDK was faced with widely scattered temporary gravesites (about 4,000) but only a few larger cemeteries, such as one erected by prisoners of war at Cervia. Many German soldiers had been buried without identification and hence, lists and records were incomplete.

The new Italian government authorized the VDK to commence activities in 1947. Under the leadership of Baron Caius von Münchhausen (a resident of Florence), a Rome office was established. About 35,000 remains were consolidated, and within a few years, 90 percent of the more than 100,000 German soldiers fallen in Italy had been identified, (reducing the percentage of unidentified from 70 to 10 percent).[27] Such logistical achievements earned the VDK domestic authority.

Authorization to establish collection cemeteries (*Sammelfriedhöfe*) in Italy was an urgent matter since, in accordance with Italian law, burial plots were to be emptied after 10 years and the bones transferred to ossuaries. Foreign soldiers' graves were exempted from the 10-year limit, but rumors circulated that individual communities were ignoring this. Such rumors disquieted bereaved German families, and the VDK petitioned the Italian authorities to intervene and to take a stand on this matter.[28] The

---

25 Fn. 21. The document speaks of "Hatred against everything German." (guidelines 1946)

26 VDK *Pomezia* brochure. Information on an engraved tablet at entrance of the cemetery. Reburial accomplished in the fall of 1947.

27 See fn. 20. With the assistance of prisoners of war and the Italian *Commissariato Generale Straordinario per le Onoranze ai caduti in guerre*, the VDK Rome had identified ca. 105,000 graves in over 4,000 locations. Ca. 35,000 were identified and rescued from having their bones exhumed and deposited in ossuaries.

28 A note from the Italian government to regional governments saying that the funeral law of Italy—burial limited to 10 years—does not apply to buried soldiers in communal cemeteries, no matter what nationality they are. The approximately 100,000

German Embassy in Rome became involved.[29] The VKD proposed that Italy erect simple, low maintenance collection cemeteries (as mandated by the Geneva Convention). Monument building was a different matter. Italy rejected the VDK's requests and appeals until an official accord between Italy and Germany had been consummated (December 1955).

Evolving New Moral Principles

An archived statement from the VDK (1949), with reference to the hundreds of cemeteries under construction in Germany, reads:

> "Germany, after two wars lost, knows how to create true soldiers' cemeteries, since the German conceives of death for his fatherland differently than do other peoples. The [German] soldier, first and foremost, is part of a whole, and thus has renounced much of his personality. The death of these soldiers is tragedy, and awareness of this as tragedy comports an appreciation of their heroism."[30]

This clearly reveals entrenched NS ideology. There is no evidence that this quotation (from Fronenberg) articulated the VDK's official stance, but the themes that German soldiers' individuality is subordinate to his group

---

German remains should be left alone until Germany (just about to be formed) will do something about it at their expense; letter from the Italian Foreign Ministry to the Ministry of Defense confirming that Italy will not permit the disinterment of German remains because it would cause the definite loss of identification. Archive of the Ministry of Foreign Affairs, *Affari Esteri 1950–1957,* busta 186, File A, (July 6, 1949); Archive of the Ministry of Foreign Affairs, *Affari Esteri 1950–1957,* busta 186, File A, (December 16, 1953).

29 An assurance from the Italian Foreign Ministry was sent to the German Embassy in response to their letter of November 13, 1951 to intervene in the disinterment of German remains. Archive of the Ministry of Foreign Affairs, *Affari Esteri 1950–1957,* busta 186 File A, (December 18, 1953).

30 From report on a meeting of the VDK with Robert Tischler in Munich. *("Deutschland, als erstes nach zwei verlorenen Kriegen schaffe echte Soldatenfriedhöfe. Der Grund dafür läge in der Tatsache, dass der Deutsche den Tod für sein Vaterland anders auffasse als alle anderen Völker. Der Soldat sei in erster Linie nur Teil eines Gesamten, und er habe auf viele Motive seiner Persönlichkeit verzichten müssen. Über den Tod dieser Soldaten breite sich Tragik, und die Kunde von den Vorgängen trage das Kennzeichen heroischen Geschehens.")* The VDK Archive, A.100–1030, (April 29, 1949).

identity, that individual loss is part of the continuum of historic events not brought about by human actions, and the use of terms such as "peoples" speaks for itself. Taking this as a starting point, the VDK's subsequent memoranda reveal evolving thought. To trace this, we begin with Dr. Werner Lindner, an architect and former NSDAP member who had prepared plans for "Germanized" settlements in Poland following victory, a commission given him by SS Chief Himmler. Lindner's 1950 essay "Caring for War Graves as a Cultural Duty" (*Kriegsgräberfürsorge als Kulturaufgabe*) initiated a process through which the VDK prepared a series of statements articulating those fundamental principles that would guide and hone its official message.[31] This series would span the next decade. The overriding goal was to fulfill the task of curating the graves of the fallen "as occidental culture's tradition was to honor them," but more important is a new sensibility expressed as such: "… for the German people striving for its moral rehabilitation, the manner in which it does this will, in high measure, contribute to the judgment of foreign nations."[32]

Lindner's 1950 essay exudes NS vernacular and notions, such as veneration of the irrational over rational and terms such as "the people's feeling" (*Volksempfinden*), "holy" (*heilig*) and "divinely gifted artists" (*Gottbegnadeter Künstler*),[33] and "…the eternal becoming and decaying of the living." The NS notion of soldiers' burial ground being a sacred realm

31 Werner Lindner, "Kriegsgräberfürsorge als Kulturaufgabe," *Der Friedhof*, 41. Year. Nr. 1. (1950). A copy can be found at the AA Archiv, Befund 92, 602–85.20 Band 271. Of note: Lindner, as well as others, such as v.Lutzau, member of VDK's directorate, continued to work for the VDK, even though both had worked for the NS regime and endorsed its ideology about soldiers' death. They belonged to the functionary elite that, because of their expertise, were able to occupy an important place in Post War German society. Most criticism of the VDK focuses on this continuity. (See fn. 9). Author takes this fact as an opportunity to measure gradual ideological changes within the VDK as evidenced in their various *richtlinien*.

32 VDK Guidelines from September 27, 1950. The VDK Archiv, A. 10–124, *("Für das deutsche Volk, das sich um seine moralische Rehabilitierung bemüht, ist die Art und Weise, in der es dieser Aufgabe* [curating the graves of the fallen] *gerecht wird, von um so größerer Bedeutung, als sie in hohem Maße zur Urteilsbildung der fremden Nationen beitragen wird.")*

33 Hitler had a list of "divinely gifted artists" (*begnadete Künstler*), and to be made part of it was the highest honor. In 1944 the *"Gottbegnadeten Liste"* listed 41 artists. <http://de.wikipedia.org/wiki/Gottbegnadeten-List>. Last accessed March 29, 2018.

(*heiliger Bereich*) flows from such ideas.[34] In the same document, within the web of past notions and values, he admits how difficult it is to free himself from the NS mindset: "We come from many decades of the most primitive non-culture, and most of us are, without noticing it and others suspecting it, still possessed by its evil spirit."[35] Lindner struggled with finding a new meaning in soldier's death, but felt more secure in his rejection of celebratory rituals, and his dismission of axial designs, elaborate entrances, and triumphal arches in evidence in British and US cemeteries (not specifying which world war), labeling such as "pathetic."[36] He asserted Germany's right to honor its dead with designs that were different from those of the past. German cemeteries must eschew triumph, as there was nothing to "boast" about (*rühmen*), and nothing "heroic" in a historical sense.[37] (Lindner became an important VDK member as a member of the Committee of Experts [*Sachverständigenbeirat*] in 1951).

Germany's first president, Theodor Heuss, in 1952, used the words "admonition to seek peace" (*Mahnung zum Frieden*): "When we stand at the crosses in silence, we hear their calm voices. Take care, you that are yet alive, that peace remains, peace between peoples, peace between nations." Thus, did Heuss articulate the VDK's mission, to use past war to assure a future peace.[38] By 1953 the VDK had firmly memorialized the notion of peace

34 Fuchs-Heinitz, *Todesbilder,* (Frankfurt, 1969), pp. 146–147. The author regards the creation of a sacred area through walls, hedges, and fences as an outcome of atavistic fear to disturb the dead. The dead might rise and take revenge.

35 See fn. 31. *("... ewiges Walten im Werden und Vergehen des Lebendigen."* and "*wir kommen aus langen Jahrzehnten gröbster Unkultur, und die Meisten unter uns sind, ohne dass sie es merken, noch ganz von ihrem Ungeist besessen.")* Linder's terminology testifies to his assessment, but author hesitates to interpret this as sign of a lingering retrograde attitude.

36 Ibid. p. 1. In his *Gestaltung von Kriegsgräberstätten, Hinweise und Richtlinien,* VDK, 1954, p. 5, he rejects axial designs.

37 *("Sechserreihen und mit Pauken und Trompeten ... pathetische Schöpfungen ... leider auch bei Stadtfriedhöfen—auf vertretbare oder unberechtigte Achsenwirkung ausgerichtet, glaubte man unlängst oft mit festlichen Zuwegen, wuchtigen Portalen, Triumphbögen und dergl. wie zum Einmarsch errichten zu müssen. Wir haben hier nicht zu marschieren und zu posaunen, sondern müssen das stille Verweilen in dem heiligen Bezirk auch schon in der Art des Zugangs vorbereiten.")* Sonderdruck of "Der Friedhof" 41. Jahrgang, Nor 1., p, 1. The AA Archive 92, 259, vol. 271.

38 Dedication speech by Theodor Heuss in Hürtgen, 1952, the largest VDK war cemetery in Germany. *("Wenn wir in der Stille an den Kreuzen stehen, vernehmen wir ihre gefasst gewordenen Stimmen: Sorgt ihr, die ihr noch im Leben steht, dass Friede*

with its slogan: "Work for Peace."[39] An interim VDK memorandum reads: "The message must stand above the debate about the war, and emphasize admonition to seek peace."[40] The VDK's continuing quest for meaning was unmistakable. By 1958 it postulated that cemetery design must be adequate to teach future generations the "magnitude of the loss." Explicitly disavowing nationalism, death of such magnitude must not be seen as sacrifice (for a wrong cause), but as tragedy. In short, the VDK attempted to design its foreign cemeteries to message a new Germany concerned with moral renewal. Established architectural forms were re-evaluated for their symbolism. Nature's role, untamed or landscaped, was heftily discussed, and old emblems and symbols were either abandoned, or their meanings redefined.[41] As family became central to the VDK's ideology and style, the object of "honor" was shifted from the collective to the family. The VDK referred to its obligation to ascertain where a family's fallen kin were buried and to honor them, as a hallmark of western civilization. The notion of understanding between peoples (*Völkerverständigung*) began to appear and the concern to be inoffensive toward other nations.[42] It is worthwhile to restate here the VDK's concern: "The German people's efforts toward moral rehabilitation by way of undertaking these obligations will contribute in high measure to esteem by foreign nations."[43] It would result in the VDK's mantra, Admonition for Peace (*Mahnung zum Frienden*) and "reconciliation over the graves" (*Versöhnung über den Gräbern*). The dove of peace became a visual logo.

---

bleibe, Friede zwischen den Menschen, Friede zwischen den Völkern!") The VDK brochure on Hürtgenwarld-Hürtgen–Kriegsgräberstätte.

39 In 1953 a Jesuit priest, Theobald Riedt from Aachen, organized the first international youth camp in Lommel/Belgium and coined these mottoes. The AA Archive, VDK's *Richtlinien für die Gestaltung der Kriegsgräberstätten des 2. Weltkrieges im Ausland,* (1958) B 92 602–85.20/271. The document *Richtlinien* is also available in the VDK Archive, A.10–124_*richtlinien–Gestaltung,* (1958) with commentaries.

40 Ibid. "The message must be beyond a debate about the war and express the admonition to peace." *("Sie [die Botschaft] muß jenseits der Auseinandersersetzung über den Krieg stehen und die Mahnung zum Frieden herausstellen.")*

41 Alex King, *Memorials of the Great War in Britain,* (Oxford: 1998). The author discusses the theory that monuments have no clearly implicit meaning, but it is given to them and that one has to make sure that it is understood. This point is emphasized also in Kossellek and Jeismann, *Der politische Totenkult,* (München: 1994). "Forms remain, meanings depend on time and place."

42 The VDK represented its efforts as building a bridge to the "*Verständigung der Völker.*"_*richtlinien* (1950). The AA Archive, Bestand 92, 602–85.20, 271.

43 Ibid. see fn. 32 for German original quote.

Fig. 7. Dove of peace. Mosaic, Costermano.

Of NS notions, "camaraderie" was one of the only to endure but it, too, underwent a shift in meaning. The notion of comradeship has pedigree in German history. The iconic song, "I Had a Comrade" (*Ich hatt einen Kameraden*) dates from the War of Liberation (*Befreiungskriege*) waged against Napoleon, during which the notion of the German nation was formed. Soldiers experienced what they had in common with other German-speakers, and the dream of a future unified nation was born.[44] (Post-war) German Basic Law begins with the sentence: "The dignity of the individual is unassailable."[45] This raised the question, which was more important, the "comrade" or "camaraderie"? Veterans wrote and urged the VDK to give expression to camaraderie, arguing that being interred together was "soldierly custom."[46] The often-expressed demand for a stronger emphasis of camaraderie (*die oft gestellte Forderung nach stärkerer Betonung der Kamaradschaft*) demanded "a dignified design

44 The text is from Ludwig Uhland (1787–1826). The lyrics of the song date from 1809. The composer Friedrich Silcher (1789–1860) set it to music in 1825. Honoring comradeship is emphasized by Lindner. See Lindner_richtlinien_ca1954_6781 pdf, p. 6. (AA Archive). In his view the sameness of the grave-markers make the soldiers' cemetery a symbol of unbreakable friendship.

45 "*Die Würde des Einzelnen ist unantastbar.*" *Das Grundgesetz für die Bundesrepublik* from May 23, 1949. (BGBI) Article 1 Basic Rights (Articles 1–19).

46 Hans Gstettner, *Deutsche Soldatenmale*, p. 14 (Berlin, 1940). Gstettner postulated that the comrades' graves represented the German spirit. He states that the individual is nothing and the folk (*Volk*) everything; Cemeteries are collective, "sacred places" of the the people. (*Heiligtümer des Volkes*), p. 32.

of comrades' graves," such as a crypt underneath the floor (*würdige Gestaltung der Kameradengräber*).[47] The comrade's grave, however, was reserved for remains that could not be separated, and individual names were engraved upon its marker. NS ideology had framed comrades as supermen,[48] while the post-WW II VDK, through this action, shifted the meaning of "comrades" to those who had suffered violent death together. In this way, "… camaraderie was no longer an ideological imperative, but rather emphasized the human factor in mechanized war."[49] (The US likewise honored camaraderie.[50])

The VDK considered their cemeteries in Italy to be works of art involving both landscape and monument, but at first, the list of potential monumental ornamentation was modest; a tall cross or group of crosses, a sarcophagus, a sculpture, a chapel-like space, a fountain, or benches.[51] Lindner's evident trepidation resulted in modestly monumentalized entryways (pioneered by Tischler for WWI cemeteries, where a monumental effect was reached inspite of small dimensions stipulated by the victors e.g. Langemark). The VDK's post WW II cemeteries were to provide information as to where battles occurred and where each soldier died and was first buried, but were to eschew glory and not be informative about campaigns and battles. Aesthetic and ideological guidelines were further articulated in 1958.[52] The right to eternal rest (*Dauerndes Ruherecht*) remained the legal and moral basis for foreign cemeteries, and they should

47 VDK's *Richtinien* (1958). Emphasis on the comrades' grave is confirmed in the guidelines from March 19, 1965. The VDK Archive, A.10–124.

48 As represented in Josef Thorak's statue of two nude oversized males, titled understatedly "Comrades" (they represent the ideal NS "*Herrenmensch*"), for the German Pavilion at the Word Exhibition in Paris, 1937.

49 George L. Mosse, *Fallen Soldiers Reshaping the Memory of the World Wars*, (Oxford University Press, 1990), p. 217, about the salvaging of the "romance of camaraderie."

50 See "Comrades in Arms," the central sculptural monument in Anzio/Nettuno.

51 See fn. 31.

52 Committee for the design of cemeteries in foreign countries. This committee had travelled to Belgium and France prior to formulating guidelines that they worked out in four sessions, which were accepted by the experts' committee *Sachverständigenbeirat* on December 17, 1957. A memorandum about this travel states that its aesthetic guidelines are for orientation purposes, not requirements. *Bg-S/Dr.P.Zi. aus dem VKD Archivbestand A.10–26 Geschäftsleitung/Vorstand/ 8 committes 8.5*. (Compare to the similar US position toward its architects.)

bear the unmistakable message that remembrance is a duty (*verpflichtendes Gedenken*).[53] The 1958 guidelines adopted the use of Christian symbolism: "In all cases Christian symbols must be used" (*In allen Fällen sind christliche Symbole zu verwenden*), framing therewith German memorialization in Christian terms.[54] Indeed, the cross is used in all German cemeteries in Italy. Although "God" is used in the preamble and articles of the German Basic Law, the German people did not become more religious after the war.[55] Christian symbolism underscored recognition of courageous heroes to peace, such as Dietrich Bonhoeffer and a host of others associated with progressive thought and pacifism, who gave their lives resisting NS brutality. Memory of their heroism caused moral ambiguity with regard to rearmament during the Cold War.

Fig. 8. La Cambe, Normandy, entrance. Note restrictive opening focuses view onto the pieta sculpture set upon distant tumulus. Tischler's motif gives form to mourning and suffering.

53 See VDK's *_richtlinien, (March 28, 1958)* p. 2., the VDK Archive, A.10–122.

54 Ibid.

55 Simone Lässig and Rainer Prätorius, *Religion: Belief and Power*, p. 41. The authors discuss religion and waning influence of established church denominations after 1945 in Germany; *Competition and Convergence,* eds. Christof Mauch and Kiran Klaus Patel, (Cambridge University Press, 2010).

As allusion to both pathos and nationalism were specifically to be absent, the appropriate display of the cross became a point of discussion.[56] During the NS period, the cross was co-opted to represent soldierly ethos.[57] It reappeared in the post WW II period in the context of mourning, for example, at the sculpture of a Pieta Merano (Italy), and as a main feature at La Cambe (Normandy).

After 1967, a tall cross in fine metalwork was erected at foreign soldiers' cemeteries, intended as a universal symbol of suffering rather than an emblem of Christianity. Nowhere does it confer ressurectionist promise. Artful craftsmanship in metal and stone beautifully expresses "earnest competence," a form of beauty prevalent in the VKD's post-war artistic repertory.

New Design Guidelines

The VDK initially favored the so-called garden setting (*Gärtnerische Anlage*) for its cemeteries, where landscaping with trees and plants predominated and architectural elements played a restricted role. This form of setting was less costly than more architectural settings, and more traditional as well. Trees would form a grove where soldiers might rest: "… an area sheltering bones as a sanctified space" (*Gebeine bergende Fläche als geheiligter Bezirk*).[58] In addition, it obviated the heretofore-unresolved issue of what, in the post-war period, constituted an appropriate monument. But due to high maintenance, this type of setting could not be seen as a permanent solution (*endgültige Lösung*).[59] For the sake of more permanence, the VDK came to favor architectural settings with stonework of

56 See Kathrin Hoffman-Curtius, "Das Kreuz als Nationaldenkmal." Deutschland 1814 und 1931, in *Zeitschrift für Kunstgeschichte*, Bd. 48, 1985, p. 16. The author theorizes that the cross was polyvalent and easily co-opted for political purposes.

57 Gstettner, *Deutsche Soldatenmale,* (Berlin: 1940) p. 14. The cross in his view stood for the will to soldierdom.

58 -richtlinien-gestaltung -1958-mit Kommentar, from January 1, 1958. The VdK Archive, A 10, 8.5.

59 Letter from Tischler to the directorate in 1957 calculated that landscape maintenance for Lommel (Belgium) and Sandweiler (Luxemburg) alone cost the VDK DM 40,000 per year. The VDK Archive, A.100–1031_tischler-symbolkreuzgruppen.pdf.

high quality. Nature would not guarantee permanence, only stonework.[60] As Robert Tischler wrote: "It must be made clear to the families how important the securing of the right to permanent rest for their dead is, and that permanence is best guaranteed through architecture."[61]

Emphasis on the individual soldier became the VDK's central ideological tenet and was applied to crypt burial as well.[62] The VDK insisted that each soldier rest in his own sarcophagus (to the extent possible). In Italy (as noted) the VDK clearly rejected mass burial, except where remains could not be separated. This was in variance to crypts built during NS times, such as at Pordoi (in the Dolomites).

The debate as to proper grave marking lingered. Tischler touted the practical value of larger, symbolic crosses placed in groups of three to five, one taller than the others and have names inscribed on small markers flush with the ground.[63]

---

60 Ibid. Tischler postulates that the cemeteries' honor site should have a limited number of graves, and be so heavily and permanently built that they could not be altered without violent disruption (*aber so schwer und dauerhaft ausgestaltet, dass sie nicht ohne Gewalt zu beseitigen sind*).

61 "Ibid. *("Den Angehörigen muss klar gemacht werden wie wichtig die Sicherung des dauernden Ruherechts ihrer Toten in Zukunft ist und dass die Sicherung in einer baulich gestalteten Anlage am besten gewährleistet ist.")*

62 Special session of the Directorate (*Vorstand*), the Experts' Committee (*Sachverständigen Beirat*), and the Adminintration, (*Bundesgeschäftsstelle*). Von Lutzau, member of the Directorate, was adamant about the veneration of the individual through grave markers. (With regard to the Futa Pass.) "In contrast to WWI, the interest in the single grave has come very much into the foreground." *("Im Gegensatz zum nach dem 1. Weltkrieg, in der das Gefühl der Gemeinschaft vorherrschte, ist nach dem 2. das Interesse am Einzelgrab sehr in den Vordergrund getreten.")* The VDK Archive, A.100–885, (March 25, 1960). This is a parallel to the Grundgesetz' beginning with the "dignity of the individual." *("Namenssteine unerlässlich, ob symbolisch Kreuzgruppe oder Einzelkreuze vorhanden ... die Ehrfurcht vor der Einzelpersönlichkeit ist von besonderer Bedeutung.")* Nevertheless Tischler's group crosses persisted for some time, as at Costermano. They disappeared completely after his death in 1959.

63 Letter, Tischler to the VDK headquarters. Thoughts about grave markers. (*Gedanken zur Grabzeichenfrage*) (November 7, 1949). The VDK Archiv, A 100–1031_tischler-symbolkreuzgruppen. While Tischler advocated for his group crosses for aesthetic and logistic reasons as well, NS ideologues had employed them to express their views about unique German soldierdom. E.g. Gstetter enthusiastically interprets Tischler's crosses as signifying a military unit: the leader and his followers. "An army unit assembled for review, eternally at attention" (*ein Heeresverband, der zur Parade, zum ewigen Appell, angetreten ist*). The cross is given an exclusively soldierly

Fig. 9. Tischler's group crosses in Consenvoye erected at the first time in 1942. (France, WWI cemetery.)

But this did not fulfill the VDK's stated need that "markers with names are indispensable, regardless of symbolic groups of crosses or singular crosses … the veneration of the individual person is of special significance."[64] The density of burials, and the requirement that names be in close vicinity to actual burial sites necessitated that these markers be engraved with multiple names (up to six), and this presented practical problems. Stones, nearly flush with the ground were difficult to see, and thus hindered grave tending. In Italy, the Costermano cemetery's grave fields were raised sufficiently for grassy ways between them. At Pomezia, smaller stones were later removed and replaced with stone crosses, large enough to bear several names of those buried in proximity, and

meaning. Quoted and set into context in Meinhard Lurz "…Ein Stück Heimat," p. 66; and see Monika Kuberek in her indispensable "Die Kriegsgräberstätten des deutschen Volksbundes Deutsche Kriegsgräberfürsorge" in *Unglücklich das Land*, pp 85–87. E.g. see Gstettner, "Die Entwicklung der Grabzeichenfrage in der Arbeit des Volksbundes," in *Kriegsgräberfürsorge* 6–7/1940. S. 76–88.

64 See fn. 62.

placed so lawnmowers could pass between. Nonetheless, each grave is that of an individual soldier (again, where possible).

The VDK's developing an image of a changed (and changing) Germany, and projecting this abroad through its soldiers' cemeteries required about a decade. This journey may be traced in the work of its architects (Tischler, Offenberg, and Oesterlen). Its largest and final cemetery at Italy's Futa Pass represents a culmination or sorts, one that shows a security, a bold newness, and a synthesis of nature, landscape, and architecture.

Robert Tischler, Architect

> "I began my work with a search of all countries' soldiers' cemeteries, sought in vain for inspirations for my new tasks, and through this process became aware that the proper grave for fighting soldiers was and remains the *Kameradengrab* (Comrades' Grave)."
>
> Robert Tischler[1]

The VDK's chief architect, Robert Tischler (1885–1959), joined the organization in 1926 and worked with great dedication until his sudden death in 1959. He was personally involved with most of the VDK's Italian projects, and participated as the lead architect on many. Such remarkable hegemony in architecture enables a visual analysis of style and messaging in the process of evolving, from Germany's pre-war into its post-war period. Removing the variable of different architects allows us to better recognize the VDK's (and Germany's) cultural development as expressed through his work.

Of Robert Tischler, British architectural historian Gavin Stamp wrote: "He rose to the terrible challenge presented by industrialized slaughter and his work deserves to be better known."[2] Indeed, Tischler's works having been associated with Nazi ideology resulted, until recently, in a "*damnatio memoriae.*" Although his creativity is in evidence in cemeteries and monuments numbering in the hundreds,[3] historians have largely limited their

1 ("*Ich habe zu Beginn meiner Tätigkeit in allen Ländern nach (solchen) Soldatenfriedhöfen vergeblich gesucht, um Anregungen für die neue Aufgabe zu finden, und bin dabei zu der Erkenntnis gekommen, dass das Grab des kämpfenden Soldaten das KAMERADENGRAB war und ist...*") Letter from Tischler to the VDK's *Vorstand* (Directorate), (January 23, 1957). VdK Archive, A. 100–1031. He was not alone in that opinion. See Gnettner, in *Soldatenmale,* see chapter II, fn. 46. In Tischler's WWI cemeteries the Comrades' grave was the ideological and spatial center. Soldiers that fight together and die together need to be buried together. This notion has a universal atavistic strength that outlasts ideological misuse.

2 Gavin Stamp, "Almost Unbearable," (London: *Apollo*, November 2, 2007). "The German war cemeteries in Belgium contain one of the greatest artistic responses to World War I. These cemeteries are also deeply moving as well as superbly designed, but in a very different way," [different from British cemeteries].

3 Tischler designed or was involved with the design of cemeteries in the hundreds in Germany by 1952. See Christian Fuhrmeister, "Robert Tischler, Chefarchitekt 1926–1959. Ein Desiderat." Kolloquium Bauaufgabe Soldatenfriedhof/Kriegsgräberstätten, Teil 1, Berlin, (February 2014).

interest to those that feature characteristically Nazi elements, such as mass graves and his large, nameless rough-hewn group crosses of natural stone. His favoring of mass graves, also for WW II cemeteries, considered by his critics as a sign of a lingering NS disregard for soldiers' individuality had aesthetic reasons. Row upon row of individual crosses covering a whole cemetery, as at the WWI Cemetery at Langemark, Belgium, would individualize mass death with a sea of crosses, but would result in a ghastly sight.[4]

Such elements have been interpreted as reflecting the Nazi emphasis on the collective (*das Volk*) more than the individual. Tischler's work is treated herein from an art-historical perspective, in disregard of his reputation as a purveyor of Nazi thought through architecture, particularly through his *Totenburg* designs. As to his vicinity to NS ideology, may it suffice simply to note facts: Tischler joined the NSDAP on March 1, 1932.[5] After the war's end, his Nazi past activities were reviewed at a judiciary court (*Spruchkammer*). He was classified as a harmless follower (*Mitläufer*).[6]

This chapter focuses on German soldiers' cemeteries in Italy, of which Tischler designed almost all, except the last to be built. Tischler's Italian work thus presents a design "continuum," a series covering both pre- and post-WW II eras. Analysis of these works reveals an inclination toward the abstract, as found in the movement named "*Neues Bauen,*" and toward modernist aesthetic principles.[7] During the period of the two countries'

---

4 See his "Thoughts about grave markers." (*Gedanken zur Grabzeichenfrage*) (November 7, 1949). The VDK Archiv, A 100–1031_tischler-symbolkreuzgruppen. That mass graves were not always a matter of choice but of given restrictions is evident in Yves Le Maner, Director of *La Coupole, Zentrum für Geschichte und Erinnerung* (center for history and remembrance) in Nord-Pas-de-Calais. No date given, <www.wegedererinnerung-nordfrankreich.com/die-hintergruende-verstehen/gedenken-an-den-ersten-weltkrieg/die-nationalen-merkmale-der-friedhoefe.html>. Last accessed March 27, 2018. The author points out that France allowed only mass graves.

5 Archive (Bundesarchiv) Berlin-Lichterfelde holds Tischler's file about his Nazi membership.

6 Per Christian Fuhrmeister. "Tischler had to go through a denazification process as all state employees and persons who had worked in public service (*Öffentlicher Dienst*). He was categorized as a '*Mitläufer,*' (literally: following along). 80 percent were categorized as such. The judiciary court was German, given their authority by the US occupation force."

7 For a note on Tischler's modernity see Christian Fuhrmeister "Die 'unsterbliche Landschaft,' der Raum des Reiches und die Toten der Nation. Die Totenburgen Bitoli (1936)

alliance (Germany and Italy), Tischler erected imposing structures. In the post-WW II period, he was obliged to downplay monumentality.

Tischler joined the VDK as chief architect (1926) and little is known about his previous work. The VDK's design and construction management activities (*Bauleitung*) were located in Munich. In the Munich workshop, modeled after medieval workshops (*Bauhüttensystem*), individual, anonymous artists and craftspeople produced the innumerable emblems, decorations, and sculptural reliefs for WWI cemeteries.[8] Tischler's archives in Munich were lost in a fire shortly after his death (in August 1959), which according to rumor, may have been deliberately set in protest against relocation of said activities from Munich to Kassel. What we do know may be gleaned from the VDK's archives in Kassel. These, however, are limited to the post-WW II period as their pre-war archives were destroyed in bombings. Hence, we rely largely on visual analysis of his designs.

Stamp's term, "industrialized slaughter" refers to fallen soldiers in the millions needing to be buried. The victorious belligerents such as the United States seized the opportunity to monumentalize and politicize soldiers' deaths. (US losses in WWI were 116,708 soldiers, small by comparison to those of the major belligerents.) Approximately one million of Germany's two million fallen soldiers lay strewn across northern France and Flanders in hundreds of *ad-hoc* cemeteries.

A violent death in the service of a nation cries out for some explanation. Some factions of Germany's "body politic" denied any meaning to such mass death,[9] while others imputed thereto a sanctification of the

---

und Quero (1939) als stategische Memorialarchitektur" p. 64 in *Kritische Berichte, Zeitschrift für Kunst-und Kulturwissenschaften,* Heft 2 Jahrgang 29, 2/2001, (Marburg: Jonas Verlag) pp. 56–70, and "Klatschmohn und Ochsenblut. Zur Ikonographie der Kriegsgräberstätten des Volksbundes Deutsche Kriegsgräberfürsorge", p. 134, in *Gartenkultur und Nationale Identität,* ed. Gert Gröning and Uwe Schneider, pp. 119–134, (Worms: Werner, 2001).

8 "Die Bauhütte des Volksbunds. Eine Kameradschaft deutscher Künstler und Handswerksmeister im Dienste der Heldenehrung" *Kriegsgräberfürsorge* XVIII Heft 1, pp. 2–12 (1938).

9 Examples are: Käthe Kollwitz' kneeling statues of herself and her husband mourning the death of their eighteen-year-old son, Peter, in Vladso, Flanders. Further, the pacifist satirist writer Erich Kästner, whose books were burned by the Nazis in 1933, wrote in bitter irony about soldiers' deaths: "*Da liegen wir/den toten Mund voll Dreck. /Und es kam anders, als wir sterbend dachten. /Wir starben. Doch wir starben ohne Zweck. /Ihr lasst euch morgen, wie wir gestern, schlachten.*" (Here we lie, our dead mouth filled with dirt/And it turned out differently than we had thought dying

nation. One such was Hans Gstettner, as his hymnic exaltation of war exemplifies.[10] Gstettner saw such carnage as a sign of greatness, specifically of the German folk, which supposedly had absolute will for "soldierdom." Soldiers' blood sanctified the ground. Though this terrible ideology appeared in writing in 1940, it was already being formulated toward the end of the Weimar Republic era.

Following WWI, Germany, in the throes of catatonic political turmoil, was neither allowed to erect monuments in France nor able to define a clear commemorative message. France and Belgium had collected most of the German soldiers' remains and erected individual wooden crosses, densely packed and marked with names (where possible) before the VDK was finally allowed into France (in 1926). These were in the process of decay. For Germans, this neglect was painful:

> "The impression of boundless poverty and abandonment that these gravesites triggered was the more shattering as the victorious states began immediately to build, with huge state means, national shrines, called Sanctification Shrines (*Weihestätte*), in most cases in the vicinity of German grave fields."[11]

That same year, Tischler began his work with the VDK. He commenced modestly with landscaping and planting, creating groves using trees treasured by Germans, such as oak and linden. These are referred to as *Heldenhaine* (heroes' groves) and when planted in rows, represented gothic cathedral pillars.[12] However, as Monika Kuberek has noted, the *Heldenhain* was not a Tischler invention, but goes back to a design

/We died. Yet without purpose/You'll let yourself be slaughtered tomorrow as we were today.) Quoted in Norbert Fischer, *Geschichte, des Todes in der Neuzeit.* Erfurt, (2001), chapter 5, "Kriegstod, Massenvernichtung und kollektives Totengedenken," pp. 73,74.

10 Hans Gstettner, *Deutsche Soldatenmale*. His book is a singular hymn on the heroic "Germanness" of soldiers' death. It is almost unberable to read, but an important source. It provides the full range of NS ideology behind its cemetery structures abroad.

11 Ibid. p. 8. ("*Der Eindruck grenzenloser Armut und Verlassenheit, den diese Totenstätten erweckten, wirkte umso erschütternder, als die Siegerstaaten von Anfang an mit gewaltigen staatlichen Mitteln daran gingen, ihre meist nahe bei den deutschen Gräberfeldern liegenden Soldatenfriedhöfe zu nationalen Weihestätten auszubauen.*") The author gives statistics: approximately 2,900 grave fields existed after WWI (in which 947,000 soldiers were buried). Belgium and France then reduced the number of grave fields to 203.

12 Meinhold Lurz, "... ein Stück Heimat", p. 66.

developed in 1915.[13] As Tacitus had observed two millennia earlier, Germanic tribes venerated nature.[14] In nature, there was to be found a primal "Germanness" (*UrDeutsch*) that was of a higher order, unfathomable, intangible, and sacred.[15] Hans Gstettner had claimed supremacy of Germany's closeness to nature over that of other nations.[16] Whether Tischler had this in mind is nowhere recorded, but one may adduce that his creative telos to express an inner monumentality played a role. One senses Tischler's urge toward monumentality throughout his life's work. (His aesthetic distinction has been largely overlooked.) Tischler might have seen in *Heldenhaine* a means to mask the horrors of mass death, as these belonged to a natural order.

Toward the end of the Weimar Republic, nature and landscaping did not suffice to express rightist German views.[17] Tischler introduced architecture. France imposed a limit of three meters in height upon structures in Germany's soldiers' cemeteries. Tischler's edifices were thus modest, but achieved impressive monumentality in spite of such small scale. His pavilion at the cemetery at Langemark (Belgium, 1932) (considered pre-fascistic by Lurz, Arts and Crafts by Stamp) displays all of his hallmark features that will later be in evidence in his cemeteries in Italy. These features include monumentality through mastery of proportion and scale, precise and tight stonework of large durable stones, and a dramatic entryway experience through a narrow passage through such stonework, as to guide the visitor from the profane to the hallowed.[18] A connection has also been made between monoliths ("Dolmen") as a theme and a Germanic mythic sense of nature, a source of inspiration for Tischler.[19]

---

13 Monika Kuberek, "Die Kriegsgräberstätten", p. 79.

14 Christof Mauch; ed. *Nature in German History*, (New York: Oxford, 2004), Introduction, p. 3.

15 Hans Gstettner. *Deutsche Soldatenmale,* p. 15. "Soldier's blood sacrifice sanctifies the site."

16 Ibid. "Strength through the landcape." He pejoratively refers to US cemeteries, with expensive marble steps and crosses, as Dollarcemeteries. P. 16.

17 See Meinhold Lurz, "…ein Stück Heimat," p. 70.

18 A prime example is Pordoi in the Dolomites, to be discussed below.

19 Brands, "From World War I" The author treats the appearance of architectural monuments by the end of the 1920's citing several of Tischler's extraordinary works such as Quero, Bitoly, and Pinzano from an aesthetic and iconographical point of view, without discussing Tischler as their creator, with the exception of Annaberg. These omissions appear to author as undeniable examples of *"damnatio memoriae,"* pp. 241–42.

Remarkably, Tischler created monumentality without resorting to classicism. His designs have distinct modernist features, blends between medieval motifs and modernist conflations of cubes, spheres, and ovals, blends that were rarely appreciated by his critics.[20] In addition, despite challenging preconditions and restrictions, each of his cemetery designs effectively fuse with its surrounding landscape. (He had been trained as a landscape architect.)

Tischler rose to prominence during the NS period when his designs included *Totenburgen* (castle-like structures sheltering the dead).[21] However, the most celebrated archetypical *Totenburg* was *Tannenberg* (1927), which was not his design (its architects were Walter and Johannes Krüger). *Tannenberg* was located in what was formerly East Prussia and became a central emblem for the NS ideology justifying the conquest of territory in the East for living space ("*Lebensraum*"). Its octagonal wall with eight square towers enclosed a large open space, especially suitable for ceremony. Its design recalls Hohenstaufen Emperor Frederick II's *Castel del Monte* in Apulia (1240's), a legendary zenith of medieval architecture. For Hitler, it bespoke Teutonic grandeur and might. Although conceived as a memorial and not as a burial ground, *Tannenberg* became iconic for the cult of the fallen soldier.[22] (It was razed after war's end.)

Through the work of Meinhold Lurz and Christian Fuhrmeister, Tischler is presently being explored. The former focuses on the continuance of NS ideology and the "castles of the dead" (*Totenburgen*) in Tischler's work. See Lurz "Ein Stück Heimat," p. 68. The latter points out the aesthetic distinctiveness of Tischler's works notwithstanding NS iconographic implications. See fn. 20.

20 Tischler's modernism in terms of abstracting tendencies in his buildings is found in Christian Fuhrmeister: "Klatschmohn und Ochsenblut…," p. 122. About his cubism derived from the tenets of the "Neues Bauen" see Fuhrmeister "Die unsterbliche Landschaft," p. 64. Tischler's aethetics would retain cubic architectural forms throughout his work, as seen in his post-war II works in Italy.

21 Brands, "From World War 1," sub-chapter titled "Untouched, Heroic and Imaginary Landscape: Kreis and the Nazi Totenburgen" (1941–43), pp. 242–256. The author discusses Kreis's designs of Totenburgen and gives extensive footnotes and bibliography and many illustrations, pp. 242–250. Regarding the subject of *Totenburgen*, see also Sergiusz Michalski, *Public Monuments: Art in Political Bondage 1870–1997*, (London: Reaktion Books, 1998), pp. 102–106. The term "*Totenburg*" denotes castle-like structures built and planned during the NS period that house crypts as burial places.

22 Tannenberg took on mythical connotations under Hitler. It became a symbol of Teutonic superiority and might. (Hindenburg's victory in WWI over two Russian armies was seen as a revenge for the defeat of the Teutonic Knights by Polish-Lithuania in

Tischler was neither a Hitler protégé nor, as often erroneously assumed, Tannenberg's designer.[23] However, he did design Totenburg-like structures during the NS regime. These were emblematic of Nazi beliefs and directives, especially the Totenburg at Annaberg (Silesia 1936–1938).[24] Its eastern location foreshadowed the imminent Nazi invasion of Eastern Europe.[25] NS propaganda praised it for incorporating the very essence of German genius. Its circular mass with affixed, squared buttresses expresses abstract qualities that testify to Tischler's creativity, and to his unique aesthetics. Situated majestically in wild nature on a steep, rocky hill (as is *Castel del Monte*), it is a vanished hallmark of these aesthetics. Its surroundings appear "natural" as if the perfect landscape had simply offered itself, but in fact, enormous engineering efforts were involved in preparing Annaberg's setting.

1410. Hitler, in 1934, ceremoniously made Tannenberg the burial place for General Feldmarshall Paul von Hindenburg and his wife. Hitler held Hindenburg, who had facilitated his rise to power by making him Chancellor, in especially high esteem. Jürgen Tietz, *Das Tannenberg-Nationaldenkamal: Architektur, Geschichte, Kontext,* Verlag Bauwesen, (1999) and see Heike Fischer. "Tannenberg-Denkmal und Hindenburgkult. Hintergründe eines Mythos," in *Unglücklich das Land…,* pp. 28–49.

23 Hitler favored Wilhelm Kreis over Tischler. Kreis was on Hitler's list of *"gottbegnadete Künstler."* See Brands, "From World War 1" fn. 21. Kreis preferred centralized monuments and mass graves. It is of note that Kreis referred to his megalomanic war memorials to be built in foreign lands—once they would be conquered—as "Kriegerehrenmal" (soldier's honor monument) and not as "*Totenburg*." Brands, "From World War I,", fn. 3., p. 216.

24 Razed after WW II. Fifty-one fallen members of the rightist *Freikorps* were buried there. For illustrations of Annaberg see Gstettner, *Deutsche Soldatenmale*, p. 4 to 7. For a hymnic description in the spirit of Nazi propaganda see p. 4.: "*Idee und Gestalt unserer Heldenmale sind so unverkennbar aus der innersten Substanz der deutschen Seele erwachsen*" (Idea and Gestalt of our heroes' monuments are unmistakenly grown from the most inner substance of the German soul). For a critical treatment of Annaberg see Korbinian Böck "Bollwerk des Deutschtums im Osten: Das Freikorpsehrenmal auf dem Annaberg/Oberschlesien." *in RIHA Journal 2017,* "War Graves."

25 This strategic destination as a defense bastion can also be found in Quero and Bitoly. See Christian Fuhrmeister "Die unsterbliche Landschaft," p. 59–66.

Fig. 10. Robert Tischler's Annaberg (now destroyed), Silesia (now Poland). Forceful architecture intended to bespeak medieval Teutonic force.

The following chapter explores Tischler's works in Italy of the NS period, as these provide the background for his post-war designs. Familiarity with these illuminates the groundbreaking shifts in ideology and aesthetics, and in the underlying urges that spawned the spirit of renewal evident in the post-war period. Comparing Tischler's post-war work with this forbidding *Totenburg* highlights a welcome new spirit.

Tischler's Work in Mussolini's Italy: Fascist Theatrics in Stone

> "… may this architecture, not only its buildings, but also its surrounding garden-like environment witness to the world that such weighty, soldierly seriousness of form and style is uniquely German."
>
> Hans Gstettner[1]

During the post-WWI years, huge monuments were erected to the fallen millions of the so-called "war to end all wars." British, French, and US monuments are in evidence on the "killing fields of Flanders." German building was not permitted before 1926, and then severely restricted, such as limiting the height of structures. Only in fascist Italy was construction allowed on a monumental scale. Again, projects in Italy present fascinating case studies, included herein in order to establish a point of departure, an ideological and aesthetic "base-line" to which subsequent post-WW II projects may be referred.

Fascist commemorative art under Mussolini was purposely theatrical and bombastic. At Redipuglia (near Venice), Mussolini's architects transformed a modest cemetery into a whole mountainside of marble steps.[2] Climbing to the top was a pompous pilgrimage. In the German/Italian alliance of 1939, Hitler's acknowledgement of Italy's annexation of South Tyrol (1918) cleared the way for commemoration of former enemies. For the VDK this meant that there was no obstacle to building massive crypt structures for re-interment, honoring WWI German and Austro-Hungarian soldiers fallen on Italian soil, even though these had been Italy's enemies.

Robert Tischler's three pre-WW II sites in Italy testify to this shared urge toward monumentality. These are at Quero above the river Piave, at Pordoi in the Dolomites, left unfinished when Italy joined the allies in

---

1 Hans Gstettner, *Deutsche Soldatenmale,* p. 15 ("… *denn diese Architektur, zu der nicht nur die Bauwerke, sondern auch die gärtnerisch gestaltete Umgebung gehört, zeigt einen unvergleichlichen, eigenen Wuchs, einen Stil, dessen schwerer, soldatischer Ernst der ganzen Welt seine deutsche Herkunft kündet.*")

2 See Hannah Malone, "Fascist Italy's Ossuaries of the First World War: Objects or Symbols?" *RIHA Journal* 2017, "War Graves."

September 1943 and finished after the 1955 treaty between West Germany (FRG) and Italy, and at Pinzano, abandoned in 1943 and never finished. All bespeak defiance and mighty confidence in the new age that the fascists believed was dawning. At Pordoi, the addition of post-WW II elements reveals the beginnings of the stylistic and ideological leap to the post-war period. These pre-war monumental cemeteries are discussed in order to illuminate this leap.

Quero: A Totenburg in Modernist Form

> "So lie we in rank and file as once in life together we stood, that the same Cross and Medallion might mark our graves. Now we rest, far from travails, awaiting eternity's consolation."
> Inscription in crypt of Quero[3]

It was Tischler's monument in Bitolji, (Macedonia, planned 1929–30 and built 1935–37) which was first labeled "*Totenburg.*"[4] Although strictly speaking not a *Totenburg* since its approximately 3,000 WWI soldiers were not buried within the edifice but rather outside in a large open space surrounded by a high massive circular wall, the term "stuck." The "*Totenburg*" became an enduring topos for Nazi veneration of soldiers' valor in the service of the nation, such as mass graves as well as castle-like structures.[5] The building displayed typical Tischler features, a stereo-metric cubical form of dense stonework, with heavy towers buttressing its main corps, and emphasis on entrance design.

---

3 ("*Wir liegen zusammen in Reih und Glied, wir standen zusammen im Leben, drum gleich Kreuz und gleicher Schmuck ward uns aufs Grab gegeben. Nun ruhen wir aus vom heißen Streit und harren getrost der Ewighkeit.*") Note the lack of Christian symbols and concepts.

4 Christian Fuhrmeister, in "Die unsterbliche Landschaft," p. 57 gives an overview about the origin of the term. The concept was developed by Tischler but he did not call his designs *Totenburgen*.

5 Between the two world wars, Tischler designed a dozen *Totenburgen.* See Meinhard Lurz, "... Ein Stück Heimat", p. 70. The author discusses the characteristics of Tischler's *Totenburgen*.

Tischler designed a *Totenburg*, erected at Quero (planned in 1934–35 and dedicated in 1939) with similar combination of a block-type structure and graves contained within peripheral stone walls. This commemorative monument was to honor the deaths of mostly Austro-Hungarian soldiers. Its site overlooks a valley on the Piave River and as with Annaberg and Bitolji, is a defensible fortress able to resist enemy attack.[6] Quero is an impressive creation of architectural clarity and ingenuity evident in its proportions and its assemblage of cubic and oval forms. The visitor confronts a massive structure, an unadorned wall, blank except for small window openings high above that recall the architecture of fortifications.[7] No ornamentation offers relief from absolute starkness. Stonework of red porphyry is of the highest quality with interstices that are practically seamless.

Fig. 11. Tischler's Quero. Bastion-like structure in mountainous landscape.

6 See Fuhrmeister, "Die Unsterbliche Landschaft," p. 65. For more information on Quero see VDK Archiv, kunsthandwerk sheets on Quero and see Fritz Kirchmeier, "*Eine Festung, die ein Friedhof ist.*" <http://www.volksbund.de/fileadmin/redaktion/BereichMeldungen/Presse/2014/Prospekt_Quero.pdf>.

7 Ibid.

The composition of the total structure is a configuration of two cubic edifices and an adjacent wall, curvilinear at one end, straight at the other. Hans Gstettner raved about the "indescribable pleasure of measure and proportion of the grounds and the architecture" (*unsagbare Wohllust der Maße und Verhältnisse … der Anlagen und Architektur*).[8] A rounded watchtower overlooking the Piave river is integrated into the wall, in keeping with the latter's height. The totality has modernist features, as Christian Fuhrmeister has pointed out.[9] Tischler was evidently familiar with non-axial access ways and stereo-metric spatial conceptions, as seen in the works of Le Corbusier and Mendelsohn. Yet, its heavy stonework, purposefully executed in the fashion of medieval craftsmanship, counters the affect of modernism.[10]

Tischler worked with contrasts to achieve dramatic results. The red porphyry of the building effectively contrasts with the green hillside. This man-made structure on a hill is juxtaposed against nature's formations. Across the Piave valley an expansive mountain cliff presents nature's opposite to the compact building complex. Such juxtapositions had great appeal for Tischler, as will be evident in his post-war design proposal for a crypt building at Monte Cassino where in his unrealized design, a castle ruin has dialogue with the monumental crypt.[11] At Quero, access to the entrance is gained via a small path. Changing directions, it quotes international modernism as it zigzags upwards, passing the main structure where a flight of stairs leads to a narrow entrance in a cubic tower.

Small entrances were a Tischler hallmark, dramatically ritualizing access to the interior, a separate experience of space. One enters into a low, windowless hall roofed by crossvaults. Tomb-like Romanesque elements, such as capitals made to look old, ornaments, and carved emblems simulate a medieval world. Examples of craftsmanship in stone and wrought iron abound, such as the exquisite gates at the crypt's entrances.

Light guides the visitor's steps toward the crypt-like space, which lies lower. Three steps lead down to it through three round-arched openings in the massive wall, the number three being of symbolic significance.[12]

8 Hans Gstettner, *Deutsche Soldatenmale*, p. 22.

9 For Tischler's modernist urges see "Klatchmohn und Oschensblut," p. 134.

10 Ibid.

11 See chapter on unrealized Cassino project.

12 Symbolism expressed by the number three is characteristic of Romanesque architecture, e.g., three portals as in the late Romanesque Abbey of *la Trinità* Caen, Normandy, France.

This sacred center is a square and is spanned by a single pair of heavy cross vaults. A mystical atmosphere is produced through a bright shaft of light entering through an oculus in its thick ceiling and piercing the dark, windowless space (recalling Rome's Pantheon). Below, a massive central block of black Swedish granite resembles a roofed sarcophagus.[13] Ancient symbols of timelessness cover its surface (by sculptor Karl Krober, Munich). This quasi-altar rests on short, fluted pillars that seem pressed into shortness, creating the impression of concentrated power.

Fig. 12. Quero, crypt with altar and frescoes of soldiers. A medievalized inner sanctum.

The altar carries an open book made of copper that lists 3,461 buried Austrian and Germans, of which 2,561 are nameless. Three sides of the crypt's walls are covered with marble mosaics, each showing four over-life-sized,

13 The space displays the full NS-sanctified repertory of artistic elements. The altar evokes the sarcophagus of Frederick II in the cathedral of Palermo thus creating a tie to the grandeur of medieval German history. See Monika Kuberek "Die Kriegsgräberstätten," p. 84. For article about Quero, see: (<http://www.volksbund.de/fileadmin/redaktion/BereichMeldungen/Presse/2014/Prospekt_Quero.pdf>).

heftily built soldiers in frontal view holding their helmets in their hands, rifles at their sides, and with their gazes directed downward (by Lois Gruber, Munich). Their large size conveys monumental importance. Beneath, a band of large-lettered inscription reads: "So lie we in rank and file as once in life together we stood, that the same Cross and Medallion might mark our graves. Now we rest, far from travails, awaiting eternity's consolation."[14] Words convey the message that these soldiers will still exist, but in a dimension beyond time. They appear substantially identical, making a clear statement that the individual is subsumed into the collective. The word "eternity" asserts that their state of soldierdom is immutable. They await no resurrection as individuals. US cemeteries, with individual grave markers, message through their art that the individual soldier is owed an afterlife by the grateful nation.[15] A mosaic over the entrance side depicts a sharp-angled eagle, emblem of the Nazi state, spreading its wings. The term "comrades' blood" (*Kameradenblut*) appears next to it, conveying the notion that soldiers' blood sanctifies the site.[16] The space delivers what was intended, a feeling of awe surrounding soldiers' deaths. Tischler's creation makes the visitor shudder.

At the end of the dark entrance hall, light pours in from the outside through the narrow (open) wrought-iron door and pools on the dark stony floor. One's steps are pulled toward it. In Tischler manner, a right-angle change in direction is necessary to pass through a modestly large exit, to the outside. (Such jarring change of direction is a design element to be embodied at the Futa Pass cemetery, to be discussed later herein.) Stepping into the open, the visitor is surprised by the sculpture of an angel's face that confronts him literally head-on. Its huge hands and long fingers, reminiscent of Assyrian art, clutch a shield that bears the VDK emblem of five crosses. The angel lies atop a massive wall, head slightly turned and looking over the land beyond, as if keeping watch. It seems as though it had grown out of the wall and will remain there forever.

14 For translation see fn. 1.

15 E.g. at Anzio/Nettuno, an individual soldier is taken to heaven. See chapter IV.

16 Gstettner, *Deutsche Soldatenmale*, p. 15.

Fig. 13. Quero. Angel in martial form with shield bearing VDK logo keeps watch overvalley below.

The sculptor Ernst Geier (Munich) blends classical with medieval canons. The angel's face reminds one of the Apollo of the temple pediment at Olympia, and of the Bamberg Knight. There is an element of softness in its expression befitting an angel. Its horizontal forms emerge from the wall, body and mighty folded wings tightly stacked above each other. A solid bank of clouds is formed of sharp (but not harsh) edges and angles. Compositional stacking of forms in relief is also seen in art eco, but the latter generally uses puffier forms. In its compactness, it answers Tischler's compact architecture. Oriented toward the visitor, it beckons him to take the narrow steps that are wedged between it and the crypt's wall.

Mounting, the visitor attains a rectangular field covered with low-growing cotoneaster, with mass grave under foot. A stone informs that us that 3,461 soldiers rest beneath. From this spot, the view over the lands (to be defended) is superb. The angel (as his shield reminds us) keeps eternal watch over the dead, and shall be ever watchful over the surrounding landscape, alert to the approach of an enemy. Gargoyles in form of toothy dragonheads on the building's wall facing the gravesite, right out of the medieval repertory of horrible creatures, contribute their threat.

A most interesting part of the site is the so-called "communication trench" (*Laufgraben*). Dug low into the ground and accessible through stairs, it winds around the mass grave and finishes in a flourishing curve that ressembles a defense bastion. Together with the trench, the depth of a man's height, framed by walls, it seems like a commemorative mise-en-scène of a bastion of defense, the trench recalling the trenches from which the last big offensive of WWI in this region was launched. As Fuhrmeister points out, Quero was indeed intended to function as a defense structure and because of its position it could have been easily defended against enemy attacks.[17]

Wall meets wall and stone meets stone in configurations ever changing with one's point of view. Harshness is inexorable. Different levels interconnect via stairs, with changes of direction. The story of the war is told through the architecture itself, abstract but at the same time it *is*, through its Gestalt, what it is intended to mean. Tischler's capability to impart an immediate impact and meaning through architectural abstraction is evident.

Mass graves have been interpreted as evidence of Tischler's Nazi ideological leanings. However, Quero's mass grave was likely a practical solution for re-burial, neither imposed by Italian authority nor the result of Tischler's (alleged) NS leanings.

Pordoi: A Totenburg in the Dolomites

> "Beyond time, beyond the mountains, may stone and star-filled heavens bring us closer to your eternity."[18]

This monumental cemetery in the Dolomites, at some 2,200 meters elevation, lies close to the WWI battlefronts where Italian, Austro-Hungarian, and German soldiers suffered in the freezing winter between bouts of slaughtering each other. Views upon the *Col de Lana*, and south onto the "sea of ice," the *Marmolate* (at 3,342 meters elevation) are awe-inspiring. The harsh mountainous landscape where heavy fighting occurred during WWI viscerally transmits the horrors of warfare in such a severely harsh environment.

17 Fuhrmeister, In "Die unsterbliche Landschaft," p. 55. With regard to the trench (*Mannshoher Laufgraben*) author quotes v.Lutzau: *"...aus dem im Oktober 1917 die deutschen Truppen zur großen Herbstoffensive vorbrachen."* (...from wich the German troups broke out to the big autumn offensive). v.Lutzau, "Deutsche Totenburgen in Italien. Sinn und Werk deutscher Heldenehrung." (Siedeln: *Wohnen 20,* JG Heft 1, January 1940). Of note: v.Lutzau refers to the Austrian troups as German.

18 So reads the inscription over the entrance to Pordoi's castle-like central structure: ("*Fernab der Zeit spüren wir über der Berge Firn nahe bei Stein und Gestirn Eure Ewigkeit.*")

Below dizzyingly high snow-covered peaks, soldiers were fighting on skis. Surrounded thus by nature's grandeur, Pordoi's circular, castle-like structure of red-brown stone with compact central tower claims man's place in nature's wildness.

Fig. 14. Tischler's Pordoi, a Totenburg in the Dolomites.

Italy's joining the Allies interrupted Pordoi's construction, but its crypt had been finished and holds the remains of 8,582 mostly Austro-Hungarian (some German) soldiers killed in the area during WWI. Burial at or near the place of death was valued, an atavistic urge that was similarly felt by the ABMC. After WW II, work resumed under Tischler (1956) and the monument was dedicated on September 19, 1959.

The monument at Pordoi is in the form of three concentric rings (the outermost 54 meters in diameter), and is surrounded by a low wall. A moat is excavated, giving the impression that the whole structure is rooted or anchored in the earth. Within lies a wall, five meters high and thirty meters in diameter that supports a paved platform reachable only from within the monument itself, thus making one's entry into the monument mandatory. It offers the viewer an entirely wild panorama where one may experience the vast grandeur of this spectacular mountain world.

Pordoi's octagonal tower (8.5 meters above grade) is covered with a flat metal roof and its tall, narrow windows are grouped three to each of the eight faces. It is footed centrally, on the monument's substantial circular stone platform whose outer walls are canted inward and crenulated to recall medieval fortification architecture.

The visitor at Pordoi breaches its outer ring, the "moat," to reach the monument. Large stone monoliths frame a small entrance, Tischler's signature. Contrasting the monument's mass with its small entry door of wrought bronze, Tischler has created an air of mystery. From afar, the dark opening seems to soar above the ground, above three rounded risers. The interior is neither visible nor imaginable from without.

Pordoi's entryway is a passage to another realty, which beckons entry. One enters a narrow, barrel-vaulted and roughly paved tunnel that cuts beneath the second ring's platform and recalls an Egyptian tomb. The tunnel narrows twice, engendering anticipation and hope for relief. On a shiny floor at the end of the sight line a massive marble altar is bathed in natural light. As if by magic, the visitor emerges into a spacious hall in twilight. The floor, covering the crypt, is of hard porphyry, assuring that the dead (from WWI) interred below may rest in peace, locked safely away.[19]

The marble altar that beckoned one to enter the space stands in the center on a platform of darker stone, echoing the building's design, a tight round core from which eight buttresses project. Their number eight, with its tower-like shape and substantiality, refer back to *Tannenberg* and *Castel del Monte*. Once, the altar carried a bowl with perpetual flame of sacrifice placed on a gilded crown (indispensable in Nazi ideological repertory) designed by Fritz Schmoll genannt Eisenwerth, with whom Tischler had collaborated in Annaberg.[20] The light from this fire was to be seen from without, through the small entrance. After WW II the sculptor replaced the bowl and flame with an inlaid cross of lead within a circle of marble. The fire that had mystified soldiers' deaths being extinguished effectively

19 Fuchs-Heinitz, *Todesbilder, p. 146*. The author postulates that soldiers are feared for aggression toward the living, because their death was caused by aggression and killing. Therefore the typical arrangement of graves ressembles a parade. Military discipline may not release the dangerous dead soldier. A crypt confines him forever.

20 An enormous sculpture titled "Germany awake" was the center-piece of Annaberg's interior. For illustration see Hans Gstettner, *Deutsche Soldatenmale,* p. 8. For a treatment of Annaberg tinged with NS ideology see Franz Hallbaum. "Das deutsche Freikorpsehrenmal auf dem Annaberg" in *Kriegsgräberfürsorge* 9–10/1943, pp 54–55.

de-sanctified the interior's "sacredness."[21] Four pairs of tall figures, soldiers hewn in limestone in half-relief, stand each on its own base, high above over a railed passageway that circles the interior. They look solemnly downward. Their uniforms are identical and the generalized facial renditions show types rather than personalities. The only element of individuation is the position of hands holding their doffed helmets. These were not altered in the post-WW II years. Their style, however, is not typical of Nazi art, as the figures lack the sharp angles and locked square jaws, and they are not bombastic.

Their sculptor, Albert Allen of Munich, had specialized in small pieces for refined upper class tastes, Art Deco figures such as nudes (e.g., a veiled Arabian woman on a camel). As his renown grew, he was propelled toward public works expressing NS ideology such as the imperial Nazi eagle and maidens for the *Nibelungenbrücke* in Regensburg.[22] The commission for the soldiers at Pordoi was issued in 1938. His Pordoi soldiers, through their smooth continuous outline, reveal the artist's Art Deco aesthetics. (The US parallel would be Paul Manship.) A large emblematic eagle with spread wings, the work of Fritz Grau of Munich, hovers over the buried soldiers from the ceiling and dominates all. In its stylization, particularly the spread of its wings, it seems to share shape and form with Italian fascist emblems.[23]

The monument's most striking features are compactness in combination with clarity of form. Its lines are sharp and precise, and the building is comprised of clear entities in harmonious proportions. The combination of round shapes with the octagonal central dome reveals a sure handling of stereo-metric forms, ratios, and proportions. In the stonework Tischler utilized a progression from very large stones in the outer wall confronting the mountains, to smaller stones in the second ring, to yet smaller and finer stones in the central octagonal. As always, his insistence on exactitude of

21 The motif of the bowl of flame was central to NS repertory. It was still the central motif in Tobruk, Lybia, built after WW II although without the "eternal flame."

22 As of this writing there is no study of the sculptor. The eagle discussed above was reestablished on a rebuilt bridge from 1950. The swastika sign under the claws of the eagle was replaced by an oak branch and as such stood there, much beloved, until a new bridge was built. The eagle and the maidens are, at the time of this writing, stored in a depot. The eagle's shapes are not as angular and harsh as those of other sculptures of NS eagles (e.g., the one in Nuremberg's Reichspartaigelände).

23 For Italian fascist eagles see: <http://en.wikipedia.org/wiki/Fascist symbolism>. accessed March 19, 2018.

stonework resulted in roughly hewn natural stone laid so precisely that just the thinnest of joints is noticeable. Repetition and intensification of a design motif, such as the round wall circling in stark opposition to the octagonal, creates drama. Such repetition with a stepwise augmentation or diminution of a motif is central to Tischler's aesthetic canon. As at all of the VDK cemeteries in Italy, handicraft in iron and bronze formed an integral part. In Lindner's words, they "… sparkle with life."[24]

Although completed in 1956, Pordoi was built according to pre-WW II plans, which included ritualized entrance, mysterious interior and NS emblems. Its massive structure seems built for eternity, its castle-like character defying the terrors and hardships of mountain warfare. Death is presented as part of nature's eternal cycle, and death's cause is not a subject. Nature and fate were pillars of NS ideology.[25] Nature meant eternity without the unpredictability, or the accountability of human history. Nature elevates the fate of the fallen soldier to nature's higher order, embedded in the eternal sequence of life and death.[26] This is reflected in the inscription above the entrance: "Beyond time, beyond the mountains, may stone and star-filled heavens bring us closer to eternity." A soldiers' death is transcendental.[27]

Two flights of broad circular stairs descend into a moat where, following WW II, soldiers were buried in individual graves. This part of the monument bears testimony to the VDK's post-WW II restraint, as well as to its absolute emphasis on individual graves. Horizontality bespeaks repose, and stands in modest contrast to the imposing monument and the majestic mountain landscape. The visitor experiences a skillful rhythmic patterning of paired votive stones, grave areas, grass, and paved pathways. Framed by cobblestone paving, red votive stones flush with the ground bear Latin crosses to which the post-WW II VDK had reverted.

24 The AA Archive, Werner Lindner, richtlinien, (May 4, 1954), p. 32.

25 For emphasis see inscription on Pordoi's entrance. Fn.18.

26 Gstettner, *Deutsche Soldatenmale*, p. 14."Closed sacred space above the comarades' grave… in order to create eternal values." ("*Geschlossener Weiheraum über dem Kameradengrab….zeitlos sein der Form…um Ewigkeitswerte zu schaffen.*")

27 For further thought on NS ideology and its views on transcendence and eternal life see Monika Kuberek, "Die Kriegsgräberstätten," p. 84.

Pinzano: Fascist Megalomania

"Spatial concepts realized in petrified monumentality."
Architect Rudolf Hillebrecht's characterization of NS architecture.[28]

Fascist regimes in Germany and Italy not only formed a military alliance, but also strove to form a closer union in which their future triumph would be shared. Germany's unfinished project at Pinzano would display unabashed grandeur. Italian and German soldiers were to be buried together, as in their universe, death in battle was a collective sacrifice, and an imposing monument thereto seemed appropriate. A large site was made available at Pinzano, in the region of Friuli (Venezia). Under construction from 1938 until Italy changed sides during WW II, it would be the "grand finale" of the VDK's NS period. (On September 8, 1943 Italy signed an armistice with Great Britain and the US. The allies, therewith, broke their pact with Stalin to make no separate peace.) Italy joined the allies in a formal military alliance, making Germany's *Wehrmacht*, and especially its SS *(Schutzstaffel)*, literally from one day to the next, foes of the Italian interim government. Pinzano, unfinished and its ruins resembling an archeological site, seems like a metaphor of this broken alliance between Germany and Italy.

Pinzano's ruins dominate a hill called Pion, near the *Ponte di Pinzano*.[29] Its ambitious design indicates that this military cemetery was to be used for theatrical fascist spectacles as well, and thus we can imagine that Tischler had a free hand to pursue his vision of grand architecture set in large spaces. Plans were finished in 1937 but construction began only after a wooden full-scale model (1938) had been built on site and approved, and soil engineers had confirmed that the ground could support such a massive edifice.[30]

28 ("*in versteinertere Monumentalität gebannte räumliche Vorstellungen.*") For the context of the quote with regard to building in the Third Reich, see Werner Durth, Deutsche Architekten: Biographische Verflechtungen 1900–1970. *Deutsche Architekten: Biographische Verflechtungen 1900–1970,* p. 374. Friedr. Vieweg & Sohn, (Braunschweig/Wiesbaden, 1986).

29 <https://it.wikipedia.org/wiki/Ossario_Militare_Germanico_di_Pinzano>. Accessed March 14, 2018. There is no detailed research available on Pinzano.

30 For an illustration of the life-size wood model see *Mitteilungen und Berichte vom Volksbund Deutsche Kriegsgräberfürsorge 19* (1939) p. 153.

Pinzano was to be more explicit about NS notions of grandeur than the Pordoi "*Totenburg*" and the VDK evidently decided not to undertake its completion after 1955 (when Germany was allowed to resume building in Italy). Similarly to Quero, and to a certain degree Pordoi, the Pinzano site was militarily significant. Italy's army fortified the unfinished edifice during the Cold War. It commands a spectacular view over the Tagliamento river, where it cuts through a narrow gap between surrounding steep hills. Its crypt was to hold the remains of more than 20,000 WWI German and Austro-Hungarian soldiers, Italy's enemies during WWI.

Italian sources refer to Pinzano as an ossuary, and rightly so, as its architectural center is a crypt framed by architectural elements. To form a mental picture of the whole from the many parts in their dilapidated state is a challenge but fortunately, a total plan and drawings are available in the VDK's archives.[31] Photographs exist of those parts that were completed: a large terrace atop its semicircular retaining wall (of local stone) and behind, rising high, the crypt's façade in eclectic Egyptian-Assyrian-European style. Its ponderous rectangular structure with four inwardly canted towers at its corners unmistakably recalls fortification architecture, echoing grand tradition and conferring additional imposing monumentality.

Recognizable Tischler elements include a terrace from which only two small doors allow access to the mighty crypt. A grave field/parade ground with a festive entrance existed on its opposite side. The crypt's high openings framed by double pillars support enormous lintels, monolithic blocks.[32] Judging from the plans, the crypt's terrace side was to have identical pillars. The rectangular ledge on which these pillars were to stand formed an empty interior, open to the sky over a crypt intended for the remains, most likely not to be buried in individual sarcophagi. Thus, the crypt would have been indeed an ossuary, a charnel house. Nowhere is the individual emphasized. Rather, pompous structure was to express haughty ambition.

31 See VDK Archive, kunsthandwerk-pinzano pdf.

32 See fn. 29. Some monolithic stones were 6.4 x 1.4 x 0.8 and weighed 18 tons. The stones were from Somplago. For image and short description also see <http://www.itinerarigrandeguerra.it/code/46166/Former-German-Ossuary-of-Pion-Hill>. Last Accessed August 3, 2018.

Fig. 15. Tischler's Pinzano, incomplete. Ruins still reflect fascist theatrical megalomania.

Tischler conceived the Pinzano site as a grand amphitheater. The site widens toward the distal end, creating a slightly rhomboid plan. Unfinished walls and tribunes are still in place, but with nature having had her way with them for the better part of a century, they might well be mistaken for Roman ruins. The VDK's available plan shows a large grave field (for WW II casualties) and the adjacent amphitheater/parade grounds. Two lions by Fritz Schmoll genannt Eisenwerth were to be placed between this grave field and the amphitheatre.[33] The latter was to be a large stadium with a semicircular end (recalling Rome's *circus maximus*), suitable for fascist parades. The stadium seems to have been conceived to seat tens of thousands. Along its side, two towering wall panels were planned with no evident purpose other than an architectural emphasis of an entrance.

The project's large size and its interplay between heavy architecture and designed open-air spaces would have been any architect's dream, and the crowning of Tischler's work in Italy during NS times. The plans show an impressive architectural unit with formal correspondences between, for example, the rounding of its terrace above the river and that of the stadium at the other end. Extensive earth works were carried out. The terrain on the hill around the site was evened to surround the complex like a broad belt tilting slightly toward the monument. In contradiction to traditional German romanticizing of nature, respecting it as a sacred force, the natural landscape was brutally bent to political purpose, as at Annaberg and at

33 Barbara Küppers, *Fritz Schmoll genannt Eisenwerth,* (Weimar VDG: Verlag und Datenband für Geisteswissenschaften, 2003). p. 348. (Werkverzeichnis VB 14).

Tannenberg. Nature at Pinzano was forced into the grand plan.[34] Pinzano demonstrates how the landscape, in NS ideology, could be altered in order to create a "*Gesamtraumkunstwerk,*" (a total work of spatial art) with connotations of mastering nature.[35]

Time and decay have revealed brick under the crypt's marble veneer, probably a cost-saving measure due to the exigencies of WW II. Tischler surely would have preferred a more solid edifice as the stage-set for marching and ceremony. Other than two rows of cypress behind the tribunes, there is no indication of mourning. Triumph held supremacy over mourning.

34 For plan see the VDK Archive, Record N. 000000450 *Friedhof (1) Pinzano, Italien, Kunst/Handwerk.* For the brutal alteration of natural topographies for Tannenberg that needed immense earthwork by landscape architect Heinrich Wiepking see Brands, "From World War I," p. 240, in *Places of Commemoration.*

35 On NS Germany's brutal interference with landscape and nature see Joachin Wolschke-Bulmahn, "All of Germany a Garden? Changing Ideas of Wilderness in German Garden Design and Landscape Architecture." The author analyzes NS plans to turn Poland's wilderness, visible proof of the Polish status of "*untermensch,*" into a German-type garden. In Mauch, ed., *Nature,* pp. 74–92, (New York: Berghahn Books, Oxford, 2004)

III. Anzio/Nettuno: Space of Triumph

> "Everything is being done to make the visitor feel that these cemeteries are not the places of the dead, but memorials to the living and indestructible spirit of the heroes whose memory we cherish.
> John Harbeson[1]

John Harbeson's Normandy American cemetery and memorial, an icon for the US military contribution to the allies' triumph over Hitler's *Wehrmacht*, was the first commissioned by the ABMC in Europe and was completed in 1956. As a work of art, it set the general tone in style and message for all US cemeteries on foreign soil.[2] The ABMC undertook this assignment with methodical, military-like organization, beginning with an organized report and a clear view of what the end result(s) would be. Design and construction were to commence as soon as possible.[3] Harbeson, the newly appointed ABMC supervising architect, pointed out that he could optimize time and expense by making initial visits to all the ABMC's fourteen foreign projects, and suggesting plans.[4] He traveled to Italy in the summer of 1947, visited the sites in the Florence area and Anzio/Nettuno, where a substantial cemetery was being executed by the AGRS, a branch of the US Army's Quartermaster Corps.

After WW II, a wave of self-confidence swept the US. One detects, in the archived correspondence of those American talents involved in the US military cemeteries to be built in Italy, an eagerness to return to the country where they had apprenticed.[5] The very prospect of receiving such

1 Quoted in Shenton, "They Will Never Be Forgotten," p. 59.

2 Its landscape architect, Markley Stevenson, became the ABMC's consulting landscape architect. The sculptor Lee Laurie, who did the "Spirit of American Youth rising from the Waves," at Coleville-Sur-Mère, became the ABMC's consulting sculptor.

3 The ABMC's report on survey in Europe regarding Anzio/Nettuno. RG 66, (August-September 1946), p. 24.

4 Letter from John Harbeson to Eric Gugler assuring that Harbeson would act as the "eyes and ears" while the architect hired for Anzio/Nettuno would follow the report's recommendation. RG 117, Entry 9, Box 122_0002, (August 18, 1947).

5 Barry Faulkner, *Sketches from an Artist's Life,* (Dublin and New York: William L. Bauhan, 1973). In his higly readable memoir Faulkner repeatedly records the close

commissions must have tantalized those selected, as all had been fellows at the American Academy in Rome (AARome) whose mission (at that time) was to sustain the preeminence of classical and Italian Renaissance style. Le Corbusier famously said: "To send architectural students to Rome is to cripple them for life."[6] A paradox resulted. While more contemporary "modern" aesthetics developed in Europe, the vision and tastes of a bygone era resulted in the imposition of classicist convention on America.[7]

American public aesthetics favored grand axial plans, classical buildings with columns, sculptural monuments, water features, and sight lines converging on a main axis in the form of a broad mall or waterway (or any combination of these). Axiality and temple facades characterized the main streets of US cities. Following this ideal emphasis on the main axis was essential in all US foreign military cemeteries. This movement had motivated architect Charles F. McKim to found the School of Classical Studies in Rome (AARome), which opened in 1894.[8] It provided scholarships for architecture graduates, later to include sculptors, painters, and landscape architects. During the first half of the twentieth century donors included the wealthiest Americans of the time (Carnegie, Morgan, Vanderbilt et al.) as well as foundations and corporations. Fellows of the AARome received full training and were to benefit from direct exposure to classical and Renaissance monuments. (Prior to 1949 the AARome did not allow for other styles.[9]) Harbeson, a Francophile, had trained in

---

ties of former AA Rome fellows, professionally and personal. He describes his friends, in particularly Paul Manship. They travelled together after WW I through France in order to study Romanesque art, and to Italy, Greece, Egypt and Constantinople, pp. 110 ff.

6 Le Corbusier: *Towards a New Architecture*, (London: John Rodker, 1927). A work of great influence through its systematic application of concepts such as mass, surface, and plan, it makes the point that industrial design, such as seen in trains and ocean liners, shares basic design characteristics with historical classical buildings. But, none of these revolutionary ideas impacted the close-knit AARome-bred supremacy of historicizing classicism of the ABMC's appointed architects. P. 173.

7 See Clarke's influence on the supremacy of classicism in the US. See Chapter I. fn.31.

8 Fikret K. Yegül, *Gentlemen of Instinct and Breeding: Architecture at the American Academy in Rome, 1894–1940*, (New York: Oxford University Press, 1991.

9 See Costanzo, "The Lessons of Rome", about attempts to modernize since 1949, pp. 36, 120, 205. The author depicts the Academy's attempt to modernize beginning in 1947. Attempts to invite Mies van der Rohe in 1948 "fell on deaf ears." A. Kimball, partner of Gugler, became the AARome's director (residing in Rome) during 1960–65 and J. Kellum Smith, designer of Impruneta, was a member of the board. The latter

Paris in Beaux-Arts aesthetics, but his classicism was in harmony with the aesthetics of the AARome's fellows. Classicism implying officialdom, authority, national ethos, and patriotism became a convention in America.[10] The public at large accepted and approved its use.

The ABMC's Anzio/Nettuno

At Anzio/Nettuno Harbeson found the AGRS cemetery under construction and its staff still busy collecting, identifying, and reburying remains. The site was utilitarian and practical, walled and organized in geometrical clarity. It occupied a small rectangle of land with narrowing sides at one end. German and American bodies were buried in separate sections. (Germans would be re-buried in Pomezia in the fall of 1947.)[11] On March 24, 1948 the AGRS cemetery was considered complete and officially dedicated. A ceremony was held, with US and Italian dignitaries in attendance. The ABMC considered the AGRS's cemetery merely provisory, to be replaced with a grandiose monumental cemetery as soon as possible. Harbeson quickly produced drawings in which the design centered about an axis, a design convention that all the ABMC-selected architects shared. French landscape architects (under AGRS supervision) had used this form exclusively for US WWI cemeteries in France and Flanders. After WWI the ABMC adopted certain standard elements (approved by

was open-minded toward allowing modernist trends in the service of artistic freedom. See his letter to Manship (7 Sept. 1950) in response to the latter's complaint that modernist sculpture was produced at the AARome. AAR Records, AAA (570:138, 141–142). As the author points out, very little was realized in Italy itself, only the US Consulate by George Howe in Naples and the Sicily-Rome military cemetery (Anzio/Nettuno). (The author fails to mention the Impruneta cemetery).

10 See Wilbur Zelinksy, *Nation into State,* Chapel Hill: The University of North Carolina Press, 1988. The author discusses the relation between classicism and patriotic values.

11 The cemetery was first opened on January 24, 1944. In 1947 it had 6,791 US servicemen and 2,723 German soldiers, who were buried in a separate fenced-in section. Two aerial views of the AGRS's cemetery show the amount of cultivated land that would be used for the ABMC's cemetery. RG 117, Box 122, RG 119, Box 122; RG 117, Entry 9, Box 122. See fig. 5

General Pershing), such as a marble cross for each grave, and standard grave dimensions.[12] WW II cemeteries followed established precedence, a common US cultural and historical orientation as seen in, for example, the Common Law tradition based on precedence.[13] The theory has been advanced that as a result of war, during which people learned teamwork, the US public at large became more conservative.[14] The US also became more religious.[15]

Harbeson's Hunger for Land

Harbeson ignored the AGRS's cemetery and called for terrain four times larger in size.[16] The AGRS's cemetery was totally subsumed in Harbeson's extensive, fan-shaped plan. A grand entrance, and a sequence of grand spaces would bespeak triumph. A large semi-circular plaza before the entrance would announce an aesthetic feast, like a herald in a theater piece.

One would then enter onto a long and broad, tree-lined mall leading to a large oval plaza with a central marker (a monument or fountain or oak grove, as noted in his second plan).[17] In fan-like shapes, four

12 Sue A. Kohler, *The Commission of Fine Arts: A Brief History 1910–1976*, (Washington, D.C.: 1966) Kohler and Grossman "Architecture" mention the binding standard in grave size: two meters between graves, three meters between rows. Grave plots would be grass.

13 Letter from North to the ABMC Paris office saying that Gugler (and others) are to be assisted in visiting some WWI cemeteries in order to get the "feel" of the treatment of the several features. RG 117, Entry 38, Box 3, (February 16, 1949).

14 See C.C. Michael Adams, *The Best War Ever – America and World War II*, (Baltimore and London: The John Hopkins University Press, 1994). "The war … nurtured conformity and an intolerant Americanism." p. 154.

15 On the US becoming more religious in the 1950s. see <http://www.npr.org/2015/03/30/396365659/how-one-nation-didnt-become-under-god-until-the-50s-religious-revival>. Last accessed Mach 29, 2018.

16 Design with topographical indications. Smithsonian Archives, Washington DC, 44 OV (2 ovs, #44, 47) dr 231, folder "Architectural drawings, renderings and blueprints, Anzio Cemetery" Eric Gugler papers, 1889–1977. 4.5: Memorials, 1929–1977.

17 Ibid.

tree-lined paths spring forward from this central plaza and enclose three sector-shaped grave fields separated by tree-lined walkways. Graves were to be arranged along radii, in continuous and visually-pleasing arcs. The memorial was indicated as a rectangle, with shallow projecting side-wings so as to create a "*court d'honneur*" on the sightline established by the axis. The existing brook running along the Nettuno road would feed a pond outside the actual premises (for both aesthetic and irrigation purposes).

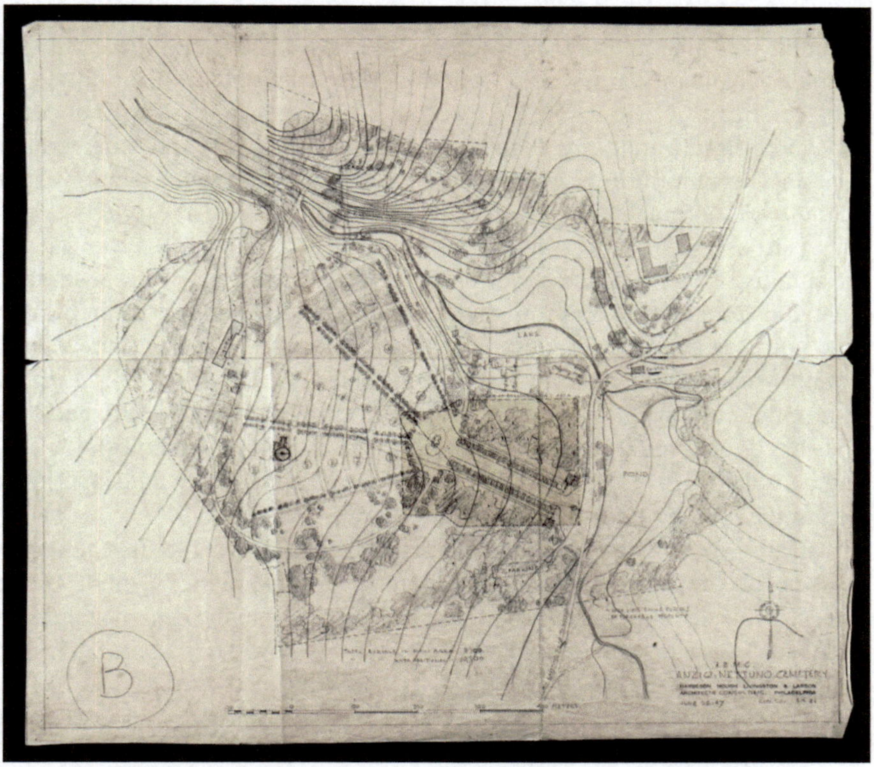

Fig. 16. Harbeson's plan for Anzio/Nettuno. Darkened area shows AGRS modest cemetery. Note the marking of the "Lone Pine" and the inclusion of pond and lake water element at the peripheries.

The pre-existing AGRS cemetery, perhaps accidentally, was oriented toward a lone pine tree that towered in the distance at its western side.[18] Harbeson marked this on his plan as "existing stone pine." For Gugler (Anzio/Nettuno future architect), this tree would become a fixed mental and aesthetic orientation point.

Harbeson's bold, triumphal plans for Anzio/Nettuno reflect the exuberance of these early post-WW II years, when everything seemed bright and possible, before budget restrictions and before Italian attitudes toward the US became more complex. Soon, the Soviet Union would become a nuclear power and Stalin would insist that the US Army leave Italy. Mao would triumph in China and a proxy war would draw the US into the cruel and thankless Korean War morass. At home, McCarthy's House Committee on Un-American Activities (HCUA) would disturb the nation's sense of security and mental tranquility. Indeed, these first exuberant plans remained but a fantasy, promptly to be dismissed by General North, with the argument that it took too much land and grass, and would be expensive to maintain.[19]

Harbeson reigned in his fantasy, returned to Anzio/Nettuno (again, in summer of 1947) and made another plan, "Plan C."[20] Its aesthetics changed from curved and rounded to straight and linear, but its ambitious size and it is appetite for land remained unchanged. The Nettuno/Velletri country road would be straightened and serve as the base line for a grand triumphal entrance sequence. A large semi-circular plaza would lead to a tree-lined mall, and then to another plaza half way to the memorial building complex. Grave fields would hug an oviform pond with an island in its center. Gugler would keep the water feature.

Harbeson stressed that a layout plan was the first submittal the architects needed to provide, to be followed by the grading plan.[21] General North was anxious to commence re-burial, but cautioned that the design should be firm, such that the graves would not have to be moved yet again later.[22] At

18 Letter, Watson (AGRS) to North. By then the "lone pine" had become the axis's point of orientation. RG 117, Entry 9, Box 122, (October 31, 1947).

19 Letter, North to Harbeson. RG 117, Entry 9, Box 122 (November, 1947).

20 Referred to as Plan C by Harbeson, (June 26, 1947). Smithsonian Archives, to be found under folder "Architectural drawings, renderings and blueprints, Anzio Cemetery" Eric Gugler papers, 1889–1977. 4.5: Memorials, 1929–1977.

21 Letter, Harbeson to Gugler. RG 117, Entry 9, Box 112, Folder 687, (8–22–47 to 12–31–48), (August 18, 1947).

22 Letter, North to Gugler, inviting him to design Anzio. RG 117, Entry 9, Box 112, Folder 687, (8–22–47 to 12–31–48), (August 18, 1947).

Anzio/Nettuno, work was to be organized by the on-site AGRS staff,[23] which grudgingly commenced to undo years of work as was required to accommodate the ABMC's new design. However, specific plans from the ABMC were not forthcoming, as needed. The AGRS had scheduled disinterment to begin on March 22, ending on May 31. Grading was to commence on June 1 and final re-burial was to begin on July 1, but they had no layout or grading plan.[24] Taking matters in their own hands, they produced three layout plans by April. These would need less land and would cost far less and be finished in little time.[25] From what one can gather from the correspondence, they would have been oriented differently. An obelisk would have constituted the commemorative feature. A good view was emphasized. One plan placed the entrance to the south. Another plan would need little cutting and grading and would take only three months to establish, and was thus favored by the AGRS's own architect.[26]

These plans were modest, but in the AGRS's opinion, adequate, patriotic, and pious: "Every effort is made by the American Graves Registrations Service to maintain the cemeteries in a state of beautification and repair befitting the resting places of our fallen heroes."[27] But the ABMC considered them commonplace, certainly not in a style that stressed grandeur, architectural distinction, and glorious landscaping. Harbeson studied and dismissed these plans, and also dismissed the fact that the AGRS had included complete cost estimates of their plans. The AGRS's initiative challenged the ABMC's claim to control, and Harbeson responded by admonishing the AGRS to occupy itself with the yet-to-be-delivered Gugler plan.[28] The ABMC's era had begun.

23 Letter, North to Gugler. RG 117, Entry 9, Box 112, Folder 687, (8–22–47 to 12–31–48), (September 7, 1948).

24 Letter, the AGRS to the QMG, Washington. RG 117, Entry 9, Box 122, Folder 687, (1–1 47 to 12–31–48) (March 17, 1948).

25 Letter, the AGRS Rome to the AMGS service Washington. Three different versions were discussed. RG 117, Entry 9, Box 112, Folder 687, (2–2–47 to 12–31–48), (April 4, 1948).

26 Letter from the AGRS's architect Lackart Roth to the AGRS MZ (Mediterranean Zone) discussing Plan 2, but favoring Plan 1 because of less cutting needed and because it could be finished in three months. RG 117, Entry 9, Box 112, Folder 687, (2–2–47 to 12–31–48), (April 3, 1948).

27 Archive of the Foreign Ministry, Rome, busta 123 F.C. P. 2., (March 18, 1948).

28 Letter from Harbeson to North. RG 117, Entry 9, Box 112, Folder 687, (2–2–47 to 12–31–48), (April 23, 1948).

Once the land was secured, Gugler would spring into action and produce a general plan and sketches. The AGRS was involved solely in digging graves and setting the crosses according to a rational, effective method where ditches accommodate footings for the crosses and anchor caskets, as well as major irrigation pipes.[29] The military's influence on the US cemeteries' style and message was substantial. When George Marshall became chairman of the ABMC he visited Anzio/Nettuno in September 1952. General North accompanied him. Marshall's stepson, Lieutenant Allen T. Brown would be buried there.[30]

Eric Gugler, Architect

The ABMC's letter inviting Eric Gugler to design the Anzio/Nettuno cemetery emphasized artistic freedom. Indeed, Eric Gugler's dedication resulting in an attractive work of art gives insights into an important side of mid-twentieth century US cultural history. Gugler was born to German immigrants, successful lithographers, in Milwaukee (Wisconsin) in 1889, a time and place were the German language was in general use. Higher education first took him to Chicago where surely he must have been exposed to Louis Sullivan's architecture. Much to Sullivan's dismay, Chicago's 1893 "Columbian Exposition" witnessed a renaissance of *Beaux-Arts* aesthetics, which would become the "gold standard" in US public architecture for generations to come. Sullivan famously pronounced that the exposition had set US architecture back 50 years. The "City Beautiful" movement went forth from this Exhibition, and imported classicism went on to prominence in Washington, D.C. and in many US cities. Symmetrical plans, buildings organized around broad axes and Greek columns, and ratios from the classical architectural cannon "popped up" all over the US.

29 "Battlefield Cemeteries Get New Marble Headstones" *Engineering News-Record,* (December 27, 1951), pp. 24–26.

30 An iconic image of this visit exists: 2018 <http://www.gettyimages.com/collaboration/boards/4FNCbKhRk0WIA0GTygyzrg?invitation_code=8dfJndPQV0KOe71_swDiZA>. Accessed March 14, 2018.

Gugler's path took him next to New York City, where he graduated from Columbia's School of Architecture in 1911. His orientation toward classicism, in particular toward ancient Greek art, predisposed him toward the tastes of the "upper classes." This, in combination with his affable personality, excellent education, and his years on a McKim stipend at the AARome (1913–15) promised to launch him to prominence and fame. It came when he shared a first prize in a competition for a World War I memorial in Chicago, a monumental classicist structure.[31] Throughout his life he maintained ties to the AARome and allegiance to classicism that, even after its popularity had peaked in the US (late 1940s),[32] would remain the dominant aesthetic canon at the AARome into the 1960s.[33] A friend's eulogy emphasized his credentials as a fully classically oriented architect: "All through his life he was in close communication with the grand style of the classical periods. He had an instinct for heroic nobility and for refinement of detail."[34]

Gugler was affluent and erudite, multilingual, well traveled through Italy, France, and Germany, nuanced in cultural history, and known as a lecturer.[35] He frequented New York's prestigious Century Club. Friendship with President Roosevelt's wife, Eleanor Roosevelt (E.R.) proved to be of great importance for his life and his work.[36] He nurtured this friendship, and her daily syndicated column "My Day" contains several references to him.[37] Her influence likely facilitated his receiving the commission to

---

31 The design was noticed by the press. E.g."Chicago Memorial in Lake," (*New York Times*, December 15, 1929). It was never executed.

32 Faulkner, *Sketches,* p. 181: Faulkner learned to appreciate the works of Calder, Leger, Juan Gris, Braque, and Solange, while Gugler remained true to his classicist beliefs (as did Harbeson). "Paul Manship and Eric Gugler looked askance at this broadening of my tastes." On the west coast, Beaux-Arts had fallen out of favor by the late 1940's.

33 Costanzo, "The Lessons of Rome." See fn. 9 about the attempts to admit modernist tendencies at the AARome.

34 Royal Cortizzon eulogizing Gugler. See orbituary of Eric Gugler, Columbia University Archives. G. Clarke Papers, (August 19, 1974).

35 Gugler gave a talk about the *Villa Bosco Parasio* that had been the seat of an academy under Queen Christina of Sweden begun in 1690, (Gugler Papers are in the possession of the Smithsonian Institute. The signatures at the time of this writing have been changed, since they by now are organized and have different call numbers).

36 Gugler's partnership with Henry J. Toombs (1902–1967) proved fortuitous. Toombs had worked with McKim, Mead & White, and began to work independently for the Roosevelt family in 1926.

37 Eleanor Roosevelt (E.R.) mentions Gugler eight times between 1938 and 1958. Gugler helped choose the White House furniture. He was the architect of E.R.'s New

build the White House's Oval Office (oval was his favorite form), followed by his alteration of its West Wing, the latter originally designed by White (of McKim, Mead & White).[38] Both commissions must have enhanced his feelings of being both important and American. However, while the Oval Office became a national icon the name of its designer sank into oblivion.

Gugler attained further prominence through a commission for New York's World Fair (1939), a treasure-trove of architectural experimentation.[39] This lead to his designing commemorative monuments for important people (albeit on a modest scale), which kept him occupied during the WW II years (and into the 1950s). These experiences lead to his being on the list of desirable architects, on the basis of experience with memorials.[40]

Although Gugler was busy during the decades preceding WW II, large projects such as Rockefeller Center, Radio City, and a skyscraper(s) did not materialize.[41] Little wonder that the offer of a commission for the Anzio/Nettuno cemetery seemed the opportunity of a lifetime. He would be able to bring his AARome sketches and drawings to bear at the cemetery, such as his romantic rendering of fountains with surrounding planting and trees, and his painstakingly copied map hall of the *Palazzo Vecchio* in Florence.[42] His Anzio/Nettuno would be decorated with maps,

---

Deal pet project Arthurdale, WV, a new homestead community for unemployed miners and their families. (The community hall has a temple façade). See Arthurdale Heritage. Org. Accessed March 29, 2018. often had lunch with her. In 1958 she referred to him and his wife, Anne, as "delightful people."

38 <http://community.seattletimes.nwsource.com/Archives/?date=19940619&Slug=1916273>. Accessed March 29, 2018.

39 Business systems and insurance building with architects John B. Slee and Robert H. Bryson.

40 Examples: monuments to Harvey Firestone in Akron, Ohio; William and Charles Mayo in Rochester, Minnesota; Waldo Hutchins in Central Park, New York. Clarke lists Gugler's memorial work in the latter's obituary. Clarke Papers, Columbia University Archives.

41 His early works are the Park Ave Squash Club, Bush House, London, circle at the Conde Nast plant, Greenwich, and the Connecticut residence of Paul Manship. When compared to the large buildings going up during the late 1920s and early 1930s in the US, these seem relatively modest structures.

42 Avery Library Archives, Columbia University holds a few of Gugler's sketchbooks (undated). Avery Library Archives, Columbia University NYC.

globes, and a zodiac ceiling. He probably would have seen his return to Rome a just requite after his ambitious design for the Spina di Borgo (the access street to St. Peter's) had fizzled.[43]

43 Gugler's attempt to leave a mark in Rome, while a fellow, through a grandiose design for the Spina di Borgo (Vatican) failed. For description and illustration of his design see Vincenzo Cazzato, *Ville e Giardini Italiani.* (Rome: Instituto Poligrafico e Zeca Dello Stato, 2004), p. 621–624.

Gugler's Anzio Design: A Vision of Grandeur

"When bereaved relatives or friends of the deceased visit the site, we hope that they will discover as they move around the building, distinguished and handsome elements to keep in mind the nobility of the intent of the Commission's and the authors' thoughts."

Eric Gugler[1]

Fig. 17. Anzio/Nettuno. A strict axial order.

Upon receipt of General North's letter of solicitation, Gugler met with Harbeson at New York's Century Club to discuss the matter.[2] Harbeson would

1 Description by Gugler of his design. RG 117, Entry 9, Box 122, Folder 687, (8–22–47 to 2–31–48), (December 17, 1947).

2 North sent the identical letter to all architects. The one to Gugler must have been written in the summer of 1947. RG 117, Entry 9, Box 122, Folder 687, (8–22–47 to 2–31–48).

provide contour maps and a design vision that Gugler should consider.[3] Harbeson advocated for his initial inspiration, a so-called "semi-arched" arrangement of graves.[4] The solicitation was divided into two stages, the first of which was a preliminary design consisting of two sketches in three dimensions. These would be judged on general proportions and not on detail. Should the design be accepted, a commission or contract would be offered, in which case the architect would be granted two trips to the site with a travel allowance of $1,250 for each trip. The architect would receive ten percent of the construction costs and four percent of said costs would go to the local representative under whose supervision the design would be executed.[5] The architect was entrusted with the site's aesthetic totality. Fencing, trees, and shrubs would isolate it from its surroundings. Large paved areas were to be avoided, as they would be too expensive to maintain. Clear specifications would firmly define those aspects where the architect could be creative.

General North had specified a "standard" for all the ABMC cemeteries to include a visitors' reception room, a chapel/memorial/museum, a wall of the missing, as well as a caretaker's house and service buildings, the designs of which must be "...complimentary to the designs of the memorial chapel and any other buildings."[6] In the preliminary first design stage, Gugler should locate these mandatory structures. Next, he would submit sketches for the commemorative center, which includes a museum and the wall of the missing (names) relating to the non-denominational chapel for visitors to pray quietly. The nation's flag was to be appropriately placed for ceremonies. Further, North wrote: "Since the cemeteries and their memorial chapels will serve as memorials to the achievements of the forces who fought in the respective region, each will include a chamber in which these achievements will be recorded by carving or in some permanent manner."[7]

---

3 Ibid. "Consulting architect shows recommendations made as a result of study of topography, aspect and prospect, including location of the main axis, areas to be used for burial, and preferred location for the chapel."

4 Letter, Harbeson to Gugler. He praises the latter's plan of radial layout of grave rows. RG 117, Entry 9, Box 122, Folder 687, (8–22–47 to 2–31–48), (October 10, 1947).

5 At that time, North estimated the construction costs at $460,000 for chapel, museum chamber, immediate surroundings, and flagpole. RG 117, Entry 9, Box 122, Folder 687, (8–22–47 to 2–31–48), (December 17, 1947). On the whole, North managed to stay within a given budget.

6 Ibid.

7 Ibid.

Thus, he thought in terms of "memorial," "armed forces," and "success," terms that indicate that the cemeteries were intended to celebrate victory.

Even though "tightly scripted," architectural solutions among the United States' fifteen cemeteries are surprisingly varied.[8] Aesthetic preferences of architects involved were able to unfold within this firm ABMC matrix of dignity and beauty. Cases in point are the two US cemeteries in Italy. The one at Impruneta is quite different from the one at Anzio/Nettuno. North stressed that the architect had stylistic liberties, but design should express national dignity. Commemorative monumentality should "beautify" the cemetery. "The Commission makes few requirements as to style other than the buildings express the dignity of their purpose, and of the government of the United States, and fit suitably into the surroundings where they are to be placed. Since there will be no other battlefield memorials in the region, it would seem proper that the chapel be somewhat monumental in aspect. At the same time, it should be simple and durable and waterproof, and contain heating and ventilation."[9] In the same letter North also passed on the consulting architect's (Harbeson's) design recommendations from his report from the summer of 1947. It suggested the location of the main axis, areas for burials, and the preferred location for the chapel, but he stressed that Gugler was not bound by Harbeson's recommendations. E.g. since the number of dead was not fixed (then estimated at 5,866), the grave plot areas could be extended, preferrably toward the outside edges, and not toward the axis.

The ABMC and the US Commission on Fine Arts (CFA) furnished Gugler with material on WWI cemeteries and provided him with a trip to view WWI cemeteries in France, so that he could "get the feel."[10]

---

8 One of the most interesting solutions is at Manila, Philippines. The enormous number of missing in action (lost at sea or in the jungles of southeast Asia) represented a great design challenge for its architect Gardner Dailey. He brilliantly created a large rounded peristyle consisting of large erect slabs covered with names. Gardner Dailey's papers are in the Archives of the College of Environmental Design, University of California, Berkeley.

9 Letter, North to Gugler. See fn. 2. For quote from Harbeson's letter see fn.3.

10 Hand-written addition by Paris officer in charge noting that Gugler and company partner Kimball were taken to Belleau (on April 8). The long unhindered vista over a densely tree-lined mall to the monument (a romanesque/gothic tower) and graves arranged in arcs likely became a source of inspiration. RG 117, Entry 38, Box 3, letter 4306, (February 16, 1949).

Given this level of practical detail, as well as design specifications, and given that Gugler's aesthetics were a known entity, there would likely be no surprises. The result would be a memorial cemetery in full conformance with the ABMC's and CFA's vision of how it should look and what it should message.

Gugler's Design

Gugler's design is the result of a process of consensus building. He accepted criticism from others without considering his artistic integrity compromised. North respected the inspiration of his architects and artists and consistently saw to it that no aesthetic decision would go against their will. He fought for the funding necessary to carry these through. Harbeson, the overseeing architect, although whole-heartedly involved in each cemetery's design process, never allowed ego to stand in the way of smoothing out personality clashes. Archives bear witness to this consensus building, that characterize a US communication style at that time.

Gugler vigorously took advantage of the design license he had been promised by orienting the whole site toward the lone stone pine, which Harbeson had noted on his initial sketch. He shifted the cemetery's plan to the south and left a wide strip of land at the north unoccupied in order to center the cemetery's axis on this tree. Having survived WW II battles as well as the AGRS' construction activities, it evidently held sacredness for both Gugler and Harbeson.

Gugler's first plan in December 1947 maintained Harbeson's original design, with modifications.[11] He oriented his design toward the lone pine.[12] Although he retained an emphasized entrance area, complete with

11 Letter, Harbeson to North telling him that he was pleased with Gugler's plan. This plan was used by the Italian land owners as the basis for their counter-design. See chapter I under "Site Acquisition at Anzio/Nettuno" p. 67. For plan and further drawings see CFA Record RR 66 NA (October 1947) and (December 10, 1947), RG 117, Entry 9, Box 122, (January 9, 1948).

12 Gugler developed a romantic fixation on this tree. It regulated his whole design.

wide gates, pillars, and the US eagle, his version had a different rhythm and sequence of spaces. He shortened and narrowed Harbeson's grandiose tree-lined entrance mall and replaced the round plaza with a large oval pool, which he moved closer to the entrance gate. He rearranged the grave areas from three to two, separated by the mall, and arranged the grave rows radially (with two center points) that seem to look toward the memorial on the knoll in the west. He kept the transition of entrance to grave areas; those closest to the entrance wrapped as semi-arches around the plaza with the pool. At its center, he placed an island, upon which was placed a prominent sarcophagus surrounded by dense trees. His axis would be uniformly wide from beginning to end. While his first layout plan lacked detailed museum and chapel, an early ink drawing testifies that he had a complete vision of the memorial area early on, including the two gardens lateral to the memorial structure.[13] Its central core would be a Doric temple façade.[14] The temple motif was common in the US at the time for civic buildings such as post offices and courthouses, connoting patriotism, republic, and national distinction. Here was his opportunity not only to express his love for ancient Greek art, but also to create a "little America" on Italian soil.[15]

By February 28, 1948, Gugler's composition of the memorial area design was in place. The requirement of a chapel and museum made it impossible to have an isolated temple structure. A model of some later date demonstrates how he solved this requirement.[16] The memorial consists of a three-partite unit of identical height with the central "temple" section and two solid, windowless attached lateral pavilions.

Photographed from above, the model makes it clear that Gugler conceived the memorial as a unified structure, albeit buttressed laterally by a chapel at its south side, (with a lowered apse in a medievalizing semi-circle plan) and by a museum at its north side. In the model, the

---

13 For ink wash by Gugler of the memorial area; RG 117, Entry 9, Box 122, (0028), undated. December 1947 seems feasible judging by correspondence.

14 Gugler made a beautiful large watercolor of his "temple" façade of the memorial. The tree is visible behind the façade. (The pine tree appears in all Gugler's ink washes). Smithsonian Institute Archives. "Architectural drawings, renderings and blueprints, Anzio Cemetery" Eric Gugler papers, 1889–1977. 4.5: Memorials, 1929–1977 ß. (Current signature).

15 See Wilbur Zelinksy, *Nation into State,* Chapel Hill: The University of North Carolina Press, 1988, p. 209.

16 RG 66, RG 117, Con 323, Box 19, (received March 9, 1950).

chapel's apse is fluted with an insinuation of columns, later to be executed as a smooth wall. A common cornice unifies the three-partite composition. The result is an eclectic creation, Gugler's brand of Greek classicism. The "temple" topos predominates. The chapel and museum, in simple cubic form with solid walls and without entrances, do not compete with the temple. Further, the façade, consisting of six Doric columns, projects forward a sizable distance, establishing itself as the capital piece. The view of the front is dominated by this impression of a temple façade. Lateral structures remaining subsumed by it contribute to the temple theme. Instead of frontal entrances, their facades are reserved for reliefs.[17] These, in a sense, complete the memorial's identity. Sculpture, traditionally found under the gable of a Doric temple, is here seen at its sides, comporting a symmetrical, classical coherence.

Fig. 18. Gugler's model of monument building, with sculpture and "lone pine" in open court.

17 Gugler's early sketches indicate two large globes, one celestial, and the other earthly. (He was fascinated with astrology, as will be seen on the chapel's ceiling). RG 117, Entry 9, Box 122, Folder 687, (8–22–47 to 12–31–48). The drawing must have been loosely added to the file, since it has no signature.

Fig. 19. Monument building. Note temple-like portico's central prominence.

In Gugler's skillful eclecticism, the temple façade is just that – a façade. Behind is an open, double-columned court with a rounded rear section projecting outward into the landscape, describing an oval, one of Gugler's aesthetic trademarks.[18] The dark and light effect achieved by the court's double peristyle of dark marble columns interacting with the trees behind is typical of Gugler's romantic play of light and shadow. In the center of the grass-covered court, a sculpture on a socle depicts two young men, later to be delivered by then-famous Paul Manship, a close friend.[19] The lone pine is on their right side, as it was in reality before it perished.

Through the museum's open door and three large vertical rectangular openings, the visitor's view is free to pass onto the sunken north garden, and to come to rest on its distal Roman-type fountain framed by background vegetation. An artfully curved staircase leads down into the garden.

The harmony the visitor senses within the court most likely results from Gugler's mastery of Vitruvian ratios and geometric shapes such as the oval

18 He used the oval for his home in Palisades, NY, and for the White House's oval office.

19 See page 175 for illustration.

(his favorite) as seen also at the White House's Oval Office. Anzio/Nettuno's plan for its memorial consists of a large square, two smaller squares and two rectangles, symbolic of life and resurrection. The basic shapes not only bring about harmonious proportions and relationship, but also have ancient symbolic meanings.[20]

Gugler, adoring Greek architecture, duly gave his columns an "entasis," or slight convex curvature.[21] He intended that the columns be an impressive 30 feet high,[22] but when advised that this would exclude allowance for sculpture and an entrance building, he unhappily responded that the column height could be reduced to 24 feet (about eight meters) but no further.[23] He accepted budgetary limitations, consoling himself with the thought that the court was a memorial structure and not a monument.[24]

Gugler's eclectic design deviates from classical aesthetics. The façade of the museum's garden side (to the north), has three large rectangular openings separated by narrow areas of wall, a prime example of "stripped-down classicism." The chapel's apse bears a band of graphic ornamentation in fine stone filigree. Deviating from classical temples, the space between the central columns is greater than the distance between

20 Per Michael Schneider, mathematician, senior adjunct professor, California College of the Arts in San Francisco. He has published about the underlying geometry of buildings and its symbolic meanings, e.g. on the underlying meaning of the Oval Office. <www.constructingtheuniverse.com/Oval%20Office.html>. Acessed March 14, 2018. As to Gugler's design his verbatim quote is, "…the symbolism he applied is clear and very beautiful and poignant: a square is a worldwide symbol of the earth (and those buried within it). A golden rectangle is a symbol of life. And most importantly, a square flanked on two sides by two small golden rectangles is known as a square-root-of-five rectangle. Since Egyptian times it has been a symbol of resurrection. The whole design shows it to be composed of a square and a root-5 rectangle. The large square has been divided into smaller squares, golden rectangles, and another root-5 rectangle." E-mail to author from May 15, 2013.

21 Barry Faulkner, *Sketches,* p. 114. He tells of Gugler's love for ancient Greek architecture. The memoir contains many interesting details about the AARome, Gugler, and Smith.

22 Memo by Harbeson. RG 117, Entry 9, Box 112, Folder 6878, (December 5, 1947).

23 Letter, Gugler to the ABMC. RG 117, Entry 9, Box 112, Folder 687, (8–22–47 to 12–31–48) (December 27, 1947); For his vision of Greek grandeur or a temple, see ink drawing in RG 119, Con 310, Box 17. But soon, once again Gugler was forced to give up his visions of classical architectural grandeur that had filled him since his Roman days as an AARome fellow (1913–15).

24 Ibid.

side columns, accentuating the axis's center while focusing the visitor's view on the sculpture erected behind, at the center of the memorial's court.

Gugler's Romantic Design Impulses

The ABMC and the CFA considered Gugler's impeccable classicism to be original, elegant, and appropriate. However, his personal vision also embraced the romantic, perhaps due to his stormy temperament. The two impulses stand in lively dialectic. This is evident in drawings in which Anzio/Nettuno's memorial is set in a staged, romantic landscape.[25]

Fig. 20. Gugler's pencil drawing, a romanticized classical vision.

The temple's façade is shown in central perspective in the distance, diminutive in its relation to the trees. The foreground is left empty and the image seems to float. Four high and irregularly shaped pines on both of its sides lead the eye to the façade. The trees seem animated, at though bending and swaying in the winds. Dark, forest-like growth behind introduces mystery.

25 Archives, Avery Architectural Library, Columbia University, Gugler papers.

Through contrasting of dark and light, emptiness and mass, the view falls on the lone pine crowning the façade's cornice. Nature's wildness makes way and welcomes the temple to its eternity. The lone pine sanctifies. Envisioning his work amid nature's forces, and given his German background, he was likely familiar with and influenced by the iconic German romantic painter, Caspar David Friedrich. At Anzio/Nettuno, Gugler's classicism forces his romantic impulses into subservience.

Ralph Griswold's "American Space"

> "It is your [Gugler] personal concept of this axial picture, your basic scheme, your building, your island, your vision. I am merely working out the details to meet the practical landscape requirement."
>
> Ralph Griswold[1]

Fig. 21. Grave fields beneath high umbrella pines.

Through what turned out to be a very fortuitous decision, Gugler sought out landscape architect Ralph Griswold (1894–1981).[2] Both had spent

1 Letter, Griswold to Gugler. Little is written about Griswold.The Cultural Landscape Foundation has a short entry undated and unsigned. Pittsburgh University holds Griswold's papers: For this letter see Pittsburgh University Archives, Ralph E. Griswold GWSM, Inc. Collection, AIS 2001.10, Box 3, Folder 3, (January 10, 1951).

2 Letter, Gugler to Griswold. Pittsburgh University Archives, AIS 2001.10, Box 13, Folder 1, (July 15, 1949).

time at the AARome and thus had first-hand experience with classical forms and ratios, and both recognized the central role of planting in the creation of space. Markley Stevenson, the ABMC's consulting landscape architect, had been involved from the project's inception, and Harbeson considered the design to be complete.[3] Gugler's firm had even completed the north garden's fountain design[4] when Gugler approached Griswold, almost apologetically, offering him a role in design even though major decisions had been made and approved, and only planting details remained.[5] In October 1949, while Griswold began thinking and writing about the design of the north and south gardens,[6] Harbeson warned of "the pervading possibility of war" and cited the army's need to "get its reburial work done as soon as possible."[7] Involvement in this project would afford Griswold the opportunity to sojourn in Rome once again. He accepted with grateful enthusiasm,[8] writing, "I feel that I can really do justice to the Anzio work in that [AARome] atmosphere."[9] As it turned out, Griswold's

3 Letter, Harbeson to Griswold: "… the general site development plan has already been done, in consultation with the commission's consulting landscape architect, who had acted in a previous capacity for the Graves Registration Service of the quartermaster general, these matters having been done under some pressure, due to the pervading possibility of war, and the army to get its reburial work done as soon as possible." Pittsburgh University Archives, AIS 2001.10, Box 13, Folder 1, (October 10, 1949).

4 Letter, Richard Kimball (member of Gugler's staff) to Griswold. Pittsburgh University Archives, AIS 2001.10, Box 13, Folder 1, (October 21, 1949).

5 Letter, Richard Kimball (member of Gugler's staff) to Griswold. The exact responsibilities would be specified later. Pittsburgh University Archives, AIS 2001.10, Box 13, Folder 1, (October 26, 1949).

6 Letter from Griswold to F. Eisemann (member of Gugler's staff). Pittsburgh University Archives, AIS 2001.10, Box 13, Folder 1, (October 9, 1949).

7 See fn. 32.

8 Griswold wanted to go to Rome. He had to "wiggle out" of the commission for St. Avold, France, saying that he wanted to complete his education. Markley Stevenson wondered about that in his letter to Donelson, since he had completed his studies at the AARome some 25 years earlier. RG 117 Entry 13 Box 143, Folder 231.24, (August 29, 1949); St. Avold's architects, Murphy & Locraft, agreed to release him (per letter) from Gugler to Donelson. Griswold stayed at the academy for five months and worked out the plans for Anzio/Nettuno. RG 117 Entry 13 Box 143, Folder 231.24, (August 30, 1949).

9 Letter, Griswold to Gugler. Griswold informs that the academy had made him an offer for a stay. Pittsburgh University Archives, AIS 2001.10, Box 13, Folder 1, (October 3, 1949).

whole-hearted involvement would be far more than miscellaneous "fixes" for leftover problems. With persistence and vision, the project became a magnificent reality with grandeur of landscape design.

Griswold gathered the plans, elevations, and sectional drawings of the monument building and set out for Rome, where he obtained exact topographical maps and surveys of boundary lines. Residing at AARome, he could concentrate on design. Photos of the site as he first saw it show a large construction project underway, with ditches for the permanent graves and irrigation systems, and wooden crosses. He noted his reaction in his journals: "The views around the cemetery are not inspiring but they are not as bad as they have been described. There is an opportunity to do something quite distinguished. Now, the problem seems real. It has three dimensions and emotional content. It can be beautiful."[10]

While Gugler and the ABMC exchanged letters, Griswold forwarded his sketches, photos, and plans to Colonel Walker (the ABMC's emissary in Rome). From there, these went to Washington, D.C., to be passed around for review and comment, and finally to be forwarded to Gugler in New York. In an era when communication was less than instantaneous, misunderstandings were inevitable.[11]

Griswold grasped the essence of Gugler's thematic central sight line.[12] He envisioned an expansive mall bordered by a path on each side, each path accompanied by a row of oaks in "… proper proportion of width and length of the mall in relation to the memorial."[13] The trunks would be pruned to a height of eight feet (2.5 meters). Hedges would frame the mall as well as define the grave fields. Hedges toward the grave fields would be higher[14] so as to deemphasize the view onto the graves, emphasizing instead the memorial building ahead. Crosses

10 Entry from Griswold's journal. Pittsburgh University Archives, Griswold papers, (December 19, 1949).

11 Letter, Eisemann to Griswold. "Not everyone sees everyone else's response, and sometimes references to each others' letters get mixed up." Pittsburgh University Archives, AIS 2002.10, Box 13, Folder 2, (March 7, 1950).

12 Pittsburgh University Archives, Griswold Papers AIS 2001.10, Section D-D, Box 9 FF4, (undated, but from other drawings can be assumed to be from January 1950).

13 Letter, Griswold to Gugler. Pittsburgh University Archives, Griswold Papers AIS 2001.10, Section D-D, Box 9 FF4, (January 4, 1950).

14 Letter, Griswold in Rome to Gugler. Pittsburgh University Archives, AIS 2002.10, Box 13, Folder 2, (January 4, 1950).

would be viewed from the so-called "curved avenues" that intersect the central mall. Griswold argued, "I cannot see how the memorial can hold its own in your scheme against an eye-level panorama of white crosses. I think that they should each be seen in their turn with their proper backgrounds."[15] This caused serious ideological controversy beginning with the ABMC's Colonel Walker, who stated he wished to be able to see the graves better from the memorial. Griswold insisted: "If the grave areas were intended to have been emphasized in the view from the portico then the mall should not have been made the dominating feature in the design."[16]

A struggle between conflicting visions ensued. The cemetery was to embody both aesthetic and ideological, or "sentimental" dimensions, but which of the two would be its main focus? Would the cemetery memorialize individual sacrifice or the armed forces' achievements? Griswold's garden design favored the latter, and protest came immediately from various sides. Markley Stevenson thought the emphasis should be on the graves: "The graves, with their crosses, are the most interesting and important element in the cemetery. The memorial building is, in a sense, a memorial to each grave … I feel that they are the very important foreground and the *raison d'être* for the memorial."[17] Harbeson shared this view: "The crosses are the reason for the being of the memorial, are an integral part of it, and are in their pattern in the grass quite as beautiful as the memorial …" and "… crosses and the memorial are an inseparable part of one idea, not two."[18] General North, diplomatically supporting his architect, as was his custom, assuaged critics, assuring them that the importance of the graves would not be diminished, and noted that topography would guarantee that one would not "look down upon the crosses," from the memorial terrace.[19]

---

15 Ibid.

16 Ibid.

17 Stevenson's comment on Griswold's letter to Gugler from January 24, 1950 and prints submitted. Pittsburgh University Archives, AIS 2001.10, Box 3, Folder 3, (February 15, 1950).

18 Letter, Harbeson to North. Pittsburgh University Archives, AIS 2001.10, Box 3, Folder 3, (February 7, 1950).

19 Letter, Eisemann (Gugler's office) to Griswold. Refers to letter from Griswold to Gugler assuring him that the vista toward the memorial would not be obstructed. Pittsburgh University Archives, AIS 20001.10, Box 14, Folder 2, (January 12, 1951); Pittsburgh University Archives, AIS 20001.10, Box 14, Folder 2, (March 4, 1950)

Griswold did reduce the height of the hedges bordering the grave fields, and at the same time prevailed in his view that the memorial dominate at first sight, thus staunchly supporting the prominence of Gugler's temple-like memorial. He convincingly maintained that he was not neglecting graves, but for aesthetic reasons they needed to be enclosed by hedges due to the flatness of the site.

"... from nowhere (a view over) the entire pattern of the design is possible. The only pattern that is recognizable in the burial areas is the area in your immediate foreground. The graves would merge into a patternless mass without hedges around the grave fields. Pattern of crosses depends on the light. Hedges accentuate the pattern in the immediate foreground of your view."[20]

In other words, Griswold argued successfully that in this case, the graves could not be properly presented without hedges. Further, he stressed that a strongly emphasized mall must be in place on such a flat site in order to "hold the design together." From a distance, the impression of the land would be difficult to assess. "... the vastness, the flatness, the brilliant light, and the wide-open spaces stretching from the mountains to the sea would create a much different feeling than looking at a small-scale plan indoors."[21] A strong, clear, and asserted vision was needed to turn the site into a space. Griswold's Pittsburg friend, while visiting the site, was shocked and exclaimed: "What the hell could be done to retrieve that God-forsaken site?"[22] Such a site needed strong planting to shape its "... uninteresting middle ground and lead the view beyond to the mountains in the distance."[23]

Thus, as the result of consideration of many factors and much diplomatic consensus building, the memorial building was given visual preeminence over the grave fields, and the debate shifted to whether the oaks accompanying the mall's two paths would hinder the view onto the memorial

20 Letter, Griswold to Gugler. Pittsburgh University Archives, AIS 20001.10, Box 3, Folder 3, (April 4, 1950).

21 Ibid.

22 Letter, Griswold to Paul Manship. The quote is by Griswold's friend Bill Block, owner of the Post Gazette, a Pittsburgh daily newspaper. Pittsburgh University Archives, AIS 20001.10, Box 14, Folder 2, (January 8, 1951).

23 Letter from Griswold to Gugler. Pittsburgh University Archives, AIS 20001.10, Box 14, Folder 2, (April 4, 1950).

building. Shaping of the mall would require time and more consensus building. Griswold specified that oaks have "architectural" trunks, and therefore would never grow into a continuous visual wall. Rather, with each step the visitor takes along a path, these oaks would obscure the view less and less.[24] Griswold referenced a famous standard landscape architectural work of French gardens with large illustrative plates.[25] (Versailles and Chantilly were the ultimate authority for his school of design.) Still, axis-obsessed Harbeson was not completely convinced and held to the notion that the oak walks should be eliminated completely.[26] Griswold retorted: "… a skeleton with nothing but ribs and no backbone, I think we need the backbone in this scheme … because the land is so flat it needs a strong order. This is another reason why it is desirable to give the mall a strong third dimension to hold the design together."[27]

North suggested the access paths be located to the outside (not between) the rows of oaks. At this point Griswold realized he needed to convince his critics visually.[28]

Griswold went to the site, drove white stakes into the ground, photographed, and then made beautiful perspectival images that combined photographs with ink wash. With oaks spaced ten meters apart, the view onto the memorial would remain in scale with the width of the open mall, and thus he demonstrated that his "oak walk" would impede neither the view onto the memorial nor onto the grave fields. His concept thus prevailed and the much-contested double row of oaks with hedges and pines along "curved avenues" (between grave fields) remained in the plan. His vision that the view from the memorial offered only glimpses of the grave fields prevailed, as did his contention that the crosses should not be seen in mass.

---

24 Ibid.

25 P. Péan, *Les Jardins de France.* A then-influential standard work with plates that was often quoted as an authority, such as flanking masses of trees and their effect, or a path with only light as its terminus.

26 Letter, Harbeson to North complaining that the main axis does feel hemmed in through the planting. Pittsburgh University Archives, AIS 20001.10, Box 14, Folder 2, (March 12, 1951).

27 Letter, Griswold to Gugler. Pittsburgh University Archives, AIS 20001.10, Box 14, Folder 2, (April 4, 1950).

28 Undated ink drawing. University of Pittsburgh Archives AIS 2001.10, Ralph E. Griswold Collection 1912–1981, Box 69, (undated, but must be from ca. April 1950).

Fig. 22. Griswold's montage demonstrating that trees will not obscure view onto monument building.

As Gugler's memorial was to be the cemetery's dominant element, Griswold created a pleasing geometric pattern before it, one that would facilitate free movement, even for larger groups, with interlocking paved and low groundcover areas. This designed space before the memorial engenders pause, for thoughts and feelings as well as for rituals.[29]

By virtue of his Beaux-Arts training and an AARome fellowship, the ease with which Griswold filled the voids in Gugler's plan is striking. The visitor senses harmonious order based on classical rules and ratios. Issues such as, for example, how many trees should surround the flagpole (it was *de riguer* that flags dominate tree crowns),[30] or how best to connect the

29 All US cemeteries have such an area of open space in front of the memorial.

30 Letter, Gugler to Griswold. General North kept watch as well. Pittsburgh University Archives, AIS 20001.10, Box 3, Folder 3, (September 5, 1950).

"oak walks" to the gardens were intensely debated, yet Griswold's solutions consistently relied on a geometric imperative, and the basic plan's self-confident *grandeur* remains unadulterated.

Gardens of Respite from Death

The cemetery's two gardens, smaller units, lie hidden behind retaining walls and rich planting, even though they are accessible from the mall. The north garden, lying lower than the museum, may also be reached over a short terrace and elegantly curved steps. Drawing from Renaissance garden illusionistic techniques, the space is made to seem larger than it is.[31] At the north garden's northern end, a semi-circular fountain of granite is set against a wall of dark shrubs and high trees.[32] Although Griswold did not design the fountain, he became involved with its placement at proper height on a terrace, with the wall of shrubbery shielding it from behind, and with the orientation of its jets so it might be seen from various parts of the cemetery.[33] Perhaps recalling the iconic gardens of Tivoli, he envisioned two smaller jets from its lower rear wall, "like candles on an altar."[34] The garden quoting directly from the Renaissance pleasure garden-with-fountain tradition and isolated from the cemetery's grave fields, creates a welcome reprieve from the confrontation with death.[35] Nevertheless, an unusual planting feature endows

31 As a fellow at the AARome, Griswold produced plans for the reconstruction of the gardens of the Villa Caprarola, as well as for other gardens of villas. See Vincenzo Cazzato, *Ville e Giardini,* p. 358. The book is a treasure trove of information with many fine illustrations of designs by AARome fellows.

32 Villa Farnesina, Villa Borghese fountain.

33 Letter, Griswold to Gugler. Pittsburgh University Archives, AIS 20001.10, Box 3, Folder 3, (January 4, 1950).

34 Ibid.

35 Letter, Griswold to Gugler. "… the granite fountain in the north garden has all the grandeur of the Renaissance, but again requires the cypress background to complete the picture." University of Pittsburgh Archives, Ralph Griswold collection, AIS 2001.10, Box 7, Folder 7, (July 13, 1955).

the garden with a military destination. Between museum and fountain, four US (five pointed) stars of white gravel fill the garden's central bed, framed in boxwood and flowerbeds. The emblem of the US Armed Forces' "stellar" accomplishments is thus set into Italian earth.

Fig. 23. North garden, view onto museum. Note the symbolic "Army Blue Stars" embedded in planting.

The south garden begins at the semicircular outer wall of the chapel's apse, where steps adjacent to a narrow terrace lead down to it. For budgetary reasons, the ambitiously curved steps conceived in the beginning were later simplified, following Griswold's suggestion.[36] A central empty lawn space is bordered by semicircular rimmed retaining walls of *tufa* stone. A row of trees and massed flowers on three levels offers scenery familiar to Americans: "… Lebanon cedars at the end and two romantically shaped,

36 Letter, Griswold to Gugler. Pittsburgh University Archives, AIS 20001.10, Box 3, Folder 3, (August 14, 1950).

heroic sized stone seats slightly baroque in character."[37] A thick layer of stone pines guarantees protected seclusion. With meadow space, flowers, shrubs, and trees recognizable by American visitors, Griswold provided a sense of wellbeing and home, a reassurance of American values. As of this writing, both these gardens conform to Griswold's original specifications with only minor modification, such as spring-blooming plants replaced with summer-blooming varieties once it was realized that Americans would mostly visit in summer.

Anzio/Nettuno was a unique opportunity for Gugler, and all evidence points to his intention that it be his masterpiece. He found the perfect partner in Ralph Griswold, a master at his craft as well as a person who seemed not to have difficulty working in the design shadows of a tempestuous architect. Gugler enthusiastically welcomed Griswold's design.[38] The latter returned the compliment: "In my estimate the beauty of the Anzio/Nettuno cemetery is attributable to a refinement of architectural details intended to satisfy a sense of design far more sensitive than mine."[39]

Gugler, presenting himself as an avowed classicist wrote: "Outside the cemetery can be hidden, but inside I would like to maintain every iota of formality for the eye of the person entering the gate and nowhere else."[40] However, the romantic in Gugler also wanted the surprise and mystery of a magical, hidden world. Griswold's superb planting design answers Gugler's vision. From outside, the cemetery should appear as if part of an Italian neighborhood: "This would make the surprise and delight of a formal site the stronger."[41] The wide and open, very formal entrance would seem to work against the hoped-for surprise.

37 Ibid.

38 Letter, Gugler to Griswold. Pittsburgh University Archives, AIS 20001.10, Box 3, Folder 3, (April 11, 1950).

39 Letter, Griswold to Gugler. Pittsburgh University Archives, AIS 2001.10, Box 14, Folder 5, (November 19, 1964).

40 Letter, Gugler to Griswold. Pittsburgh University Archives, AIS 2001.10, Box 3, Folder 3, (April 11, 1950).

41 Ibid.

Fig. 24. Ornate entrance with US emblem guides view onto distant monument building.

However, drama begins once the visitor enters and encounters a sequence of strong visual contrasts. Mediterranean light is set against the darkness of a grove. An oversized sarcophagus of bright marble on the pool's island is made more prominent by framing with the dark green of cypress. Waters make intriguing reflections of island and sky. There are stands of oak, an emerald green "manicured" mall lined with trees, and further on, the bright memorial where stone pines stretching skyward hover above Doric columns. Sequencing contrasts of light and dark, openness and closure, all within the clarity of its formal site plan produce visual drama that seem to militate against classicism's "calm grandeur." Griswold made a reality of Gugler's dreamy vision, where pine crowns sway in wind while intimate shrubbery embraces their trunks. Mystery and wildness dialogue with the temple's classical repose.[42]

42 Unsigned and undated drawing discusssd above. See Gugler Papers Avery Library Archives, Columbia University.

The mysterious darkness of trees in the Italian sunlight suggests that Gugler was familiar with Arnold Böcklin's "*Toteninsel* (Island of the dead)."[43] Trees on the island would have a *slita* (slot) five meters wide to guide the view.[44] Gugler very specifically described the *slita* as elliptical in form, open to the sky, thus to create a sightline over the pool, from entrance to memorial: "… the only axis experience will be through the slot, that should be narrower than the distance between the surrounding trees."[45] Gugler's "aesthetics of control"[46] was thus executed with traditional clarity and resulted in a narrowed view.[47] Gugler asked for Griswold's support for his eccentric romanticized vision,[48] and Griswold responded with respect and modesty: "It is your personal concept of this axial picture, your basic scheme, your building, your island, your vision. I am merely working out the details to meet the practical landscape requirement."[49]

---

43 Letter, Larry White (partner of McKim, Mead & White) to J. Kellum Smith "island à la Boecklin."The NYHS McKim, Mead & White collection. PR 042, Box 452, (May 6, 1964). Perhaps, as Dr. Fuhrmeister suggests, Rousseau's Island in Geneva could have been a source of inspiration as well.

44 Unsigned and undated watercolor. Gugler's handwriting identifies it as his own. Furthermore, a total view of the site shows Griswold's design process, conceiving of plant massing to establish structure. Pittsburgh University Archives, Griswold Papers, (August 25, 1950).

45 Letter, Griswold to Gugler. Pittsburgh University Archives, AIS 2001.10, Box 3, Folder 3, (July 21, 1950).

46 Campanella, Thomas J. used the term. See "American Curves – The Public Works Legacy of Gilmore D. Clarke and Michael Rapuano." *Harvard Design Magazine.* Summer 1997.

47 One is tempted to see an influence of the view through a keyhole of a monastery door on the Aventine hill, which frames a famous view of St. Peter's Cathedral.

48 Letter, Gugler to Griswold, "We will have to go to bat about this one" (clipping trees around pool for slot view). Pittsburgh University Archives, AIS 2001.10, Box 14, Folder 2, (January 8, 1951).

49 Letter, Griswold to Gugler. Pittsburgh University Archives, AIS 2001.10, Box 14, Folder 2, (January 10, 1951).

Fig. 25. Gugler's watercolor reveals romanticized "Island of the Dead." Its waters reflect a cypress-bedecked sanctuary with shrine (later morphed into sarcophagus).[50]

The Saga of the Lone Pine

Griswold did his best to accommodate Gugler's attachment to the lone pine, sole survivor among comrades that once lived upon the site and were destroyed during battle. So enraptured with this tree did Gugler become, that it figured prominently in all his plans and drawings, and as mentioned, he re-oriented the entire site plan so as to cause it to shade the memorial's major sculptural element. The importance it held for Gugler was obvious from the beginning. When, in 1948, AGRS surveyors had mistakenly located it on a contour map (it actually lay 6.27 meters to the south of Gugler's axial

50 The watercolor displays Gugler's artistry also as a watercolorist. RG 117, Con 322, Box 19.

centerline),[51] it was to be transplanted.[52] (It would have interfered with the chapel's roof). In the winter of 1948 its roots were carefully pruned and it was replanted "so that branches hover over the figures and the taller ones embrace the court at the most attractive angle."[53] But alas, it refused to thrive. Gugler lamented: "The pine tree in the meadow was the original source of inspiration for our whole building and now there's a threat to remove it. It adds too much to the setting of the sculpture and Manship is mad as hops too."[54] Drainage proved to be ineffective. North consulted an Italian agricultural specialist on the pathology of pine trees (Prof. Rosa). A Prof. Sibilia suggested draining its roots. Griswold suggested raising the tree 18 inches and planting ivy around it.[55] Despite all efforts, by 1965 it had perished and a replacement was planted, obviously according to Gugler's wishes.[56] It too, as if cursed, failed to thrive and presented the danger that it might topple in a windstorm and destroy the chapel. Griswold wrote to North that Gugler would be "heart-broken" and that he, too, would hate to see it go.[57] For Gugler, the tree would root his work in Italian soil, but it was not to be. The new tree perished as well.

Griswold's deference to Gugler is noteworthy, but he would claim and defend his own contribution as well. For example, when hostilities erupted on the Korean peninsula and wider war in Asia loomed, fear of budgetary cuts plagued all those involved with Anzio/Nettuno. By the end of

51 ABMC Walker in Rome thinks this error will cause an "unbalanced effect to the entire design." Gugler handled it by giving center space to Manship's statue and having the tree next to it. RG 117, Entry 8, Box 122, (September 1, 1948).

52 Letter, North to the ABMC Rome, a qualified nursery needs to be involved in the moving. RG 117, Entry 9, Box 122, (October 21, 1948).

53 Letter, Eiseman to North. RG 117, Entry 9, Box 122, (October 20, 1948).

54 Letter, Gugler to Griswold. Pittsburgh University Archives, AIS 20001.10, Box 14, Folder 4, (March 28, 1961).

55 Various letters describing the problems in the Pittsburgh University Archives' Griswold collection.

56 Letter, North to Gugler telling him that the new tree is thriving. Pittsburgh University Archives AIS 2000.10, Box 14, Folder 6, (October 14, 1965).

57 Letter, Griswold to North, cc to Gugler. "I do remember that Eric located the main axis of his design to take advantage of the beautiful old pine that was growing in this approximate location … I know Eric would be broken-hearted if the pine tree had to be given up. I, too, feel that it is an integral part of his design and would hate to see it go." Pittsburgh University Archives, AIS 2001.10, Box 14, Folder 5, (April 10, 1967).

1950, this became a reality.[58] Gugler wanted to simplify the area around the memorial (that Griswold had designed with such care) and Griswold protested, writing: "From a design point of view most of the changes you propose are over-simplification, reducing the landscape to a state of mediocrity completely out of character with the architecture."[59] Later, Gugler made the suggestion to "… replace the hedge on the long, broad mall by planting globes of boxwood at intervals…." Griswold countered: "It would confuse, not relieve, the long vista and completely destroy its beauty. Soon all will reveal its grandeur of scale, simplicity and dignity."[60] The idea was duly dropped. When, in 1956, General North made known his intention to replace Griswold's carefully selected annuals with perennials as a measure to save costs,[61] he once more successfully defended his design: "Gardens were designed to be in character with the classical style of the memorial building and the best traditions of Italian gardening."[62]

Transcending the Classicist Model

Griswold's spatial vision was indispensable to Anzio/Nettuno's unique beauty. His drawings, his remarks and our own critical views give evidence that he thought in terms of light and shade, and of masses and

58 General North informs Gugler that the President's budget would defer ABMC's program. In his report on Anzio/Nettuno from July 1953, Griswold writes that everybody fears Washington might curtail funding. Pittsburgh University Archives, AIS 2001.10, Box 13, Folder 2, (December 21, 1950).

59 Letter, Griswold to Gugler. Pittsburgh University Archives, AIS 2001.10, Box 14, Folder 2, (January 28, 1951).

60 Letter, Griswold to Gugler. Pittsburgh University Archives, AIS 2001.10, Box 14, Folder 4, (December 16, 1954). J.J. Winckelmann's "noble simplicity and calm grandeur" seems to have been absorbed into the aesthetic principles of American classicists.

61 Letter, Harbeson to Griswold saying that the change from annuals to perennials was a cost matter and budget authorities ultimately "hold the destinies of our work in their hands." Pittsburgh University Archives, AIS 2001.10, Box 14, Folder 4, (September 17, 1956).

62 Letter, Griswold to Walker. Pittsburgh University Archives, AIS 2001.10, Box 14, Folder 4, (August 29, 1956).

generous free spaces.[63] A beautiful drawing in green pencil of the north garden's fountain shows this. Another drawing of the same scene, in black pencil alone, suggests that masses were Griswold's design starting points. His drawings reveal a sense of varying heights, depths, and shapes, with Anzio/Nettuno's stone pine canopies hovering above formally arranged mid-sized trees, and ubiquitous thick hedges and fountains.[64] His formative exposure to Italian gardens is recognizable.[65] His photographs reveal an eye trained in compositional patterns of shapes and texture. The Italian Renaissance's concept of designed space remained his constant frame of reference.

To Gugler's formulaic indication(s) of planting, Griswold rendered specificity. Actual designs were realized with great security, for example, with double hedges along the trunks of the stone pines to balance their mightiness. However, Anzio/Nettuno's landscape grandeur is of a different sort than that of much-admired models, such as the famous Tivoli gardens. Through mastery of proportion and various levels of height, the cemetery site's flatness is transformed into a space that breathes openness. Griswold was aware of the psychological effect of various types of trees: "Tall, magnificent umbrella (stone) pines dialog with the heavens, subtly bring them into consciousness, into our sphere, inviting them to work with the cemetery's sculpture."[66]

After Anzio/Nettuno, Griswold went on to design the grounds of the Agora in Athens, and often stopped over in Rome to check on the cemetery.

---

63 Letter, Gugler praising the effect of light at the end of a path as found in Versailles park. He wanted to build up four terraces at the pine grove behind the memorial because "the light and shade effect would be great." Pittsburgh University Archives, AIS 20001.10, Box 3, Folder 3, (January 4, 1949).

64 Letter, Griswold to Gugler, mentions planting bamboo thickets with groups of range-stemmed willows and calla lilies along the stream bank (fosco dei Tinozzi). He had seen the combination at Prince Gaetani's Ninfa Garden (half an hour away). Gugler's reaction was, "You are a great landscaper … the orange stemmed willow pruned back until a plume of twigs burst out of an old stump like a flame from a forge. Rows of these willows will cheer the saddest soul. I read it to Barry [Faulkner] at dinner tonight." RG 117, Box 13, Folder 2, (March 8, 1950).

65 Letter, Kimball to Griswold. "I hope you are wallowing in Italian gardens." Pittsburgh University Archives, AIS 20001.10, Box 3, Folder 3, (December 30, 1949).

66 Letter, Griswold to Gugler gives the insight that he made aesthetic decisions, such as choosing high cypresses also for psychological reasons. His reference to "sculpture" presumably refers to the man-made elements such as architecture. Pittsburgh University Archives, AIS 20001.10, Box 13, Folder 3, (January 11, 1950).

He advocated tirelessly for the correct planting: "I have a sentimental interest in the success of this landscape and will gladly contribute my services without further financial obligation."[67] Personally committed to the fulfillment of his vision, he continued his work "pro bono" (as did Gugler who on his celestial ceiling, to be discussed).

Griswold, as Pittsburgh's chief landscape architect, designed Point State Park, a flat site wedged between steep gorges.[68] There, the Renaissance was certainly a recognizable source of inspiration, but not an obligatory command. The park's site fronts on the confluence of mighty rivers, their waters racing through the mountains toward the promise of Ohio's vast open lands beyond. At Anzio/Nettuno, one may sense a similar promise of openness. The visitor strides freely. Paths are wide and shaded through canopies of trees, and views seem unencumbered. With each step, gleaming marble crosses re-arranged themselves into new constellations, tempering grief with a quickening experience of the kinetic in which spatial order frames and reassures one's emotions.[69] Anzio/Nettuno represents a triumph of American landscape architecture. Ironically, his contribution, according such little importance at the project's outset, became essential to its success. Griswold, considered a strict "traditionalist" by the AARome critics, burst the boundaries of the AARome's classicist canon.[70] The cemetery is an idealized "little America" on foreign soil, an American space, generous and free, benefitting from Renaissance patrimony but transcending it. It recalls the experience of the wide-open American midwest of Griswold's youth. It is not a space of triumph, but of freedom.[71]

---

67 Letter, Griswold to Harbeson. Pittsburgh University Archives, AIS 2001.10, Box 4, Folder 4, (April 5, 1965).

68 The park has been rebuilt again since Griswold's work, and he is not mentioned as its original designer in Wikipedia Foundation entry under Point State Park (state of April 2015).

69 Journal entry: "Now the problem seems real; it has three dimensions and emotional content. It can be beautiful." Later, he put gladiolas down on the graves of fallen soldiers from Pittsburgh. Pittsburgh University Archives, Griswold Papers, Griswold's Journal, (December 19, 1949).

70 Costanzo, "The Lessons of Rome." The author points out that Griswold together with Gugler and Manship were considered outdated old guard in the 1960's. see chapter III, fn. 9.

71 Letter, Griswold to Gugler saying all questions raised about the design in Anzio/Nettuno are based on scale. University of Pittsburgh Archives, Griswold Papers Collection, AIS 2001.10, Box 13, Folder 3, (March 13, 1950).

IV. The Best of American Art? Paul Manship and his Sculptures at Anzio/Nettuno

"To the Ultimate do we pursue the Ideal"
Paul Manship[1]

A memorial cemetery's architect chose the artistic staff, and Eric Gugler selected fellow Minnesota native Paul Manship (1885–1966) for sculpture at Anzio/Nettuno.[2] The two had been at the American Academy in Rome (AARome) together and had become great friends.[3] Both came of age during the period between the two world wars and each achieved high repute in his respective circle. Both continued working in the style and artistic idioms developed earlier in their careers. One might well wonder whether they felt marginalized in light of the enormous changes that were occurring in the art world of the 1940s (and later). Manship's work at Anzio/Nettuno, derived from European art's safe and venerated traditions, nourished by patriotism, optimism, and rectitude, constituted "the best of American art" according to General North.

Young Manship lived in Paris when Rodin's fame was at its peak. He had his "Rodin period," during which his sculpture became noticed.

---

1 The inscription is on a portrait coin Manship made of the AARome co-fellow (1909–1912) and friend Barry Faulkner, Smithsonian Institute Collection of Paul Manship's works.

2 Bibliographic sources for Paul Manship: Paul Vitry, *Paul Manship, Sculpteur Americain*, (Pairs: Editions de la Gazette des Beaux-Arts, 1927). Herbert Kammerer, "Paul Manship," *National Sculpture Review*, vol. 15, (fall 1966). Edwin Murtha, *Paul Manship*, (New York: MacMillan, 1957). Harry Rand, *Paul Manship*, (Washington, DC: The National Museum of American Art by Smithsonian Institute Press, 1989). John Manship, *Paul Manship*, (New York: 1989). Donald Martin Reynolds, *Masters of American Sculpture: The Figurative Tradition from the American Renaissance to the Millennium*, (New York: Abbeville Press, 1993). Wayne Craven, *Sculpture in America*, (New York: Thomas Y. Crowell Co, 1968).

3 Fikret K. Yegül, *Gentlemen of Instinct and Breeding: Architecture at the American Academy in Rome, 1894–1940*, (Oxford University Press: New York, 1991), p. 48–49. And see Faulkner, *Sketches,* P. 116, where he mentions that Manship admired the Romanesque frescoes of St. Savin in Poitiers.

In 1909, he was awarded a fellowship to the AARome, and this supported his sojourn there for three years, during which he traveled extensively throughout Italy and Greece. He settled in the New York area and remained a hard-working artist. As his career unfolded, he became an AARome trustee (1942), a member of the American Commission on Fine Arts (CFA) and was feted with numerous other prestigious titles and accolades. He lived the genteel life of a correct, educated, and involved gentleman, a member of New York's Century Club who sought to enhance life through beauty: "He did not deal with society's problems but glanced back to a golden age."[4]

Manship's refined sculptural groups in small scale often use the female figure to express idealized subject matter removed from everyday life.[5] One recognizes the influence of French art, such as "La Source" (Ingres), in various stylistic iterations. Often, he idealized Native American (American Indian) virtue with small figurines hunting or running, or represented classical mythology. Technically perfect, these small-scale sculptures were much admired. His styles echo details seen in ancient and Pompeian decorative art (also archaic statues, Assyrian, Egyptian and Minoan art) and were as eclectic as his subjects. He consistently presented the human body, or animals, in a blend of classical canon, *art-nouveau* flow, and decorative patterning reminiscent of French Romanesque. To this mix, he added exotic flourishes from Indian and Buddhist art. His forms' smooth elegance fit the taste for Art Deco aesthetics, which had become "all the rage" in the US during the late 1920s and 1930s. It has been noted, however, that their stylistic smoothness was not matched with depth of expression.[6]

Manship today is primarily known through his sculpture, "Prometheus," at New York's Rockefeller Center (1934), a project that showcased the work of America's artistic elite. Prometheus, on an abstracted

4 William S. Stott article in conjunction with exhibition from May 19 to August 18, 1985. Minnesota Museum of Art Landmark Center, (Saint Paul: Minnesota Museum of Art, 1985).

5 Most of his work is in the Smithsonian Institute, Washington, D.C.

6 David H. Wright, "The First World War Memorial in the Cortile of the American Academy in Rome 2000" in *Rome and Her Monuments: Essays on the City and Literature of Rome in Honor of Katherine A. Geffcken,* eds. Sheila K. Dickinson and Judith P. Hallett, pp. 57–80. (Wauconda, IL; Bolchazy-Carducci, 2000). "His real inspiration was the archaistic decorative work to be seen in Rome and Naples 'Pompeian kitsch,' exaggerating expressive effects of a few crucial details, eyes, mouth, and turn of head." p. 63.

low hill covered in shining gold, drifts sideways with outstretched arms and parted legs, as if borne by the winds. He holds a bundle of fire in one hand. The celestial context is shown through the encircling ring of the zodiac. The form and curve of the body pre-figure his later compositions and are recognized at Anzio/Nettuno. Albeit eclectic, the Prometheus figure works artistically and remains on display as of this writing. His Art Deco Ur-Man and Ur-Woman were originally part of the group and later were separated for independent display. Their style, smooth and eclectic with Art Deco patterning and immediacy of movement, took New York by storm.

Prometheus launched Manship in the field of public art and it is no surprise that he was given a major commission for New York's World Exhibition of 1939, a showcase for prominent US architects and sculptors.[7] Manship created a monumental sundial for its entrance, 25 meters tall, decorated with the three parses from Greek mythology. It also displayed figures depicting stages of the human life cycle and man's place in the universe. Their titles are: "Zodiacs Cycle of the Year," the "Compass of Life," and "The Garden of Eden." The "comprehensiveness of life" that reached beyond the human sphere to the stars, for Manship, was a major theme, as it was for US artists of his generation. Gugler shared this inspiration, as discussed above, and found its expression at Anzio/Nettuno's chapel. This urge to represent the totality of human life in art, and especially to represent the concept of progress, became a powerful cultural phenomenon. It witnessed America's claim to a special place in the cosmology of human events. It set the stage for the US's post-war mindset, a self-confidence clearly and unambiguously on display at its cemeteries in Italy.

Modernism in America began its ascent to prominence at the New York Armory Show of 1913 and by the 1940s, New York had begun its meteoric rise to become a center of abstract art, an art that challenged the patriotic and positivist view of human life that was so central to the ABMC, to the CFA, and to their chosen architects and artists.[8] When the proposal was made at the AARome (1949) to include modernist artists, Manship protested.[9] Again

7 *New York's World's Fair*, 1939/1940. Photographs by Richard Wurts and others, (New York: Dover Publications, Inc, 1977).

8 On the cultural significance of modern art see Peter Selz, *Art in Our Times, A Pictorial History, 1890–1980*, (New York: Abrams, 1981).

9 Costanzo, "The Lessons of Rome." Notes on letters from P. Manship to J.K, Smith. (September 1, 1950) and J.K. Smith to P. Manship (September 7, 1950).

in 1961, Manship defended his style against modernism's encroachment when Isamu Noguchi, Theodore Roszak, and Jack Zajac were nominated to be CFA trustees. Evidently, modernist sculpture was for him (and his coterie) anathematic. His battle was unwinnable. Although "well-connected," by 1950 his reputation was waning and his art was seen as merely conventional.

A great admirer of President Dwight Eisenhower, Manship was subpoenaed to appear before Congress's infamous House Committee on Un-American Activities (HUAC).[10] In 1952, he was forced to give up his splendid studio on New York's upper east side.[11] Soon after, both the artist and his work were all but forgotten. Much later, after his "lifetime of celebrity" and his death, his art was described as an "... eclipse into the shadow of derision."[12]

For the Anzio/Nettuno cemetery, the ABMC originally planned on doing without sculpture, but Gugler maintained that the cemetery needed it.[13] At the end of 1947 he wrote: "This building is a memorial structure in a cemetery rather than a battle monument and we feel that such elements as sculpture, painting, and inscriptions are unusually important in this particular design."[14] He had wanted Manship to be involved from the project's beginning.[15] The two men's friendship aside, his reputation as a sculptor of monuments and medallions commemorating the US involvement WWI had been firmly established. Stylistically and icographically, his earlier works pre-figure his work at Anzio/Nettuno. This comes as no surprise

10 It is difficult by today's standards to see any modernism in Manship's work. But according to Jane de Hart Mathews, modernism was associated with the international communist movement in the mindset of some influential political circles the the early 1950's in the US. A witch-hunt mentality obviously blurred the ability to discern between styles. See "Art and Politics in Cold War Armerica", *American Historical Review,* Vol. 81, No 4 (Oxford University Press, Oct. 1.,1976), pp. 762 to 787.

11 Barry Faulkner, *Sketches*, p. 179.

12 Richard Paul, "Art: The Empty Mastery of Paul Manship at the Smithsonian, an Unconvincing Reevaluation of the Sculptor," (*The Washington Post,* February 21, 1989). Quote is attributed to Harry Rand, curator of the exhibition of Paul Manship's work at the Smithsonian in 1989.

13 Letter, North to Gugler. The CFA in principle wanted sculpture at Anzio/Nettuno. RG 117, Entry 9, Box 122, Folder 687, Anzio (8–22–47 to 12–31–48), (December 17, 1947).

14 Letter, Gugler to North. RG 117, Entry 9, Box 122, Folder 687, Anzio (8–22–47 to 12–31–48), (June 23, 1948).

15 Letter, Gugler to Manship. RG 117, Entry 9, Box 122, Folder 687, Anzio (8–22–47 to 12–31–48), (December 27, 1947).

as the US, on the victorious side in both world wars, experienced neither historical nor moral catastrophe of a magnitude that would compel such public art to strive for newness. Continuity is seen between Manship's' commemorative work of WWI and WW II. For example, his "Detroit Soldiers Memorial Medal" (1919) pre-figures Anzio/Nettuno's winged female figure. The medal's angel, bearing a wreath, hovers above a cross, beneath which a soldier kneels. The angel's garb blends antiquity and Romanesque, and the medal bears the Latin inscription: "Lived, Conquered, Shall Live," suggesting resurrection of some sort. On its other side, fallen soldiers are rendered in naturalist figural style and full uniform, a common convention to indicate outstanding virtue. At the US's WWI cemetery at *Thiaucourt* in France, Manship sculpted a marble monument depicting a doughboy standing with helmet in hand, the other hand resting on his ammunition pack. Facial features are generalized to render the handsome, clean, angular, tight face of an archetypical Caucasian American "lad" that was loved and missed. (This idealized face reappears in his sculpture group at Anzio/ Nettuno). Above the doughboy, sculpted in relief and disconnected from earth and earthly reality, an angel fleeting forward glances back with heraldic trumpet at the ready, as if beaconing "follow onward to victory."[16] Her position above, her surety, and her movement bespeak passion and triumph in counterpoint to the soldier's more stolid earthbound determination.

At the AARome's court in Rome, Manship rendered crouching doughboys with naturalistic generalized facial features for a memorial (1925), designed by Gugler as a bench surmounted by a Renaissance tympanum in red granite.[17] They are garbed in detailed uniform with cartridge belts and kitbags (about three-quarters life sized) gazing downward.Their generalized features have been described as clumsy, rigid, and "probably executed by an Italian stone carver."[18] A relief from Manship's hand on the

16 "Blessed are they that long for home for they shall go home" is inscribed as the cemetery's motto. Meanwhile the fallen "sleep far from home in the sweet soil of France." (*Il Dort Loin des Siens dans la Douce Terre de France*). The inscription is behind the face.

17 The monument was commissioned in honor of an academy fellow who had fallen in the war. For details and critical treatment see David H.Wright, The First World War Memorial," in *Rome and her Monuments.* Two fellows had died in the war, sculptor Harry Thrasher and architect Walter Ward.

18 Ibid. p. 59. Wright's opinion of the value of Manship's crouching doughboys is less that flattering: "...the idea of this honor guard is dumb." and summarily concludes that Manship "...never [did] penetrating character studies...", p. 70.

monument's base depicts four doughboys attacking German soldiers. Citing Greco-Roman tradition from antique sarcophagi and Roman columns, an enemy soldier lies on the ground as angels cheer the victorious doughboys onward, making clear which side carries the angels' blessing. Manship used the terms "noble" and "strong" in this context.[19] In word and work, he reflects the US processing its WWI experience as heroic.[20]

The tympanum frames a fresco titled "Youth Sailing into the Sea" by Barry Faulkner, who would later work at the Florence (Impruneta) cemetery. (Collaboration between Manship, Gugler, and Faulkner would continue throughout their lives). Their post-WWI, AARome-bred classicist orientation delivered the style and iconography desired for the ABMC's WW II cemeteries. Faulkner's fresco, in Greek vase-painting style, represents the AARome fellow as a classical nude at the rudder of a small sailboat on a stylized sea, confidently making toward the zodiac in the heavens. A poem completes the message.[21] The fresco's style and message represent an idealization of violent death in war through (eternal) Greek art, which Clarke held to be supreme. Idealization takes violent death out of its historical context and entrusts it to the heavens. Heroic

19 A topos for pictorializing WWI victory in the US was representing the enemy as a fallen warrior in Greek-Roman style. This appeared on victory medals by various artists, including Manship. One such medal (1919) titled "French Heroes" depicts two "doughboys" charging with fixed bayonets, as a German soldier lies dead or dying. This topos is rarer in WW II pictorialization.

20 "With each war comes the collection of war memorials, generally as horrible as war itself. This is a subject for the idealists, and lends itself to the expression of all that is noble and strong. But posterity will remember us by the stock granite or bronze soldier with his little gun." Paul Manship: Art and Archeology, *The Archeological Institute of America,* Concord, New Hampshire or Washington DC Vol. 19 (1925) p. 81.

21 Poem by John Masefield reads:

"Man with his burning soul
Has but an hour for breath
To build a ship of truth
In which his soul may sail;
Sail on the sea of death,
For death takes toll
Of beauty, courage, youth,
Of all but truth."

The poem is quoted here in full because it reveals the attempt to generalize and eternalize a specific historical event in US history. Such detachment from the earthly horror of death in war is also found in German soldiers' cemeteries built during the NS regime, as will be discussed below.

death in war is presented as a pathway (in this case, a waterway) to eternity. Christian emblems strengthen this. Two angels hovering above the fresco hold a cross, which is surrounded by a funerary wreath. Blending antique and Christian messages, Faulkner detaches death from the human realm, where inconsolable despair at fate's cruelty rules. Such detachment is central to the ABMC cemeteries' heroic message of hope and eternity.

It is generally acknowledged that, among WWI belligerents, the US was by far the most fortunate, its losses relatively light and its industry profiting greatly prior to its entering the war shortly before the Armistice. The nation commemorated the experience with monuments large and small in practically every city and town. Generally, the US's commemorative art of the inter-war period sought transcendence from reality to the heavens, from whence comes rectitude, honor, consolation, and even resurrection. In Manship's examples, naturalist doughboys contrast with a non-historical universe, but there is no indication of further critical, or nuanced questioning. Manship's commemorative work for WW II, however, betrays a subtle shift in message. While his WWI works portray the Germanic enemy as the brutish barbarian, his WW II works at Anzio/Nettuno are without enemy and convey a sense of tranquility and reconciliation attained through remembrance. Such may be said as well for the US cemetery at Impruneta (Florence). The elimination of the "enemy" topos in the form of a German soldier might well reveal recognition that by the 1950s, (West) Germany had become an economic dynamo in its own right as well as a staunch US cold-war ally.

Manship and Gugler planned to travel to Italy in September 1948.[22] General North and Harbeson corresponded on the matter and wondered whether they shouldn't try to discourage Manship.[23] Their correspondence reveals that it was a matter of policy that a site, while under construction, was "off limits" to visitors.[24] Furthermore, there should be no *reportage*

22 Gugler and Manship, the latter on his own costs, plan to travel to Italy in September 1948, supposedly to design the map. Letter from Gugler to North. RG 117, Entry 9, Box 122, Folder 687, Anzio (8–22–47 to 12–31–48), (June 23, 1948).

23 Letter from Harbeson to North. RG 117, Entry 9, Box 122, Folder 687, Anzio (8–22–47 to 12–31–48), (June 23, 1948).

24 Letter, North to Harbeson saying that they cannot really prevent Manship from visiting Anzio, but Colonel Walker should accompany him. Letter indicates that relatives of men buried were not allowed to visit during that stage. RG 117, Entry 9, Box 122, Folder 687, Anzio (8–22–47 to 12–31–48), (July 20, 1948).

concerning the state of construction.[25] Wariness about Manship's visit was likely due to the fact that the decision to omit sculpture had already been made.[26] However, there was a clause in the ABMC's mandate that allowed for exceptions: "… because of the individual and exceptional merit of a particular artist's talents."[27] A few months later correspondence began to revolve around costs. Manship had proposed a program that would amount to 15 percent of the entire cost for construction, an amount exceeding the limit the ABMC's committee on sculpture (of which Manship was a member) had planned to allow for any of its overseas cemeteries. Harbeson wanted this reduced[28] and suggested eliminating the proposed sundial, writing: "Somehow, in this day of mechanical timekeepers a sundial seems no more appropriate today than an hour glass."[29] Manship was allowed to go forward with his other suggestions; a two-figure group ("Comrades in Arms"), two large reliefs, and a triptych for the chapel. The "Comrades in Arms" sculpture group would cost $20,000, and the two relief panels (19 x 15 feet) would be reduced to (only) $25,000 if quarter-scale models were produced and shipped to the site where Italian craftsmen would execute them in full scale. No additional sculpture was authorized.[30] Designs needed CFA approval, which was issued on July 20, 1949, with the stipulation that reliefs would not project from the façade (they were consequently recessed). There was subsequent discussion as to whether the bronze statue group should be gilded, but Manship was authorized to proceed.[31] Gugler and Manship were indeed "comrades in arms" in their arduous campaign to safeguard their vision against financial hindrances and an intrusion of new tastes.

---

25 North, in a document marked "Confidential" adds (alluding to Manship): "Please make sure that there is no talk about the state of the cemetery." RG 117, Entry 9, Box 122, Folder 687, Anzio (8–22–47 to 12–31–48), (July 20, 1948).

26 Letter, Harbeson to Eisemann (in Gugler's office), which clearly states that the CFA and the President had approved the design without sculpture. RG 117, Entry 9, Box 122, (April 11, 1949).

27 Ibid. Harbeson quotes Gugler.

28 Letter. Harbeson to North. RG 117, Entry, 9 Box 122, Folder 687, Anzio (1949), (June 4, 1949).

29 Ibid.

30 Letter, Harbeson to North. RG 117, Entry, 9 Box 122, Folder 687, Anzio (1949), (June 10, 1949).

31 Letter, Harbeson to Gugler attesting that the contract was signed. RG 117, Entry, 9 Box 122, Folder 687, Anzio (1949), (September 23, 1949).

Fig. 26. Manship's "Comrades in Arms."

Manship's design featured two young men with bare torsos on a socle, striding forward, each with an arm over the other's shoulder. The Commission was told that the figures were "… intended to represent American fighting men, shoulder-to-shoulder in comradeship, fearlessly fighting the enemy."[32] The ABMC's General Markey (who had been part of the search party for suitable sites near Florence) thought that the figures should wear complete uniforms. In response, Harbeson offered: "I think it will be unfortunate if we should upset the enthusiasm of a sculptor by an argument about clothes."[33] One ABMC staff member in Rome commented that the

32 Letter, the ABMC Paris to the ABMC Washington, D.C. RG 117, Entry, 9 Box 122, Folder 687, Anzio (1949), (August 5, 1949).

33 Letter, Harbeson to the ABMC's Shaw. RG 117, Entry, 9 Box 122, Folder 687, Anzio (1949), (October 19, 1949).

two figures did not look like Americans and did not adequately reflect "… the sacrifice of soldiers dying in honor for the glory of victory in arms for their country." He would prefer a squad-sized group and went on to describe his vision, complete with the title, "The Last Patrol."[34]

Despite controversy, Manship went forward with his vision, producing quarter-scale plaster models that were shipped to Italy. The "Comrades in Arms" figures were cast in Milan, and the two reliefs cut in marble by Italian craftsmen. In a hand-written letter Manship explained the meaning of his work: "The 'Comrades in Arms' were intended to convey a mood of youth, with faith in right, marching forward, courageous in freedom." The reliefs were intended to show "… memory placing a wreath on a gravestone and a soldier carried to immortality."[35]

"Comrades in Arms" was designed to be viewed from all sides, a rarity in Manship's work. Most of his works were to be viewed from one point of view. Two young men, soldier and sailor, one slightly taller, stride with right leg forward. The forms of their lower bodies and their shoes are almost identical. Each has one arm over the other's shoulder while the other swings freely in cadence to their stepping. Their free hands form fists. Their torsos are nude except for "dog tags," identification tags worn around their necks. In form and content the sculpture is characteristic of Manship's work. Its craftsmanship is impeccable, its naturalist style coherent with details such as belt and pocket slits. The nude torsos show generalized well-trained young male bodies in late pre-classical/early classical Greek style, fashionable in the 1930s. At the same time, they have a realistic presence, an immediacy, as though they had just finished washing and shaving.

As in all Manship sculpture, the expressions of the two faces are generalized rather than individual.[36] They show a distinctively male "look" characterized by even features, high foreheads, and hair combed neatly

---

34 Letter, R.C. Brown, officer in charge in Rome to North. "A bronze group in field uniform of a soldier guarding, an officer crouching, and pointing, issuing an order from a map to an NCO saluting, acknowledging 'the orders' the title could be 'Last Patrol.'" RG 117, Entry, 9 Box 122, Folder 687, Anzio, (November 18, 1953).

35 Handwritten note, Manship to North in preparation for the Commission's meeting. RG 119, Entry 9, Box 122, (March 8, 1950).

36 Manship's bust of J.D. Rockefeller (Smithsonian Institute) is a penetrating work of portraiture. The subject's impenetrable nature and severity are effectively rendered in a terse style. By contrast, his small equestrian statue of Civil War General (Stonewall) Jackson presents a face without expression and quotes late archaic Greek art.

and parted at the side, a recognizably American beauty ideal from the 1940s. In this sense, they are an idealization that parallels the fascist ideal, but lacks the latter's exaggerated musculature, furrowed brow, and set jaw meant to convey harsh determination. By comparison, their physiques are less powerful and their faces more "pretty."

The choice of a nude torso raised uncomfortable ideological questions, as men with bare torso and long trousers belong to the iconographic repertory, and used to represent the working class in socialist as well as Italian fascist art circles.[37] As Italian cemeteries are treasure troves of sculpture, Manship most likely would have seen such figures of males with nude torsos on funeral reliefs in Rome's Verano cemetery (signifying the working class). When Leo Friedlander (Manship's student and later an equally famous sculptor) created reliefs of bare-chested, long-trousered male figures for Rockefeller Center, he was accused of creating "socialist art."[38]

Manship's figures delighted Gugler, and Griswold as well. After a visit to New York City the latter wrote in his journal that he found the figures "… truly inspired."[39] Aesthetically, Gugler wrote, the sculpture fits the dimensions of the court very well. He lauded Manship's sense for size and scale and likely admired his mastery of representing bodies in motion.[40]

37 William M. Stott, the author of an article for a Manship exhibition from May 19 to August 18, 1985, Minnesota Museum of Art Landmark Center, Saint Paul, Minnesota, tells that Manship went to Italy in September of 1948, supposedly to work on the design of the future maps for the museum. The author comments that the streamlined style of Mussolini, Hitler, and Stalin art was influential on American "public" art. He considers the "Comrades in Arms" "the best instance of combining streamlining with European-style social realism" that the author calls "an aesthetics of control." He comments on the strange phenomenon that this style in the US was stripped of its ideological goals: "It sold toothpaste, cars, and the New Deal."

38 Interview with a guide at Rockefeller Center, February 2013.

39 Griswold's journal entry. Pittsburgh University Archives, (November 1949).

40 Letter of condolence, Gugler to John Manship (son) after Manship's death in 1966. Gugler praised the right size of the sculpture group for the courtyard in Anzio/Nettuno. "There is something about size that Paul understood so well." Also, a letter of condolence from a person named Cecil to Gugler after Manship's death saying that he left a lasting legacy of beauty to America. Lawrence White, J.K. Smith's partner, after a visit to Anzio/Nettuno found the group should have been a bit larger. See his letter: Smithsonian Archive, Gugler papers (November 9, 1967); The NYHS McKim, Mead & White collection; Florence cemetery, file no. 7, January 1, (May 6, 1954).

Critics, however, did not share this enthusiasm. Manship's work, albeit well executed and artful in a decorative sense, lacked the depth needed to address the new hunger for expression. In the new art "scene," a lust for destruction of academic tradition was afield. Manship exhibited his "Comrades in Arms" at New York's Metropolitan Museum in 1951 and Emily Genauer wrote in the New York Herald Tribune: "It's so bad it's downright embarrassing. It depicts two bare-chested American doughboys as expressionless and about as sculptural as a pair of department store dummies."[41] Howard Devree in The New York Times described the sculpture as "… monstrous as figures in that clothing advertisement display in Times Square."[42]

Although Manship's work at the Anzio/Nettuno cemetery recalled a bygone era, Manship himself, along with General North, Harbeson, and Gugler, held fast to their universe. The two large reliefs for the lateral parts of the monument are titled "Remembrance" and "Resurrection."

Fig. 27. Manship's "Remembrance" at chapel's exterior.

41 Quoted by Stott, see fn. 37.
42 Ibid.

Fig. 28. Manship's "Resurrection" at museum's exterior.

"Remembrance" is a smooth synthesis of recognizable art-historical styles. A soaring female figure ("angel" in the ABMC brochure) combines Romanesque cloth patterning with classical traditions in facial features, figure, and gesture. She lays a funerary oak wreath upon decoratively arranged crosses, a Star of David among them, of differing heights. The crosses do not stand straight, but appear askew, as if bent by time, and the ground upon which they stand is gently convulsed and flowers sprout. This motif belonged to the funerary repertory developed by the AARome fellows in the 1920s.[43] The image exudes gentleness and kindness and might well have had a consoling effect on viewers. In "Resurrection," a female figure (again, "angel" in the ABMC brochure) soars above a long bilious cloud as she bears a fallen soldier in her arms toward the heavens. Wind and the ether billow her shawl and gown as she ascends. He seems small in relation to her large size. His torso is nude and his trousers are depicted

43 This motif was developed in AARome circles. An unsigned and truncated image from a fellow's exhibition recalls Faulkner's work from the early 1920s. It shows a female figure hovering over crosses. Manship used a similar formula for representing clothes moved by air at Anzio/Nettuno. Reproduced in Cazzato, *Ville et Gardini.*, p. 201.

in sharp naturalist detail. His hair is cropped, as in Indian or classic statuary. The young man's boots are outsized and his dangling legs suggest a sleeping, trusting child. The image might be called *Kitsch* were it not for its superb craftsmanship, its elegant composition that works with counterpoint, its tender gestures, and the calm sentiment it imparts. This is the entire cemetery's only depiction of death, here softened, as if a caring mother were tenderly taking her sleeping child to bed. The ABMC brochure reads: "... borne to his reward by a guardian angel." Death, in the United States military commemorative repertory, is never ugly. Here, life and its beauty accompany the dead, and in spite of the potpourri of styles, the mood is coherent and there is "beauty." One senses the work of a serious and mature artist.

The ABMC's emphasis on beauty is consistent with its policy that relatives of those buried were not allowed to visit their loved ones' graves while a cemetery was under construction.[44] The sight of open grave ditches and coffins would surely have brought death's banality to the fore and marred the experience intended to be affected by the manicured, park-like cemetery. Evidently, the ABMC's telos had an aesthetic as well as a political component.

44 This policy can be deduced from the letter from North to Harbeson cited in fn. 24.

Reaching to the Heavens: The Chapel

> "The spacious firmament on high, with all the blue ethereal sky, and spangled heaven, a shining frame, their great Original proclaims an expression of immortality."
> Joseph Addison (1672–1719)

The Anzio/Nettuno cemetery's messaging did not end with Paul Manship's reliefs and sculpture. The chapel, expressly non-denominational, is utilized fully for ideological statements. From the open oval court, two pillars adorn its entrance and large bronze doors pierce its solid wall (cast by the Marinelli foundries in Florence). Next to the doors we find the dedication that is present in all US foreign military cemeteries: "In proud remembrance of the achievements of her sons and in humble tribute to their sacrifices, this memorial has been erected by the United States of America."

The chapel's apse serves as the obligatory wall of the missing. Engraved are 3,095 names complete with rank, organization, and state of entry into the military service.[1] Above is inscribed: "Here are recorded the names of the Americans who gave their lives in the service of their country and who sleep in unknown graves." Harbeson insisted that the names be engraved rather than presented in bronze elevated lettering because the latter would likely be removed and taken as souvenirs.[2]

At first view, precious materials indicate that Eric Gugler intended the chapel to be a "focal point." The walls are of white Carrara marble, the floor of *Rosso Levanto* marble, the pews of walnut. The chapel's altar is of white *Serravezzo* marble from Carrara, designed by Manship (in spite of the ABMC's decision not to invest in any further sculpture, Gugler and Manship prevailed).[3] It includes a marble triptych. In its entirety, the chapel presents a programmed commemoration experience drawing from biblical,

---

1 A soldier was declared missing after a year and a day. Names of those who were discovered later have a rosette next to them. See the ABMC brochure for the Sicily-Rome cemetery.

2 Letter, Harbeson to Gugler. RG 117, Entry 9, Box 112, Folder 687, (8–22–47 to 12–31–48) (January 27, 1948).

3 Letter, Gugler to North. "It is essential to the dignity and meaning of the chapel, a focal point of spiritual significance to which the relatives and friends of lost ones might turn in silent prayer." RG 117, Entry 9, Box 122, 0016.

mythological, astrological, textual, aesthetic, and iconographic sources, all put to the purpose of anaesthetizing the visitor to death's sting. It would elevate the soldiers' death to a noble level, transcendental, removed from its historical context.

Manship's angels (on the triptych) descend from the heavens toward inspirational texts inscribed in gilded letters. On the left, a text from Simonides evokes ancient Greek stoic virtue, eulogizing the soldiers who sacrificed themselves at Thermopylae:

> "Nobly they ended, high their destination. Beneath an altar laid, no more a tomb, where none with pity comes or lamentations but praise and memory, a splendor of oblation who left behind a gem-like heritage of courage and renown, a name that shall go down from age to age."[4]

On the right is a quote from the Christian Bible's eighth psalm, imploring God's almightiness to transcend personal grief and loss:

> "When I consider thy heavens, the works of thy fingers, the moon and the stars, which thou hast ordained, what is man, that thou art mindful of him? And the son of man, that thou wasteth him for thou hast made him a little lower than the angels and has crowned him with glory and honor."

On the triptych's central panel Saint Michael, armor-clad with helmet and sword, soars above the four archangels who sound trumpets to announce victory, seconded by angels on the side panels holding palm fronds.[5] The "angel of peace" has a firm place in US military cemetery iconography, but the choice of St. Michael to allegorize peace lends ambivalence. St. Michael, avenger of evil, destroyer of heretics and restorer of good, is a key icon of the counter-reformation. Here, he signifies that US valor and victory were heaven-sent and that the US's Italian campaign is sanctioned and rewarded by the heavens. He holds his sword horizontally and at the bottom of the relief, a message is inscribed: "Peace on Earth. Good Will Among Men." While the inscription honors peace, the sculptural message is an unmistakable glorification of war.

---

4 *The Greek Dead at Thermopylae.* Famous verses by Simonides, ancient Greek poet (556–468 BC) often quoted as exemplary for the soldiers' duty to hold a post in war to the last man.

5 Handwritten letter, Manship to North in which he explains the meaning of his composition. RG 117, Anzio 1–1–57 to 27–38–59, (February 16, 1957).

Gugler was apparently driven to commemorate US military achievements through all available means, one such being six very large flags to frame the altar. Flags with the insignia of artillery or infantry did not exist in such size. Stiff resistance therefore came from the ABMC's historical department, but in the end, it had the flags fabricated.

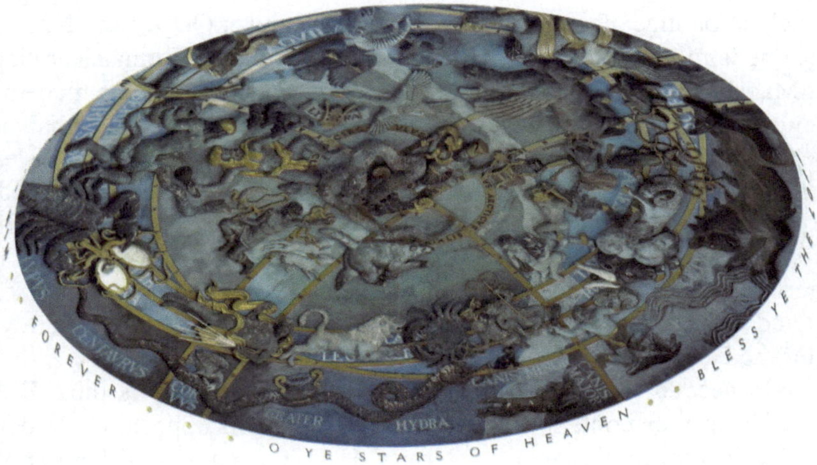

Fig. 29. Gugler's celestial ceiling in the chapel.

The Celestial Ceiling. Victory Eternalized in the Stars

The ceiling of the chapel, a flattened dome, bears a large round relief of the heaven's stars and planets. Star maps surround the space in heavenly orbs. Following tradition, planets are rendered in various degrees of relief. The constellations show the precise positions of stars and planets at the moment the Allies landed on Anzio's beaches. Thus, Gugler linked the US's Italian military campaign to astrology, and to a higher order. Using the chapel's ceiling to elevate military victory to the realm of the universe might strike visitors as an unseemly statement of the ABMC's nationalistic self-importance. This was not the case, but rather was the result of Gugler's personal vision and persistence.

Both Gugler and his friend Paul Manship shared a passion for astrology. In the late 1920s Gugler brought a five-foot (1.75 meter) etched glass celestial sphere with stars forming mythological figures from Germany.[6] Manship, for amusement, copied the images and then rendered some in high relief. In so doing, he became smitten with the representation of mythology and undertook the study of "astronomy" (probably meaning "positional astronomy," now called astrometry).[7] The two, Gugler and Manship became familiar with Assyrian, Babylonian, Greek, and Roman star maps. At Manship's home one would dine at a large table with its top made of a single piece of polished wood, its surface decorated with the zodiac and the seasons.[8] Subsequently, he produced several celestial spheres. His earliest, cast in Germany in 1934, depicts stars of the first four magnitudes in 66 constellations, and it rotated on its polar axis.[9] Later, for New York's World Exhibition in 1939, he reached a wide, approving public with his enormous celestial globe.[10] (Though this globe was an intricate work of art and craftsmanship, it was rendered in perishable plaster and destroyed after the Fair.)

He needed to advocate passionately for his vision, as the ABMC would have been content without it.[11] It would be his opportunity to commit US history and victory to eternity. Harbeson supported Gugler and urged the ABMC committee members, before deciding the matter, to view Gugler's celestial ceiling at the Harrisburg education building. This

6 This first globe sold for $100,000 in 2006. <http://www.invaluable.com/artist/manship-paul-howard-l5sqra47zg/sold-at-auction-prices>. Last Accesses Aopril 11, 2018.

7 Rand, *Paul Manship*, p. 125. He quotes Manship and his striving to represent constellations correctly.

8 Faulkner, *sketches* p. 135.

9 This globe was 16 inches high and stood on the figure of a blindfolded woman representing night. For description and image see <http://www.christies.com/lotfinder/lot/paul-manship-celestial-sphere-1792128-details.aspx?intObjectID=1792128>. Last accessed May 10, 2015.

10 E.g: an armillary/globe sphere for the Palais des Nations, Geneva.It stands 13 feet on four cast bronze turtles ad was cast by Bruno Bearzi in 1928. A large globe at the World War I Aero Memorial in 1950 in Philadelphia (five meters in diameter). The last large one was for the World Fair of 1964–65.

11 Letter, Harbeson to Gugler supporting his vision. RG 117, Entry 9, Box 112, Folder 687, (22–47–12 to 12–31–48), (March 1, 1948).

ambitious undertaking had cost four million dollars (in the middle of the Great Depression).[12] Responding to a US cultural urge at the time, its decorative program sought to locate the US in civilization's "progress."[13] Notably, large-scale maps were used for such didactic and commemorative purposes.[14] Maps, at the time, were also used as private memorials, and would play a central ideological role in all the ABMC's cemeteries.

Gugler advocated for his celestial ceiling on the basis of spiritual quality: "… the stars and planets to be presented in the exact position as at the time of the landings at Anzio, at 0200 hours on January 22, 1944." (Jupiter, Saturn, and Mars dominated the skies, at that moment.) It would take another 44,000 years before that exact constellation would reappear. Impressed, the commission acceded, but found the symbols to be aesthetically too dominant, and Harbeson agreed with them.[15] Gugler persisted.[16] Accepting the elimination of a proposed sundial (by Manship) that Harbeson considered too anachronistic, Gugler announced his plans to sail to Europe in September (with model maker Tagliabue, who had made a model of the dome using "plasticine").[17] There, he would rendezvous with Manship who was residing in Florence at that time. Manship would finish the model and add color. Gugler and Manship would work at their own expense.

A stop-work order was issued in late October of that year (1953) when the commissioners (ABMC) proposed eliminating the ceiling sculptural

---

12 Letter, Harbeson to Gugler warning him his "firmament" might not find support within the ABMC. RG 117, Entry 9, Box 112, Folder 687, 12–22–47 to 12–31–48, (November 3, 1948).

13 The walls had a mural painting with chronological tables and maps and depicted the history of the world from the ancient civilization of the Orient and Occident down to the beginning of WWI in 1914.

14 Faulkner, *Sketches,* p. 110. Faulkner did four large charts showing commercial sea routes for the great hall of the new Cunard building at Broadway and the Battery. "It was an all-academy in Rome effort. Ezra Winter painted, John Gregory sculpted." This is a good example of the persistence of collaboration of former AARome fellows in the US.

15 Report of the ABMC meeting to Harbeson. RG 117, Anzio, Entry 687, Anzio (6–16–53 to 8–15–53), (July 7, 1953).

16 Letter, Harbeson to North reminding him that Gugler is waiting for the official "go ahead." (Costs estimated at $24,000.) RG 117, Entry 9, Box 122, 0016, (June 4, 1949).

17 Letter, North to the ABMC in Rome and to the ABMC, Washington, D.C. Ciampaglia is to be used for the execution of the model and would be paid $10,000 for his work. RG 117, Entry 122, Folder 687, Anzio (5–16–54 to 10–20–54), (July 28, 1953).

figures altogether.[18] However, Gugler's intentions were unstoppable. A scale model, made in Rome, was shipped to New York[19] and in July 1954 was shipped back to Rome, where Bruno Bearzi, a sculptor based in Florence, was to execute the design.[20] An accompanying text listed a number of small corrections needed before the ceiling could be painted, and concerned details such as the location, form, and size of the stars and planets[21] as well as instructions such as "reduce lion's paws in Hercules" and the placement of the small lights that would dot the ceiling. The constellations had to be correct. Lettering had to be fixed. In January of 1955 Gugler finalized his ceiling design and had full-scale models made of its Jupiter, Saturn, and Mars, the three key gods that witnessed the Anzio landing. These, he displayed at his New York office.[22]

Gugler returned to Anzio/Nettuno several times to work hands-on with the ceiling,[23] and it was rumored that he appeared even after the cemetery had been officially dedicated in 1956. He offered to collaborate with Italian artist Carlo Ciampaglia in repainting the ceiling, at his own expense, "to finish and perfect it."[24] Manship, still in Florence, also traveled

---

18 Memo, ABMC's Lt.Colonel Shaw to Gugler. RG 117, Box 115, Folder 687, Anzio 8–16–53 to 1–15–54, (October 29, 1953).

19 ABMC's Shaw to the custom's office asking for free duty of the model. Angelo Tagliabue prepared the model. The letter mentions that an Italian artist commissioned to carry out the model. For photo of model see RG 117, Box 115, Folder 687, Anzio 8–16–53 to 1–15–54, (November 16, 1953).

20 Bearzi, a bronze caster in Florence would carry out the work as he had done the central map in the museum. Bearzi became an honorary member of the American Institute of Architects on August 24, 1964. Smithsonian Archive, Gugler papers, memorandum of July 1955.

21 Comments and suggestions of Mr. Neeley (an expert) for the Anzio dome about what needed to be changed in order for it to be correct. Smithsonian Archive, Gugler papers, (February 5, 1954).

22 Letter, Gugler to Colonel R. Brown, Smithsonian Archive, Gugler papers, (January 5, 1955); Letter, Gugler to North also says that in the summer of 1955 Gugler was in Anzio to supervise the progress of the work, but had to return to New York abruptly because of illness in the family. RG 117 Entry 687 Folder Anzio, 4–14–55 to 10–31–55, (September 20, 1955).

23 Griswold reports that Gugler wrote him that he would be in Nettuno in September to work on the ceiling. Pittsburgh University Archives, Griswold Collection, (July 15, 1955).

24 Letter, North to Gugler received April 11, 1956. Gugler could continue working until and after the dedication ceremony. RG 117, File 687, Anzio (4–6–56 to 12–31–56), (April 11, 1956).

to Rome and called at Nettuno to check correctness of the colors.[25] He likely found these to be not up to expected standards because a few months later Gugler beseeched North to have the ceiling retouched: "The painting of this dome means a great deal to me, more than I know how to say, for it is the last touch."[26] Following an inspection tour in August of 1957, Harbeson described the chapel as "… completely satisfying except the horns on the dragon's head on the star ceiling should be painted darker and not be shiny."[27] Ciampaglia traveled to Rome and repainted the ceiling on the basis of the colored model.[28] We know that Gugler joined him because Griswold noted in his journal that he visited the two at the site.[29] The result of the collaboration was colors masterfully applied in different hues of blue, and the Milky Way suggesting a fine veil weaving through the skies.[30] Gold highlights the figures' attributes as well as planetary courses.[31] North, inspecting the dome later that year, sent Gugler eight color photos and in a report from Rome described the coloring as "excellent."[32]

Although sculptural figures vary in both size and depth of relief, with some protruding almost to full sculpture and others just slightly insinuated, and although the space available seems less than generous for this level of sculptural and decorative complexity (about seven meters in diameter),

25 Letter, Harbeson to Walker (ABMC Rome), saying that Manship would be in Rome on May 7, 8, and 9. RG 117, File 470, 46 Boxes 115–122 (92 Stack 39), (May 2, 1957).

26 Letter, Gugler to North reminding North that there was indeed a budget for the completion of the work. A movable scaffold and drop cloth would be needed. RG 117, File 687, Anzio 1–1–57 to 12–30–59, (July 29, 1957).

27 Report, Harbeson to North referring to this task as the "culmination" of the chapel. RG 117, Entry 687 Anzio folders (August 24, 1959) North shared this opinion, referring to it as a unique feature only to be found at that cemetery. RG 117, file 687 Anzio folder 1–1–57—12–30–59, (September 9, 1957).

28 Letter, ABMC Rome to the ABMC Washington. The ABMC Rome reporting that the ceiling was finished. RG 117, entry 13, folder 687, Sicily-Rome cemetery (1–1–57 to 12–30–59), (November 26, 1957).

29 Pittsburgh University Archives, Griswold papers, Griswold's journal, (October 3, 1957).

30 Report, ABMC Rome to North praising the ceiling. RG 117, Entry 13, File 687, (1–1–57 to 12–30–59) (October 14, 1957). RG 117, entry 13, file 687, Anzio (1–1–57 to 12–30–59), (October 14, 1957).

31 Report, ABMC Rome after North's visit with the officer in charge, Aliotta, who wrote the report. RG 117 entry 92 folder 39 1–1–57–12–30–59, (December 1957).

32 Letter, North to Gugler with eight color photos praising the repainting. RG 117, File 687, (1–1–57 to 12–30–59), (December 4, 1957).

Gugler's tenacity resulted in a composition with an aesthetic totality, less the ABMC's message and more Gugler's personal vision.

The surrounding text reads: "O YE STARS OF HEAVEN BLESS YE THE LORD PRAISE HIM AND MAGNIFY HIM FOREVER." Clearly, Gugler's intent was to capture the heavens' mysterious infinity, hypothecate these as creations of a monotheistic Lord God and in so doing, liberate glories won by the US Armed Forces from their historical reality. His celestial ceiling spans the names of the missing and the triptych as well, messaging transcendence to heavenly spheres.

The Glory of the Armed Forces

> "The American Battle Memorial Monuments Commission was created by the Congress for the purpose of commemoration of the services of the American Armed Forces abroad during World Wars One and Two. This duty includes the beautifying of our overseas military cemeteries as well as commemorating the achievements of our forces by battlefield monuments."
>
> General Thomas North[1]

The ABMC had decided that there would be no memorials to specific battles or campaigns in WW II cemeteries.[2] North instructed: "Since the cemeteries and their memorial chapels will serve as memorials to the achievements of the forces who fought in the respective region(s), each chapel will include a chamber in which these achievements will be recorded by carving or in some permanent manner."[3] It was his idea that the permanent records would take the form of large, elaborate, didactic maps. These would be beautifully executed as frescoes, mosaics, or intricately inlaid stonework. A text record was to be included in upper case letters, in both English and in the language of the host country, on "Allied or co-belligerent soil."[4] Maps had figured prominently in North's military experience. During WWI in France, he produced day-to-day situation maps under Paul Cret.[5] During WW II North

1 Undated (assumed September 1947) letter from North inviting Gugler to design the cemetery at Anzio. Identical letter, mostly penned by Harbeson, was sent out to all architects who designed ABMC's cemeteries. RG 117, Entry 9, Box 122, Folder 687, (8–22–47 to 12–31–48).

2 Letter, North to Gugler. "[For WWI], museum chambers are smaller and less of a historical record than for WW II." This was because WWI cemeteries "were duplicated by historical monuments upon the battlefields close by." RG 117, Entry 9, Box 122, (till end of 1949), (February 6, 1948).

3 North, "Our Nation's Tribute," *The Army Digest*, (Official U.S. Army Magazine), vol. 19, no. 11, (November 1964), pp. 42–46.

4 Ibid. p. 45.

5 Shenton, "They will Never Be Forgotten", p. 59. (Shenton made maps for St. James, Brittany and shared North's focus on maps as a central commemorative element.)

prepared the day-by-day military maps for briefings with President Roosevelt and Army Chief of Staff General George Marshall.[6]

US WW II cemeteries abroad include maps, wall-size and smaller. The smaller maps show the yearly diminishing landmasses held by the Axis powers, both in the European theatre and the Asian theatre, and thus afford the viewer a progressive worldwide perspective.[7] Representing WW II's progress in understandable graphic form from regional, theatre-wide, and worldwide perspectives, such maps would contextualize local events. Information was needed to be complete. Symbols and emblems were to be used. Army divisional insignias were to be included as well as self-evident representations for air and naval operations. The accompanying text was to be "… accurate and eloquent prose to stir the imagination of the visitor and create a vivid picture of the battle condition."[8] Lieutenant Colonel Mitchell, chief of the ABMC's historical division, wrote the text for Anzio/Nettuno, assisted by North and approved by General Marshall. North adds these maps constitute the battle monuments, being of permanent materials such as bronze, marble, mosaic, engraved stone, and granite. (No two cemeteries alike.) According to North together with accompanying inscriptions, maps were intended to commemorate, without vainglory, the achievement of the US Armed Forces, but he reminds that triumph exacted sacrifice: "Each with its chapel forms an integral memorial to the achievements and the sacrifices of our Armed Services."[9] In each of them the visitor finds an accurate, compact record of the war in which all allied participants are commemorated.

For North, these maps were much more than didactic tools in the cartographic military tradition. They were works of art: "It is my belief that maps have an inherent, dramatic quality which is equal in interest to the finest pictorial art."[10] (As a young man North had taken art courses at Christ's hospital and King's College in London.)[11] Each cemetery would employ different techniques and use different media. The requirement was simply that each be executed in durable material that required little maintenance.[12]

---

6 Smithsonian Institute Archives, File "Anzio." Then signature in Febr. 1013 was Gugler MSS AAA 23/12/12.

7 These smaller maps are identical in all cemeteries.

8 See Shenton, "They will Never Be Forgotten", p. 59.

9 See North, "Our Nation's Tribute."

10 Shenton, "They Will Never Be Forgotten," p. 59.

11 Ibid.

12 Ibid.

At Anzio/Nettuno, the memorial building has, in its center, a floor-mounted round bronze map of Italy, set on a low pedestal with various colored marble tiles set into it.[13] On the north-facing wall the Italian campaign is depicted in fresco with easily read representations. Three continuous maps cover the west-facing wall's whole width. They depict the "Capture of Sicily," "The Strategic Air Assaults," and the "Naples-Foggia Campaign." Beneath are small maps: "The War Against Germany" and "The War Against Japan." Its east-facing wall mounts the map depicting "Landing at Anzio and the Capture of Rome."[14]

Fig. 30. Anzio/Nettuno's museum, depicting "path to victory" with symbols and arrows representing campaigns moving inexorably forward.

The ABMC commissioned New Jersey based Italian painter Carlo Ciampaglia to create Anzio/Nettuno's maps, credentialed by his having been an AARome fellow (1922). North himself would watch over every detail during the creative and executive process.[15] Outlined in correct geographical dimension, the detailing of Italy's mountains, valleys,

13 Executed by Bruno Bearzi in Florence.

14 The ABMC booklet, "Sicily-Rome Cemetery."

15 Memorandum, North to Shaw expressing his discontent with coloring. RG 117, Folder 687, Anzio, (5–16–54), (received August 3, 1953).

rivers, plains, and the surrounding seas is beautifully rendered. Arrows show the movements of infantry and bear the names of the army corps involved, their widths and lengths indicating the strength of the attacks. British and other allied forces are likewise shown, progressing northward. Supply ships are shown unloading in harbors while warships cruise and patrol along the coastlines of Sicily, Naples, and Anzio. Aircraft are shown flying toward Germany.

European map-making traditions include symbols, allegories, emblems, and emblematic landmarks that reflected the politics and worldview of period and place. Ciampaglia, familiar with these, proposed thick clouds in Art Deco style as a means of imparting a celestial perspective. Female figures would float among the clouds, as seen in the works of Manship and Faulkner, in the Romanesque/Art Deco eclectic style nurtured at the AARome. They would brandish traditional emblems such as flower garlands, an olive branch of peace, or the trumpet of fame. North would have none of this. It was, after all, a military map. He dismissed such embellishments as unnecessary and eschewed their formulaic meaning. He wanted thick clouds deemphasized. The ladies, in his view, had nothing to do with the bombing of Germany and therefore he wanted them just "soaring."[16] Renaissance style sea-monsters were out of the question: "… the literalness of the relief model, to my mind, does not invite sea animals or fish in the sea."[17] All of the above were banished, leaving Ciampaglia's subtle coloring and the composition of lines to make the message, of which North approved. The result was maps with, to repeat North's words, "an inherent dramatic quality."

Emblems were used to represent important cities. North advised strongly against including Monte Cassino, which had been destroyed by Allied bombing. Although at that time its wanton destruction was defended as a "military necessity" there was concern that placing it on the map might offend Italian sensibilities.[18] The Vatican needed to be excluded as

16 Letter, ABMC Rome to North agreeing with him that "ladies" should not "participate" in the bombing missions, and raises the question of what relation there was between the bombing of Germany and the Anzio/Nettuno memorial. (Bombing missions started from Foggio/Naples). RG 117, Anzio, 1953, (July 9, 1953).

17 Memorandum, North to Shaw. RG 117, Anzio, (July 28, 1953).

18 Letter, J. B. Mitchell, (ABMC's chief of the memorial division) to Gugler about not representing Monte Cassino. RG 117, Anzio, 1953, (July 2, 1953).

well as it had not been a battleground. Instead, an image of the Coliseum was used to represent Rome. The image of the *Forte di Anzio* was used for Anzio as Gugler had expressed his liking the choice.[19]

Texts aided in understanding these maps. The ABMC's carefully prepared story of the campaign was inscribed at the sides of the maps in large letters, in English and in Italian. The central map indicates the Balkans, Austria, and parts of Germany and indicates the bombing of Germany, represented as rays radiating from Naples-Foggia air base in the south. "The Strategic Air Assault" text reads: "Our bombers and fighters attacked incessantly …" causing "… a progressive dislocation and destruction of the enemy's military and economic systems."[20] North was critical of this because they originated well behind the battle lines in the north.[21]

The museum maps' didactic mission was undertaken in decidedly military spirit. The prominent representation of bombings beyond the Italian campaign's boundaries, and the ABMC's initial decision to include a representation of Monte Cassino, betray an acceptance of war's horrors and link such "achievements" with post-war's "Pax America," the latter resting upon the former's foundations.

19 Letter, Mitchell to Harbeson reporting a search of a photo of Anzio, of interest because Gugler is quoted as the ultimate decision maker. RG 117, Folder 687, (6–16–9 to 1–5–53), (July 17, 1953).

20 The ABMC Sicily-Rome booklet gives full text informing that the 15th Air Force attacked aircraft factories in Regensburg and Budapest, oil refineries in Ploesti and Brasow, enemy airfields in Northern Italy, transportations systems in Munich, Vienna and Budapest, and as far away as Berlin.

21 North' attitude may be gleaned from ABMC Rome's letter responding to North's criticism. RG 117, ABMC Holdings, (Anzio 1953), (July 9, 1953).

A Legacy to Civilization: The Statue of Orpheus

"Art follows War."
Paul Manship[1]

Upon entering the south garden at Anzio/Nettuno the visitor's attention will be immediately drawn to a bronze sculpture of Orpheus next to a bed of red roses. More than life-size, his nude body (cloth or hide at the waist) is manly, youthful, and graceful. His gaze is skyward. Legs bend at the knees with feet set firm on the stone base, as to convey energy, as if pushing his torso out of its prone repose. He is neither playing nor singing, but imploring. His hands offer his lyre to the heavens. A large armillary sphere and a sundial surround him. While his body is classicist-derived, his hair quotes from Indian sculpture. Stylistically and thematically, the figure is characteristically Paul Manship.

How the sculpture came to be at this military cemetery is (so far) not clear.[2] Did Manship bequeath it to the cemetery? Had it been in Eric Gugler's possession? In 1977 Gugler designed its base.[3] The armillary sphere is most likely by Gugler as well. Was it he who donated it to the cemetery, perhaps following his late friend Manship's wishes?[4] Was it the statement or the wish, or the bequest of both artists, that civilization might overcome war?

For Manship the figure of Orpheus was an allegory of art. He had made drawings and models of Orpheus over the years (also with his beloved Eurydice).[5] Such sketches for funerary monuments may be found

1 Struck on coin by Manship, bearing image of Faulkner, (Smithsonian Institute). See Faulkner *sketches,* p. 63.

2 ABMC's information on Anzio/Nettuno becomes scanty in the NARA Archives around 1958. The sculpture is not mentioned by Griswold or others. Manship died in 1966.

3 Smithsonian Institute Archive of American Art, Eric Gugler papers, 1889–1977, (June 29, 1977). A blueprint of a design for the statue's pedestal by Eric Gugler is dated 1977 (the year of his death), under the title: "Eric Gugler plans for Orpheus sculpture and Armillary Sphere at Sicily Rome Memorial Cemetery."

4 Ibid.

5 Orpheus and his meaning for civilization was one of Manship's key themes. He made several sketches of Orpheus showing him kneeling, sitting, or prone. In bronze, statues

dating back to as early as 1912. A sculpted Orpheus, the appeaser of wild animals, was used for the Rainey Memorial Gateway, New York Zoological Park (1929–1933). A small gilded bronze statuette of a standing Orpheus dates from 1954 and a pencil drawing from 1966 (the year of Manship's death) shows this version. On the backside of his medal of Faulker's profile, mentioned above, Orpheus kneels, lyre in one hand, looking upward toward a winged genius that arches over him, both in ideal nudity. The inscription of "to the ultimum do we pursue the ideal" at the coin's rim sends the clear message. It is a beautiful composition and sums up the prolific sculptor's intentions.[6]

Throughout the cultural history of the western world Orpheus has represented the forces of civilization.[7] His brave journey into the realm of the dead to save his beloved, with music to sooth the "dogs of hell" represents civilizing forces overcoming chaos and war, (attractive as well for artists in fascist Italy).[8] Orpheus returning to earth prefigures Christ and more generally, the many ressurectionist cults the mediterranean world has known through its ages. This classical icon of humanity became an important harbinger of enlightenment during the eighteenth century.[9] For the twentieth century avant-garde artist, Orpheus became a complex, often-contradictory figure betraying the artist's self-view and his struggles.[10]

Manship's Orpheus seems settled and distanced from inner torment, eerily so, in mild and conventional form synthesizing Greek and Indian sculpture. Hardly memorializing, he messages little about honor, glory,

---

of Orpheus as a lyric muse in (1912), and on a marble base (1927); with Eurydice (1935); two models of Orpheus with Pegasus (11.5 inches) (ca. 1930) for a monument. The Paul Manship Bequest, Smithonian Institute.

6 For Illustration see Faulker *Sketches,* p. 63.

7 Judith E Bernstoc, *Under the Spell of Orpheus: The Persistence of a Myth in Twentieth Century Art*, (Southern Illinois University Press, 1991).

8 Manship was not alone in his interest in Orpheus and its meaning as an expression of civilization. An Orpheus by Libero Andreotti from 1931 was known and exhibited at the 1934 Venice Biennale. Exhibition catalog, "The Thirties: The Arts in Italy Beyond Fascism." Palazzo Strozzi, Florence, Tuscany, (September 2012). The influence of art under Mussolini on the AARome warrants further study.

9 Johan van der Zande, "Orpheus in Berlin: A Reappraisal of Johann Georg Sulzer's Theory of the Polite Arts in Central European History," (*Atlantic Highlands, NJ: Humanities Press International, Inc.* 1995 vol. 28) pp. 175–208.

10 Bernstoc, *Under the Spell,* Introduction by Peter Selz: "He [Orpheus] reflects the modern artist's acknowledgement of feelings of inadequacy, projects his [the artist's] own fear that his imagination of an ideal world might deceive him." P. XXV.

victory, death, or resurrection. Nor is he funerary, messaging neither grief nor remorse. Rather, he seems to belong to a higher realm, gained when wars' horrors are banished from consciousness.

That the sculpture came to be at Anzio/Nettuno in 1977 may be read as a terminus of sorts, where Manship's quest to express the essence of human life is completed. It also is a testimony to his friend Gugler's continuing dedication to the work of his lifetime.

Fig. 31. Manship's Orpheus.

Gugler's (Unrealized) Monumental Dreams

"… we are the dreamers of dreams …
For each age is a dream that is dying,
Or one that is coming to birth."
Arthur O'Shaughnessy, poet (1844–1881)

Eric Gugler's monumental art at Anzio/Nettuno may be seen as an attempt to capture America's paradigm of patriotism and exceptionalism during a brief post-war period, and to fuse it with thousands of years of European history recorded through art. So formative was his experience of designing Anzio/Nettuno that he envisioned a future for himself as an ordained master of grandiose public historical monuments. A short exploration of his projects following Anzio/Nettuno is presented here because these reveal US culture and tastes moving onward, leaving Gugler and his dreams behind.

Hall of History

He proposed a gargantuan monument to be named "Hall of History," to be located on a 2,000- acre site bequeathed to the state of Georgia by late President Roosevelt (FDR).[1] It would measure 75 by 125 meters. Eleanor Roosevelt (E.R.) described the project and "the famous Eric Gugler" in her syndicated column *My Day* (August 15, 1953). The façade was to mimic Gugler's prize-winning proposal for Chicago's World War One (WWI) memorial (more than 30 years earlier and never realized).[2] The

1 Eleanor Roosevelt, "History of Nation to be Sculptured in Shrine on a Mountain in Georgia," *New York Times*, (August 7, 1953), a long article with images. See also *William and Mary Quarterly*, 3d series, vol. 12, no. 3 (July 1955) pp. 475–80.

2 For a short description and two illustrations of the monuments' plans see *The Western Architect*, (January 1930), vol. XXIX, p. 9; Wright, "The First World War Memorial," p. 61: "Mercifully it was not built."

court would accommodate thousands, and earphones would guide visitors in five languages. As Gugler explained: "The walls will have portrayed on their granite surfaces impressive inscriptions of great distinction, imperishable phrases chosen from our past and in high relief groups of figures, episodes, and events of our history."[3] Scenes were to be "dramatic" and would begin with the Age of Discovery and finish at the eve of WWI. There would be sculptures of "… eminent educators, statesmen, leaders in the fields of religion, labor, industry, the arts, and the military." Great documents would be displayed, "America's big story book" (ER's words). The entire site would be surrounded by pine forest, and Gugler's sketches show boughs swaying in the winds. Anzio/Nettuno had been budgeted at about $600,000, but the Hall of History was estimated to cost $25 million, to be raised through public subscription. Although E.R. (and other VIPs) advocated for this project, their influence was apparently not sufficient to secure a funding guarantee by Georgia's legislature. The project was never realized.

"Shrine to Freedom"

In 1960 Gugler proposed a project called "Shrine to Freedom," to be erected on federally owned land along Washington's Potomac River, at an estimated cost of $24 million (1960 dollars). Again, the model showed a huge frontispiece with pillars (100 meters long, 62 meters wide and 21 meters high) flanked by two lateral pavilions slightly set back, recalling Anzio/Nettuno.[4] These were to be covered with reliefs by Paul Manship,

3 Quoted in *The New York Times*, "History of Nation to be sculptured in Shrine on a Mountain in Georgia," (August 7, 1953). The author gives information on the pictorial program. "The Era of Discovery." (Leif Ericson's ship. Columbus, De Stoto and others.) Especially dramatic would be Daniel Webster delivering his "Union Forever" address to the United States Senate. Near it would be a sculpture of Abraham Lincoln, the 21-year old sitting under an apple tree in Illinois and reading the address that influenced him in later life. Shown also would be Lewis and Clark on their journey and covered wagons journeying westward.

4 Ada Louise Huxtable, "$24,000,000 Shrine is Proposed for Capital," *The New York Times*, (June 3, 1960). The article shows the frontal view of model. Information about the project drawn from this article.

with didactic inscriptions on the subject of US democracy. A tree-lined alley with fountains would lead the eye to the Washington Monument, the Capitol Dome, and the Jefferson Memorial. Two American eagles on high pedestals at its entrance, wings spread wide, would proclaim the monument's national destination.

In 1954 Congress created a National Monuments Commission (NMC). Gugler's project was reviewed and approved. The NMC addressed a letter to President Eisenhower stating that it had fulfilled its duty. But for Ada Louise Huxtable of the New York Times, the project might have gone forward. Huxtable wrote a scathing article damning its aesthetics as "tired cliché" and "… a flaccid, watered down classic style, innocuous and uninspired." The sculpture was to be executed in "… that peculiarly artificial and specialized manner that might be called 'historical realism' which relies heavily on groups of explorers and Indians balanced with mechanically-molded trees, cotton-ball clouds, and overly familiar symbols." Further, Huxtable wrote: "Its empty pomposity is a poor substitute for one's visit to the original documents of freedom." Huxtable, in the same article, exposed the fact that three NMC consultants, whose verdict had been decisive for the project's approval, were none other than John Harbeson, Gilmore Clarke (co-designer for the project), and Gugler himself. Thus, the designers served as their own jury. A veritable "old boys' network" had set out to impose their vision onto the US capitol at taxpayers' expense. Huxtable's criticism proved devastating. The project was abandoned, and the "old boys" subsequently came to refer to Huxtable as "that dreadful woman."[5]

Undeterred, Gugler then proposed for a memorial to Franklin Delano Roosevelt (FDR).[6] A large Greek temple upon hill would dominate a forest surrounding two ponds.[7] However, it was discovered that FDR had recorded his wish that any monument to him "… should consist of a block about the size of this desk." In accordance with his wishes, a votive granite stone was so set in Washington. Gugler did get to design the modest stone and the engraving, stating simply: "MEMORY OF FRANKLIN DELANO ROOSEVELT 1882–1945." This meek footnote to Gugler's grandiloquent vision was matched by his memorial for his benefactress Eleanor Roosevelt, which

5 Hand-written at the side of an article by Huxtable on a Beaux-Arts show in New York, sent to Clarke. Clarke papers, Columiba University Archive.

6 An ink wash exists of this ambitious project. Avery Architectural Library Archives, Columbia University, NYC, Gugler papers. Washington History vol. 10, no. 2 p. 72.

7 Avery Architectural Library Archives, Gugler papers.

is set in the United Nations gardens in New York (1966) and consists of a simple granite bench/exedra on which is engraved: "1884 – Anna Eleanor Roosevelt – 1962." Opposite stands a slanted stone stele bearing a flame in bas-relief and below, the inscriptions: "She would rather light a candle than curse the darkness" and "Her glow has warmed the world." Some puffy Art Deco clouds are squeezed in but there is no statue. The ground between exedra and stele has irregular paving stones, and no mention of Gugler is to be found in guidebooks. Gugler kept other "irons in the fire." His monument to Theodore Roosevelt in Washington (1967) includes a five-meter-high bronze statue of Roosevelt with one arm raised in rhetorical gesture by Paul Manship.[8] Its size and almost painful naturalism remind one of Socialist Realism.

Academic naturalistic style in sculpture continued to be sanctioned by conservative public agencies such as the Commission on Fine Arts, but even gigantism could not hold back the modernist trends that both Gugler and Manship had struggled against throughout their careers.[9] Gugler et al. evidently quested to create an American iconology, a recognizable and permanent visual artistic record of its history, but it was not to be. The historical period during which the US public at large would accept the artistic visions of the well-connected elite, evidently, had passed.

8 David H. Wright, "The First World War Memorial" p. 76. He refers to the statue as if it "… might have been one of his doughboys standing up and waving his right arm in a politician's rhetorical gesture."

9 Gugler's Spark Center at Wabash College (Indiana) main building has a tri-partite composition similar to the Anzio/Nettuno's memorial, complete with temple façade of the middle tract (1954). It is the closest US version of Anzio/Nettuno's memorial design.

V. Impruneta, American Modern Classicism: Grandeur and Intimacy

"J.K. was a classicist to the bone, but his adaptation of tradition to modern needs flowered in many places…"
Barry Faulkner[1]

Fig. 32. Impruneta, officially called "Florence Cemetery" by the ABMC. Note insertion into surrounding forest and co-ordination of flag-pole and pylon.

Like Gugler, John Kellum Smith (1893–1961) was a life-long adherent to classicist design principles. His classicist leanings were groomed during a three-year fellowship at the AARome, which he was awarded upon winning the "Rome prize" in 1923. His ties to the AARome remained strong. He served as its president (in New York) between 1937 and 1958. He belonged to a group of prominent and prestigious designers, architects, and artists who maintained cordial relations with one another. In 1929, he

1 Faulkner, *Sketches*, p. 120.

became a full partner at the prestigious and influential New York architectural firm, McKim, Mead & White, on Park Avenue. College buildings became his specialty. From 1930 until his death he was the consulting architect for his alma mater, Amherst College.

Smith and Gugler were friends, and even had the same prestigious business address on Park Avenue in New York. Remarkably, despite all the commonalities they shared, including adherence to the standards of the ABMC and the need for CFA approval for their designs, their cemeteries in Italy differ substantially. Comparing these, we see the results of different personalities. Landscape architect Ralph Griswold once visiting Gugler's office, looked in on Smith and noted in his journal; "… (this) afternoon went to Smith's office to see his model for the cemetery near Florence. It is altogether different from ours."[2]

Smith's creation, his "appropriate architectural expression,"[3] resulted in an original work in "stripped-down classicism," in accordance with the aesthetic principles of the venerable Paul Cret, who wrote:

> "What is to be substituted for the proven efficacy of training the eye to proportion, to rhythm, to composition, is not as yet divulged, and those who condemn them as stifling originality forget that originality so easily stifled is not very robust."[4]

Despite Cret's authority, Beaux-Arts Classicism for non-governmental edifices had been dismissed outright by the time these two cemeteries were designed. Beginning in the early 1930s, architectural Modernism had largely replaced it in the US.[5] For its adherents, however, the aesthetic canon was irreplaceable. Seen today, their works seem unaffected by Expressionism, Cubism, Futurism, Surrealism, International Modernism, the Bauhaus, and Le Corbusier's Modernism. However, in Smith's design for Impruneta, one senses that modernist impulses find subtle expression in the cubic, abstract quality of his structures and in the geometric matrix that

2 Journal entry, Griswold. Pittsburgh University Archives, Griswold papers, (November 11, 1949).

3 Letter, Smith to North. The NYHS McKim, Mead & White collection, PR 042 (correspondence files), Box 452, (September 1, 1947).

4 For full quote and source see Chapter I fn. 33.

5 The Museum of Modern Art, New York (MOMA) held an exhibition titled Modern Architecture: International Exhibition. 1932.

underlies the total design.[6] This indeed testifies that within the parameters of Cret's vision, a work of art could be created that did not seem outdated in style.

The automobile traveller of the 1950s following on the old *Via Cassia* connecting Florence and Sienna would see the Impruneta cemetery "bursting" into view. From today's newer improved roadway, it remains visible but from a speeding automobile, the view is both fleeting and shocking. One glimpses something obviously alien in the Tuscan landscape, something nestled into a natural amphitheater beyond a river, surrounded by woods. One notices rows of gleaming white crosses on lush meadows that flank a mall of well manicured grass, widening as it leads to the low, broad memorial below the hill's crest. A sturdy, triangular pylon (25 meters high) is set in front of it and a large, amorphous sculpture seems to ooze from its pinnacle. A closer look reveals this pylon in central sightline with a 30-meters-high flagpole at the cemetery's entrance flying a large US flag, topped by a gilded American eagle, the venerated totem of the American nation.[7] The way these two bold verticals work in counterpoint with the architecture's horizontality to create a dynamic feel captures the visitor's interest at once. Smith's keen design brilliance conveys an immediate albeit foreign presence.

It has been noted that the flag plays a central role in American life, public and private. "Is the United States unique among the countries of the world in its obsession with its national flag, and, specifically, in the propensity to display it in public places?"[8] The flag held a central place in the ABMC's ideological program. General North stressed that a prominent place for the flagpole must be provided since it was of paramount importance.[9] There are many archived instances in which North reminded his architects of flags and flagpoles, the design of the latter being intensively discussed.

6 On Smith's open attitude toward modernist impulses see Yegül, *Gentlemen of Instinct,* p. 110; on modernism see Henry Russell Hitchcock, *Architecture: Nineteenth and Twentieth Centuries*, (Penguin Books, 1958).

7 Wilbur Zelinsky, *Nation into State,* (Chapel Hill: The University of North Carolina Press, 1988), pp. 198–202. The author discusses the role and symbolic power the American Eagle in detail.

8 Ibid. Under the headline of "Flag and Eagle" pp. 202–208. (Author notes that in eastern states, the flag is ubiquitous and even flies over drive-in hamburger restaurants).

9 See North's letter of invitation to Smith stresses that the flagpole was important for ceremonies. RG 117, Box 72, (August 26, 1947).

The Impruneta site is accessible via a broad gate, complete with emblems, though not as ornate as those at Anzio/Nettuno. The large tract of land reaching to the highway is left with its natural contours and is cared for. The river Greve, straightened and its banks paved, its waters moving peacefully (from south to north), separates this area from the cemetery proper. Two mirror-image travertine buildings in cubic form (rigorously kept pristine) host a pleasant reception office, replete with seating and fireplace, and a visitor's building.

Fig. 33. Reception room in form of idealized American "front parlor."

They frame and border a bridge that leads across the river. Crossing, one feels subliminally reminded of the river Styx. Once beyond, the visitor has the impression of being in a different reality, onstage in an amphitheater formed by the hillside's natural embrace.

Smith's memorial design is indeed unlike Gugler's "temple." Two rectangular identical projecting porticoes, standing against the wooded hillside, frame a court of honor. The bold pylon erected on an outcropping of the middle terrace stands in central line to the "wall of the missing." The latter serves as the pylon's backdrop and at the same time, is enhanced by the pylon's affect. The resulting dynamic interplay between horizontality and verticality quickens awareness of the fates of 1,409 individual names of those whose bodies were never recovered. Through its darker entablatures in rectangular stone arranged in several continuous rows, it acts as the visual and ideological focal point of the whole cemetery. Architectural homage is thus paid to the sacrifice of life.

Fig. 34. The memorial with the pylon.

The two lateral projecting porticoes feature vertical "cut-outs" from their walls, and these appear as planar pillars. Perfect ratios produce an effect so pleasant and calming that any supplemental ornamentation, it seems, would only detract. Without ornamentation or projecting details, the pavilions have an abstract quality sought in Modernist architecture of the time. It is quite obvious that geometric clarity and simplicity, where rectangle, square, and triangle rule, are the keys to Smith's aesthetics. All of the cemetery's forms relate to one another through a consistent geometric matrix, yielding wholeness that conveys a sense of wellbeing and calm.

The memorial's soft creamy travertine, the hillside's dark green forestation, and the emerald green of the mall make a contrast that feels grounded and harmonious as though Smith wanted to create a monument to peace. From every point in the cemetery, lightness embraces the gleaming white of marble crosses. The site's gentle curvature engenders a sense of "uplifting." One feels a subtle spiritual dimension.

Access to the memorial is provided via two tree-shaded pathways, each aligned with the central opening of an atrium. In beautiful profile, three terraces slightly wider than the court of honor and of generous frontage dimension, foot and stage the memorial. Its two porticoes are atria, as in classical Roman architecture, open to the sky. Large rectangular basins in their centers have beautiful profiles. From a central spout, crystal clear water splashes lavishly into each's basin onto large grey pebble stones, producing a continuous and soothing audible dynamic. General North's wish to have flowing water here was more than fulfilled.[10] The motif of the atrium with basin presents life and plenty.

Fig. 35. North atrium's water basin and view on "wall of the missing."

The south atrium leads to the chapel, fitted between atrium and the hillside such that identical structures are seen at both of the memorial's sides. In the south atrium recessed wall sections present inspirational quotes, to be discussed further below. The north atrium serves as the museum where the obligatory large military operations map (or "battle map") is mounted on its west wall. General North had made it clear that the map should indicate progress of Allied Forces, but needed to give precedence to the US operations: "... on the map, the American effort must be the dominant one,

10 Letter, North to Smith. He imagined a fountain that would use little water. His letter included photos of a fountain at Villa d'Este, near lake Como (Italy). RG 117, Entry 9, Box 72, Folder 687, (6–1–48 to 3–31–49).

primarily telling the story of our own troops."[11] The sidewall (of *baveno* granite) has texts inscribed in English and in Italian, telling the course of the campaign and naming ground and air battles in which US forces participated. The ABMC attached great importance to the map's didactic destination, detailing campaigns and operations in Italy. Its topography was to be correct, and on its upper left corner were to appear twelve shields in four rows of three, showing the insignias of US ground and air units that had been involved in the fighting.[12] The AMBC reviewed and corrected preliminary designs that came from Smith's office.[13] Arrows signifying the military would vary in length and width.[14] Representational content was the ABMC historical committee's responsibility. Smith's task was not the map's content, but its aesthetics and placement.[15]

North insisted that these battle maps to be works of art. The Florentine sculptor Bruno Bearzi, often employed by the ABMC, was commissioned to produce Impruneta's large battle map using a traditional Florentine craft called *Intarsia*. The technique produces a mosaic of colored marble, and demanded great skill to produce beautiful results. A smaller map of fine handicraft is inserted into the larger map, below the shields.[16] Six key maps, identical to those at Anzio/Nettuno, show the total forward progress of war against fascism and the collapse of the Axis powers worldwide, year by year.

The vaulted chapel adjacent to the south atrium displays, in its apse, a mosaic. Designed by prominent artist Barry Faulkner, this fills the wall space above the altar and is discussed in detail below.

---

11 Letter, North to Smith. The NYHS McKim, Mead & White collection, correspondence files, PR 042, Box 451, (January 20, 1956).

12 Letter, Harbeson to Smith informing Smith that the ABMC's historical division, in charge of maps and historical decorations, needed about six more months to prepare the material for the map. RG 117, Entry 9, Box 72, Folder 687, Florence, (4–1–49 to 3–31–50), (November 25, 1949).

13 Drafts of maps containing corrections in red as to length and width of arrows (showing the progession of the of campaign). The NYHS McKim, Mead & White collection, PR 042, Box 453, (Undated).

14 Harbeson, the ABMC's Paul Mitchell, and the firm's partners White and Vegezzi consult over the map. Corrections in style followed later. The NYHS McKim, Mead & White collection, PR 042, Box 451, (August 11, 1952).

15 Letter, Smith to Harbeson requesting guidance in battle plans, inscriptions, etc. RG 117, Entry 9, Box 72, Folder 687, (8–26–47 to 5–31–48), (February 26, 1948).

16 Executed in local *scagliola* technique by Emilio Martelli, Florence. The ABMC visitor's booklet on Impruneta.

Smith, a passionate Italophile, must have understood Italian social structures and must have had some inkling of the social dislocations that locating such a cemetery on rich farmland would visit upon those affected. He had witnessed some of their protestations in person. But Smith was also a patriot and had participated in WWI as an aviator and in WW II with the rank of a lieutenant colonel in the Army Air Force. It seems evident that he viewed the US's post-war presence as an assurance of peace and stability, and his mission was to contribute to this effort through beauty, which he understood as: "All elements of the design are bold and simple, qualities characteristic of the best memorial architecture from the most ancient times to the most modern … the whole composition should provide for the graves a setting of simple, ordered beauty, intimately related to the surrounding landscape."[17] In his vision of "Pax Americana," there was room for his genuine and heart-felt love and respect for Italy and its heritage. As an example, he wished to maintain the site's three ancient stone farmhouses, arguing: "They add a certain amount of agreeable Tuscan flavor to the scene."[18] They could be rebuilt and used as superintendents' housing. Closer inspection, however, revealed they were unsuitable for such as there would be animal stench after so many centuries of animal husbandry, and probably vermin.[19] Nevertheless, further thoughts went into how the buildings could be saved, but in the end, they were demolished.[20]

17 "Description of the American military cemetery, Florence, Italy" by J. K. Smith. The NYHS McKim, Mead & White collection, PR 042, Box 452, (Undated). Impruneta is officially called "Florence Cemetery" by the ABMC.

18 Letter, Harbeson to North (September 1947) supporting Smith's wishes to keep the ancient farm buildings. RG 117, Entry 9, Box 72, Box 687, Florence, (4–1–49 to 3–31–50), (September 28, 1949).

19 Letter, ABMC Rome (Lt. Col. Strandberg) to the ABMC Washington, D.C. inquiring about the future of the farm buildings. RG 117, Entry 9, Box 72, Box 687, Florence, (4–1–49 to 3–31–50), (August 31, 1950).

20 Memorandum from North to Shaw. RG 117, Entry 9, Box 72, Box 687, Florence, (4–1–49 to 3–31–50), (September 18, 1950).

Designing Impruneta's Memorial: A Labor of Love and Dedication

"My connection with Italy has been so long, and my handling of this particular project so personal that it seems too bad to have to forgo any direct credit."
James Kellun Smith[1]

Impruneta is a splendid work of art. Smith's dedication was absolute, but there is no mention of him at the cemetery. The ABMC's convention was to display the name of the architectural firm on each foreign cemetery and the forceful Gugler managed to have his own name displayed at Anzio/Nettuno. Smith, evidently more modest, when offered the choice between McKim, Mead & White and his own name, chose the firm's name, but not without regret.[2]

Harbeson had invited Smith to consider designing a future monument near Florence on June 26, 1947.[3] North wrote to Smith from the ABMC's Paris office offering him the commission.[4] Residing at the AARome when he received North's letter, Smith immediately undertook to visit site E (in the future to be referred to as "Impruneta") and concurred that it was very suitable and beautiful. North then sent out an official commission.[5] Smith,

1 Letter, Smith to North in response to the latter's asking him to either credit his name or that of the firm McKim, Mead & White as the cemetery's designer. Smith, in modesty and with deep regret renounced his claim to fame. The NYHS McKim, Mead & White collection PR 042, Box 451, (October 19, 1955). (In German cemeteries only the name of the VDK was indicated. Letter, VdK to architect Dieter Oesterlen. The AdK Archive, the Dieter-Oesterlen Archive, Büro-Ordner AR-Nr. 67. However, the latter did manage to have his name engraved at the Futa-Pass Cemetery. See Chaper XI).

2 For said letter from North to Smith see the NYHS McKim, Mead & White collection, PR 042, Box 451, (October 19, 1955).

3 Letter, Harbeson to Smith inviting him to design the Florence (Impruneta) cemetery. The NYHC McKim, Mead & White collection, PR 042, Box 452, (June 26, 1947).

4 Letter, North to Smith. RG 117, Entry 9, Box 72, Folder 687, (2–6–47 to 5–31–48), (August 26, 1947).

5 RG 117, Entry 9. Box 72, Folder 687, (8-26-47 to 5-31-48) (Sept.1, 1947).

an Italophile and a veteran, accepted without hesitation, writing that he was "deeply touched."[6]

Harbeson's "short list" of suitable architects, prepared for North, included Smith's name as one who had some experience in monument design. For his alma mater, Amherst College, Smith created a memorial field with a war memorial (dedicated in June of 1946) in remembrance of those young men from the college lost in both World Wars.[7] This memorial commemorates using architectonic elements and landscaping, but does not use sculpture. A stone amphitheater is built into a hillside, and has a raised granite circle at its center. Stone seats mark its periphery. Interaction of topography and building pre-figures the work in Impruneta, where he took full advantage of the site's natural amphitheater-like curvature. Smith's service in the Armed Forces in both World Wars as a field grade officer was an important attribute for General John Markey.[8] In addition, Smith's personality was said to be pleasant.[9] By all accounts Smith was a well-educated gentleman, sensitive, with refined tastes, and polite to all.

Evolving Design of the Memorial

The seeming ease of the memorial's coherent design was the product of Smith's having experimented with different concepts. Smith was commissioned to provide two different designs. His innumerable sketches all put architectural emphasis on the center of the monument. They reveal his

6 Letter, Smith to North. The NYHS McKim, Mead &White collection, PR 042, Box 452, (September 1, 1947).

7 Webpage: HCAP Council of Independent Colleges, Historic Campus Architecture Project, Memorial Field and War Memorial. It gives a short description and illustration of the memorial.

8 Letter, North to General Markey telling him who, among their commissioned architects, had served in the war. He added the working members of the ABMC's architects' offices that were war veterans. The letter shows signs of having been crumpled and smoothed out again. One can imagine an irritated North responding to Markey's repeated inquiry. The latter likely favored Smith because he had served in both World Wars. RG 117, Entry 13, Box 143, Folder 231.24, (December 8, 1948).

9 Letter, Clarke to North. "He has a pleasant personality, gets along well with others..." RG 117, Entry 13, Box 143, Folder 231.24, (April 29, 1946).

initial orientation toward Paul Cret's WWI memorial at *Chateau-Thierry* by whose façade consists of an uninterrupted row of "wall pillars."[10] As he sought to accommodate a chapel and a museum while maintaining coherency in the cemetery's overall design, he experimented with centrality and horizontality, both features characterizing his final design. Cret's iconic monument in France remained an important inspiration. What Smith obviously took from *Chateau Thierry* was the memorial being set on a raised terrace and accessible via broad steps.[11]

Smith gradually distanced himself from Cret's authority, beginning with a central structure that would interrupt the façade's row of pillars.[12] His various experiments with centrality would finally result in the court of honor formed between two pavilions projecting forward from the façade, but the design underwent many iterations.[13] However, he maintained Cret-like central structures albeit varying in plan and elevation, and with the museum and chapel attached to the rear of the memorial's pillared façade in line with the cemetery's axis. In some sketches, a large central door would provide access from the front. In other sketches, a central structure is behind the memorial unseen from the front, and still others, emphasize it as a marked central design feature interrupting the façade's row of pillars.[14] These sketches witness Smith's distancing his designs from Cret's precedent, and movement toward framing the court of honor by means of lateral tracts projecting forward from a façade.[15]

---

10 Sketches are archived in the NYHS McKim, Mead & White collection, roll container NS1 N -139(4), NS1 N-137 (6) and NS1–138 (71).

11 *Chateau Thierry* was a standard for North, by which he evaluated Smith's design. He found that his terraces exceeded the formers' in height – it "upstaged" the venerated Cret – and therefore should be lowered. He drew Smith's monument over a photo of Chateau Thierry to prove his point (Financial reasons likely played a part). For image see RG 117, Entry 9, Box 122 and see fn.17.

12 This design phase based on Plan B's centrality had several versions. The NYHS McKim, Mead & White collection, roll container NG 1 N139 (4), (February 4, 1948).

13 For example: A design where lateral tracts project front and rear with central structure behind the pillared façade. The NYHS McKim, Mead & White collection, roll container NG 1 N139 (4), (January 27, 1948). Smith's designs were developed prior to site acquisition.

14 RG 117-Con records, (February 4, 1948).

15 Design with lateral tracts projecting front and rear, still with the central structure behind a pillared façade (Plan B) (and a row of trees, a feature he kept). RG 117-Con records, (January 27, 1948).

Smith, in parallel, experimented with the idea of adjunct buildings behind the façade's ends and a very narrow court of honor. Until the fall of 1949, this concept remained under consideration.[16] The CFA liked it and Smith elaborated on it, beginning with narrow rectangular lateral buildings pillared toward the court of honor. Merging the two concepts, these two pavilions became atria complete with water basins, (should water be available), and the court of honor widened. This concept did not change in its essential features, although Smith had to reduce the memorial's size for financial reasons.[17] Consequently, he altered the width of the mall in order to preserve the hard-won inter-relatiship between memorial and mall.[18] This trial and error exploration exemplifies creativity hampered by an overwhelming repertory of the classical past.

The Pylon, Object of Controversy

The cemetery's vertical element appeared early in Smith's conception.[19] This evolved from a central tower in front of a crescent-shaped memorial with heavy projecting buttresses.[20] Once its tapered form had evolved, Smith experimented with several solutions for its exact position,[21] finally designing a rhomboid base that projects from the middle terrace onto

16 Design from January 27, 1948 shows rear buildings and chapel with some adjunct structures at both ends per Harbeson's original sketch in RG 117, 1947, RG 117 Entry 9, Box 72, Folder 687, (8–26–47 to 5–31–48).

17 Letter, Harbeson to North. "S. has to shrink design in width and height, but will make design fit the countryside." RG 117, Entry 9, Box 72, Folder 687, Florence, (4–1–49 to 3–31–50), (March 27, 1950).

18 Letter, Smith to Rossi saying that the overall sizes of the memorial and pylon had to be reduced. The NYHS McKim, Mead & White collection, PR 042 Box 452, (March 27, 1950).

19 "Its principal feature will consist of a large triangular pylon rising from a terrace shaded by plane trees that would be clearly visible from the Via Cassia," Smith explained. "Description of the American Military Cemetery, Florence, Italy" by J. K. Smith, the NYHS McKim, Mead & White collection, PR 042, Box 452, (Undated.)

20 This drawing can be found in RG117, Entry 91 Box 72, Folder 687, Florence (8–26–47 to 5–31–48).

21 See sketch on "buffy paper" at the NYHS McKim, Mead & White collection, roll container NS1 ND-137.

the lowest.[22] Without its dominant presence rising skyward, the memorial complex would lack impact, but with it, the resulting composition has unique dynamism. Smith's pylon quickens his memorial's calming horizontality.[23] Elegantly tapered, the pylon stands tall upon its pedestal, a massive yet graceful vertical against the horizontal memorial with the horizon rising behind. (See Fig. 34) The sculpture atop, difficult to make out from a distance, depicts the American eagle accompanying a flying allegorical female, the two watching over while consoling those buried beneath. (Sculptor Sidney Waugh and his Art Deco style will be discussed in more detail below.) The sculpture being displayed as if airborne likely emanates from Smith's having served in the Army Air Corps (later to become the Air Force) in both World Wars. Not all shared Smith's enthusiasm for the pylon. He had to defend his vision. Having been assured artistic freedom, he prevailed on the subject of the pylon, as later he would on the subject of flagpole placement after Harbeson claimed it might disturb the view onto the monument.[24] Inspiration for Impruneta's pylon may well have been the sensational Trylon (triangular pylon) at the 1939 World Fair in New York, a huge tapered needle abstraction pointing skyward. (Together with the Perisphere and Helicline, it became the fair's logo.) Harbeson stood by his fellow architect when North wanted it omitted for financial reasons, arguing that the site required something dramatic, and hence it would be better to economize at the atria.[25] The CFA also objected to the pylon, but Smith, classically educated, countered that the pylon had been used at ancient Olympia, as documented by Pausanias. (Smith even gave reference to a

22 Pylon on a high triangular base attached to building façade, depicting a mighty structure was abandoned. The NYHS McKim, Mead & White collection, roll container NS1 N–137, (March 11, 1947).

23 Blueprint emphasizes the axial relation between flagpole and pylon. Their verticals work in counterpoint to the sight's horizontality. The NYHS McKim, Mead & White collection, role container NG139 (4), (January 15, 1948).

24 Letter, Smith to North. The NYHS McKim, Mead & White collection PR 042, Box 451, File Nr. 9, (September 4, 1956).

25 Letter, North to Harbeson. North found the whole chapel combination unusually large and suggested that Harbeson reduce size and omit the column. Harbeson replied: "The site requires something dramatic, and the column, I feel, is an essential part of the scheme as seen in the perspective. I'd rather reduce the size of the atria and fill the open spaces between the pillars." RG 117, Entry 9, Box 72, Folder 687, (8–26–47 to 5–31–48), (December 30, 1948).

standard work by the German archaeologist, Dörpfeld.)[26] This is a striking example of the CFA's classicist mindset, where in the end, antiquity's authority prevailed over their own aesthetic preference.

Unlike Gugler, who hoped the Anzio/Nettuno project would establish his credential as a master architect for large monumental projects of the future, Smith's Impruneta project did not seem to affect his modest sense of self-importance. His subsequent projects centered on buildings for college campuses and his designs in the Georgian/Federal tradition often had no columns. Where classicism was called for, he continued his restrained "stripped-down" style that emphasized a clarity of composition, pure shapes, simple lines, elegant proportions, and a preference for no columns. Although Smith never became involved with commercial buildings of note, toward the end of his career he was commissioned to design the National Museum of American History and Technology (later to become the Museum of American History), a large building on the Washington mall. (He did not live to see its completion.) Ada Louise Huxtable, The New York Times architectural critic and Beaux-Arts lovers' nemesis, did not consider it a success: "An awkward attempt to marry the classical and the modern, the building is legitimately neither; it is a monstrous and meaningless misalliance."[27]

---

26 Smith probably referred to the work of Wilhelm Dörpfeld. *Alt-Olympia, Untersuchungen und Ausgrabungen zur Geschichte des ältestes Heiligtums von Olympia und der älterenen griechischen Kunst.* Berlin: E.S. Mittler & Sohn, 1835.

27 Louise Huxtable, "Blending the Classical and the Modern: The Museum of History and Technology of Smithsonian Opens Doors Today," *The New York Times,* (January 23, 1964).

Quest for the Perfect Form

> "My emotions have been so intimately involved in the effort and sacrifice of our forces that I could have no personal ambition greater than that of succeeding in giving the appropriate architectural expression. The fact that the memorial is to lie in the Italian countryside, the beauty of which I have long known and admired, adds to the challenge."
>
> John Kellum Smith[1]

Impruneta's seemingly finished perfection was actually achieved through compromise, in response to circumstances beyond Smith's control. As previously pointed out, the large parcel of land between the Greve river and the Florence/Sienna state road, which had not been part of the original site requisition, was eventually included therein and acquired by the ABMC.

Harbeson, on his scouting tour, visited site E and soon after (August 15) refined his rough layout plan that fixed the axis, producing a cemetery design plan that included all the requisite elements.[2] (Smith, expected to arrive on-site shortly, could have done his own reconnoitering.) Harbeson's plan shows that he thought in the terms of a grandiose totality, as had his plan for Anzio/Nettuno. With the open land between the river and the state road now available, a "*Via Triumphalis*" was imagined, which would be an expression of triumph and importance. The triumphal entry would begin with a round plaza adjacent to the road followed by a broad mall, and following that would span the empty tract, cross the river, and continue to the memorial. Graves were to be placed laterally to the mall inside the cemetery on large rectangular fields, curving toward the monument.[3]

This standard Beaux-Arts vocabulary presented no challenge to Smith. Nevertheless, during the approximately one year's time between Smith having first seen the site in the early fall of 1947, and his submission of detailed

---

1 Letter, Smith to North. The NYHS McKim, Mead & White collection, PR 042, Box 452, (September 1, 1947).

2 Harbeson through this sketching, addressed the technical difficulties that needed to be addressed as a first step. RG 117, Entry 9, Box 72, Folder 687 (8–26–47 to 5–31–48).

3 Harbeson's survey plan from RG 117, Entry 9, Box 72, (8–26–47 to 5–31–48) (August 10, 1947).

grading and layout plans in December 1948[4] (finalized in early 1949),[5] his design for the total site underwent a great many versions. A large number of colored sketches on semi-transparent paper, unsigned and undated, suggesting they were part of his design process. They took on a playful quality at times.[6] (The fact that these were done before the site was secured [October 18, 1948] testifies to Smith's confidence that the AMBC would succeed in securing the site.) His vision oscillated between a statement of "triumph" and one of "restraint." Geometry and symmetry, with architecture and landscaping in a tight totality unify the many variations. Distinct, basic geometric shapes such as rectangular, square, and trapezoidal, determined all elements of the total site, including planters. In one, he envisioned a sequence of huge square planters filling the mall while in others, the rectangle controlled the plan. Quoting royal Versailles park, Impruneta's axis remained the supreme regulative.

Smith's extensive experimenation quotes the whole repertory of Beaux-Arts motifs. One sketch includes a cross-axis Latin cross arrangement. Another, particularly striking, features trees in rows continuing over the bridge, thus unifying the open land and the cemetery. Yet another shows colonnades in different length and form (straight, oblong, and Bernini-like). Others present a square, then a rectangular pool, or a terrace containing a pool imposed on the river's course in sightline with the axis.

Design considerations had to include practical necessities, such as parking and toilet facilities. Parking was initially envisioned to be as remote as possible from the entrance. Gradually, sketches show that parking space was moved closer, finally to the immediate vicinity of the entryway. Likewise, the administrative office and reception buildings were first imagined to be remote from the cemetery itself so that the view onto the memorial monument might be unimpeded. The latter was to be separate from the grave areas, as well as from the surrounding world, thus visually locked in space and time and imbued with "timelessness." A lengthened entrance section, accessible by foot, would enhance the experience of approaching

4 The first plan was altered several times until April 24, 1949. For various plans showing alterations see the NYHS McKim, Mead & White collection, roll container NS1 N–314 (4), (December 17, 1947).

5 Site plan. The NYHS McKim, Mead & White collection, container NS 1 N-139 (4), (January 29, 1948).

6 These sketches are to be found in the NYHS McKim, Mead & White collection, roll container NS1 N -139 (4), NS1 N-1376 and NS1–138 (71).

a special space, in similar fashion to crossing a plaza in order to reach a temple.

The cemetery's central mall underwent variations. One plan envisioned a continuous green mall unitizing both sides of the river. This plan was abandoned and focus went to the smaller mall on the cemetery side of the river. The mall's width fluctuated with that of the memorial to which it leads, first wider at the entrance, then evenly wide.

The plan for grave fields underwent only minor variations. Individual grave markers describe arcs, their (virtual) centers behind the memorial such that the pattern of markers splays outward from it. Pathways between were to be grass-covered. Framed by trees, two access pathways at the mall's sides would lead up to the memorial. As at Anzio/Nettuno, the issue of trees accompanying the pathways was carefully considered, as views onto the memorial should not be impeded by trunks or branches as the visitor approached on foot.

The space in front of the memorial underwent many design iterations with paved terraces and grass areas alternating in various constellations and dimensions. Again, geometry and symmetry would characterize Smith's designs. Shrubs, groundcovers, and trees would delineate and confirm the monument's architectural elements (such as walls, planters, and stairs). The area it occupied would be large, terraced, and composed. One drawing shows a crescent of terraces supported by an outline of trees. Another recalls a Bernini baroque ensemble with pillared portico, which would have answered the crescent of the entrance zone (before the latter was profaned by including a parking lot). Drawings show marble-rimmed planters bedding roses, and cypress (the classical Italian funerary tree) planted on both sides of the bridge. Near the river, shrubs and trees in various densities were to be planted as compositional elements, to frame open spaces and pathways.

In the end, the grandiose entrance space (Harbeson's and Smith's *Via Triumphalis*) shrank to a short tree-flanked path, leading in from a simple semi-circular tarred, two-lane roadway. For both practical and financial reasons, grandiose ideas remained phantasmagoria (even though the natural course of the river was straightened and its banks strengthened, giving it the character of a canal, in response to catastrophic flooding of the fall of 1949 that destroyed graves). A remnant of Smith's initial grandiose vision is seen in small areas of ordered planting on both sides of the river. Financial constraints became urgent by 1952 and this lead to a focus on the cemetery proper. The open land remained largely unused and undesigned, but for the simple semicircular access road (with parking), which spans the width of the cemetery's grave fields.

Smith's search for form and plan reveals that he, from the outset, realized the importance of a skilled landscape architect. With none of his more grandiose designs being implemented and the focus falling almost exclusively on the cemetery proper, the hand of a landscape architect experienced in "special aesthetics" became even more essential.

Michael Rapuano's Brilliant Contribution

> "His [Rapuano's] ability and experience are of the highest order. The fact that he is a fellow of the AAR (AARome) and therefore knows Italian landscape and conditions well, seems very important to me."
>
> John Kellum Smith[1]

Both Smith and Rapuano had been fellows at the AARome (albeit at different times). Rapuano (1904–1975) was a prominent figure in his field.[2] At the AARome (1927–1930) he studied plans of ancient cities.[3] Later, under Robert Moses, as mentioned above in the context of Gilmore Clarke, he and his business partner Clarke created park-like systems in New York City (1934–35) for urban transportation arteries, such as the Garden State Parkway. These large projects earned Rapuano his reputation as a master of "spatial aesthetics."[4] Rapuano was also lauded for his collaboration on the Beaux-Arts design for New York's Bryant Park and later, for the grounds of the United Nation headquarters.

Smith and Rapuano were well acquainted, having worked together on the design for the grounds of New York's 1939 World's Fair.[5] Smith consulted

---

1 Letter, Smith to Harbeson, RG117, Entry 9, Box 72, Folder 687, (6–1–48 to 3–31–49), (December 28, 1948).

2 For a short bio see tcif.org/pioneer/michael-rapuano — The Cultural Landscape Foundation stand of October 26, 2014; Bradford M. Greene "Michael Rapuano" in *Pioneers of American Landscape Design*, Charles A. Birnbaum and Robin Carson, ed, (New York: McGraw-Hill, 2000).

3 Vincenzo Cazzato, *Ville e Giardini,* pp. 211 & 285 on Michael Rapuano.

4 Campanella, in "American Curves," theorizes that Rapuano's studies at the American Academy in Rome in the late 1920s enabled him to develop a lean and economical "public-works Baroque," spatial aesthetic that complemented the Anglo-Romantic heritage of the Olmsted era and became a signature of both the Moses era and the city (New York) itself. Campanella discusses the pivotal roles that Rapuano and Clarke played in shaping the designed landscape of the United States as a whole in the twentieth century.

5 The NYHS library blog "From the Stacks, written by Sue on April 2012" about Clarke and Rapuano.

with him informally from the outset on the Impruneta project.[6] It was noted that different surveys showed contour discrepancies. Rapuano was able to resolve the issue and produce an exact grading plan from which further designs would proceed.[7] Smith's vision demanded that Impruneta's natural amphitheater-like contour be perfected. This required cut-and-fill in order to affect a slight indentation at its center, and the erection of a substantial reinforced concrete retaining wall along the hillside to secure the site's uphill side against landslide and erosion. The wall was rendered practically unnoticeable by means of cladding in dark stone, and skillful use of planting. On the basis of Rapuano's proven skills, Smith argued to the CFA that Rapuano should be officially commissioned. Rapuano's vision was indispensable to him: "His ability and experience are of the highest order. The fact he is a fellow of the AAR (AARome) and therefore knows Italian landscape and conditions well, seems very important to me."[8] CFA policy limited a designer to a single cemetery project and Rapuano was already engaged in designing the grounds for the US cemetery at Margraten (Netherlands). The CFA's restriction was apparently waived in this instance. As Rapuano was the AARome's president in New York, he probably enjoyed certain professional latitude.[9] Also, Rapuano's intimate knowledge of Italian Renaissance aesthetics likely facilitated the CFA's making of an exception to their "one site per designer" rule, as likely did the fact that his business partner, Gilmore Clarke, chaired the CFA.[10]

---

6 Letter, Harbeson to North about Rapuano meeting at Smith's office. Talks about percentage of slope, and the bowl shape to be preserved. RG 117, Entry 9, Box 72, Folder 687, Florence (8–26–47 to 5–31–48), (May 5, 1948).

7 Letter, Smith to North about discrepancies between the new and old survey and that Rapuano would work immediately on a new grading plan. Rapuano proved himself indispensable. RG 117, Entry 9, Box 72, Folder 687, (8–26–47 to 5–31–48), (December 10, 1948).

8 Letter, Smith to Harbeson. RG 117, Entry 9, Box 72, Folder 687, (6–1–48 to 3–31–49), (December 28, 1948).

9 Rapuano was trustee and president of the AARome and is honored with a bust at the AARome library in Rome.

10 Letter, Harbeson to Smith. Rapuano approved through Clarke (his business partner) at the CFA. Because of Rapuano's knowledge of the Italian landscape and language, Rapuano asked to be paid $10,000. See his September 5, 1950 letter to Smith. He reminds Smith that these professional fees were extremely low. RG 117, Entry 9, Box 72, Folder 687, 6–1–48 to 3–31–49, 0089, (January 13, 1949).

At the AARome, one of Rapuano's assignments had been to create a restoration plan for Rome's *Villa D'Este* at Tivoli, the *non-plus-ultra* of Italian Renaissance garden design. His plan's compositional essence consisted of squares, rectangles, and triangles arranged around a central axial sightline toward the villa situated above.[11] Such were the very elements that Smith experimented with in order to develop Impruneta's integrated totality.

There were initial details to be dealt with, such as Italian law requiring a wall two meters high at the property line. A waiver was obtained and a wire fence was established around the border, to be surrounded by thick planting.[12] Such planting encloses an earlier Rapuano model, made before the site was secured.[13] The cemetery was conceived as triangular, it's broad base against the up-hill retaining wall. Grave fields were divided into three sectors. Access pathways were to be tree-lined, those leading to the monument with four rows of trees. The memorial building was to be nestled into a horseshoe-shaped cutout from the forest, and consisted of an arc of pillars.

Obviously, this model did not please Smith. Neither did Smith approve of Rapuano's suggestion to locate the memorial at the entrance, looking toward the grave fields that would fan out behind toward the hill's crest, thus embedding the cemetery completely into its own separate environment.[14] (A chapel at a cemetery's entrance is an Italian convention, as in the Italian plan for the Anzio/Nettuno cemetery.[15]) Rapuano then produced a splendid design plan (undated), which justifies Smith's advocacy of having him at his side. This plan shows Smith's final design with two projecting pavilions, a high vertical column before it, and the requisite reception and administrative buildings framing the bridge.

Rapuano's plan presents a landscape design with a secure sense of the site's totality, and would undergo only changes in its details. It reveals a vision of openness where nothing is hidden. The whole is visible from

---

11 See Cazzato, Vincenzo, *Ville e Giardini,* p. 347 for Rapuano's restoration plan.

12 Letter, Shaw to Smith. RG 117, Entry 9, Box 71, Folder 687, (4–1–50 to 11–30–50), (May 9, 1950).

13 Photographs of the construction and development of WWI and WW II cemeteries and monuments 1920–1960. RG 117-Con 100, records of the ABMC Box 5 of 33.

14 Letter, Harbeson to North. The January 14, 1948 first complete layout plan has trees bordering the path only toward the grave field side. RG 117, Entry 9, Box 72, Folder 687, (2–6–47 To 5–31–48), (January 14, 1948).

15 For Italian plans for Anzio/Nettuno acquisition, see above chapter: "The Siting Process."

every point. Rapuano's trapezoidal plan integrates all parts of the cemetery. The mall's widening toward the hill makes counterpoint with the cemetery's narrowing.[16] The mall's trapezoidal form renders the two lateral grave fields also trapezoidal, and this invention enables Smith's entire design to "work" with the site's amphitheater-like hillside. The resulting subtle and intriguing interplay of geometric shapes effectively realizes Smith's own vision while fitting into Rapuano's perfected amphitheater, where grave rows seemingly embrace the hillside memorial. The completed cemetery blends calm with a subtle dynamic quality, and Rapuano was pleased with the result. In his words: "The graves are arranged on either side of this axis in areas shaped like the pages of an open book."[17]

Fig. 36. Rapuano's plan. Note clarity and harmony, the widening mall, total enclosure through vegetation, and the plan's craftsmanship.

16 Several sketches illustrate how dull a straight mall would have rendered the site. The NYHS McKim, Mead & White collection, roll container NSI N-137 (1).

17 Letter by Smith describing Rapuano's vision for the cemetery (not dated). The NYHS McKim, Mead & White collection, PR 042, Box 452.

Planting was indispensable to Smith's vision as evidenced in his many sketches, and yet is restrained at Impruneta. Bareness and horizontality, the design's elements, are presented starkly. Neither trees nor hedges parse and manage the experience of individual grave fields. The open green expanse renders an unhindered view onto a sea of individual gleaming white grave markers, engendering a sense of quietness and peace. With every step the visitor takes, rows of grave markers re-arrange themselves to render a kinetic experience. At Anzio/Nettuno, hedges, sunken gardens and classical architecture manage the visitor's experience and give visual precedence to the memorial. At Inpruneta, one cannot escape awareness of the fallen.

Two paths that lead to the monument are aligned toward the center of the atria. Rapuano's plan envisioned one row of trees planted on the side of the grave area, but this was changed to two rows framing each path. Rapuano located the trees inward, into the mall, and on their outer sides, outward onto the grave zones, aligned with the atria. Their trunks cast intriguing shadows on the greens and the paths, changing with the time of the day. They also create a sense of airiness, a quality that matches the mood for the whole cemetery. At the memorial, four large plane trees shade the "wall of the missing." The shadows of their leaves interplay, bringing attention to the engraved names, making one aware that these were once full human beings whose resting places are "known only to God" as the inscription on the wall says.

Shielding the cemetery from view from without had been the ABMC's conception at the outset.[18] Rapuano's also intended that it be so. Rapuano drew a thick girdle of trees and shrubs bordering the cemetery laterally, and along the sides of the entrance. The forestation on the uphill side would remain in its natural state. Rapuano's plan suggests dense tree planting in the area that now constitutes the cemetery's entrance area. It seems Rapuano intended that the cemetery be sheltered from view from the highway as well. It is noteworthy that in 1950 both Harbeson and North argued, unsuccessfully, that the existing agriculture remain so as to provide additional screening. When, in 1956, the ABMC decided to fell the trees between the highway and the cemetery, Rapuano protested: "The philosophy of the design was to create a cemetery contained within itself, a peaceful,

18 Commissioning letter, North to Smith mentions this requirement. RG 117, Entry 9, Box 72, Folder 687, (2–6–47 to 5–31–48) (August 26, 1947).

quiet surrounding expressed with dignity."[19] Smith was horrified as well.[20] The composition would be a failure if exposed: "Entrance buildings would appear too small and out of scale. The contrasts of shade and sunlight, which are so important a part of our conception, will be lost, and the whole thing to my mind will be reduced to the commonplace."[21] His memorial, he argued "…because of its height, would be visible from the highway, but its aesthetic effect would need, at least, a screen of trees."[22] Harbeson, who in 1948 had promised a magnificent view from the highway, obviously did not have a bare piece of land in mind.[23] He attempted to convince the ABMC to change its decision, arguing: "Successive views, seen between groups of trees, are more attractive, truly more memorial in character than a wide expanse of nothingness."[24] The ABMC, however, ordered that the trees be felled. Rapuano acceded to authority but nonetheless insisted that his landscaping design continue to be followed to the letter.[25]

The cemetery became and remains fully visible from the highway, contrary to its designers' preference, and this is why, for today's traveler on the *Via Cassia*, the entire gem-like presence literally "bursts" into view with a disturbing suddenness. One sees a foreignness abruptly inserted into the surrounding Tuscan landscape. For the visitor, the message is gratitude for sacrifice, pride in victory, and an innuendo of triumphalism. At the same time, the unhindered view of the crosses serves as a reminder of the US's sacrifice of life for Italy's liberation. Perhaps this explains why the ABMC chose to expose it.

---

19 Letter of protest, Rapuano and Clarke to Smith. The NYHS McKim, Mead & White collection, PR 042, Box 453, (February 27, 1956).

20 Letter, Smith to North. The NYHS McKim, Mead & White collection, PR 042, Box 451, (February 8, 1956).

21 Ibid.

22 Ibid.

23 Letter, Harbeson to Smith. RG 117, Entry 9, Box 72, Folder 687, (6–1–48 to 3–31–49), (December 21, 1948).

24 Letter, Harbeson to North. The NYHS McKim, Mead & White collection, PR 042, Box 453 M, File No. 9, (January 1, 1956).

25 Letter, Rapuano to Smith' office (Mr. Vegezzi). Rapuano insists that his plant choices be followed. The NYHS McKim, Mead & White collection, PR 042, Box 453M, File No. 9, (June 21, 1956).

VI. Impruneta's Patriotic Embellishment: Message and "Beauty"

"I am placing more than the usual dependence upon the painting and sculpture elements and have subordinated all other decorative elements of the design to them."
John Kellum Smith[1]

Smith secured the collaboration of muralist Barry Faulkner and sculptor Sidney Waugh for the Impruneta cemetery project and in their work there, one sees General North's vision of "the best in American art." Faulkner and Waugh, former AARome fellows, drew eclectically from European traditions of Romanesque and Renaissance style and iconography in order to create a recognizably and distinctly American artistic iconology.[2] Steeped in Art Deco, this "American idiom" (accepted as such by the AARome) used figural allegories borrowed from European artistic tradition to message intangibles such as "thought" and "progress," as well as "poverty" and "ignorance," notions of central importance to the American mindset oriented toward progress. Their work attempted to grasp and encapsulate American life, and to express it through recognizable emblems and symbols. This was meant to give form to US identity, to its accomplishments, and to its beliefs. This idiom, created after WWI, was no longer considered *avant-garde* after WW II.

1 Letter, Smith to Harbeson. RG 117, Entry 9, Folder 657, (4–1–50 to 11–30–50) (September 6, 1950).

2 The term was coined by Erwin Panofsky, *Studies in Iconology*, (New York: University Press, 1939). It is used herein to comport the notion that art in America in the 1930s was occupied with giving consistent meaning to the American experience, its view of life and establishing its place in the world.

Faulkner's work during the 1930s included mosaics for the RCA building at New York's Rockefeller Center entitled "Intelligence Awakening Mankind." Stylistically eclectic, drawn mainly from Romanesque sculpture and Umbrian Early Renaissance with touches of Art Deco, it expresses America's hope and faith in progress. A central large allegorical frontal figure, "Thought," is Christ-like, rising and blessing. The facial features suggest a Madonna. Smaller-scale figures at its lower sides look upward adoringly, as though being blessed. These represent the "spoken word" and the "written word." These reprentations of the common man are clad in short tunics of the sort found in Romanesque and classical art, and in Pentecostal tunic. The message of the common man being blessed through knowledge is clear. All figures stand on clouds in the heavens, framed by the celestial sphere. Romanesque-like angels float at the sides proclaiming good news via radio waves, with waves of inspiration for couples with children at their sides. Thus, families are the beneficiaries of enlightening progress transmitted through the invention of the radio. The horrors of the Last Judgment scenes, quoted from medieval art history, are transmuted to represent the depravities of "poverty" and "ignorance." Green figures fall into the fearsome fires of hell. Thus, Faulkner handily yoked the medieval tradition of damnation to the nation's service. Hell is invalidated through optimism, democracy, and belief in progress, the contemporary equivalents of the medieval heaven.

Faulkner not only enjoyed high fees for his work, but garnered lasting fame.[3] In 1933, thanks to the AARome ties, he received his most important public commission, one that would assure his work a place in America's pictorial identity.[4] He secured a commission for two large frescos in

3 The center panel of the ceiling in Mortensen Hall (at Bushnell Center, Hartford, Connecticut, 1931) is the largest hand-painted ceiling mural in the US. Entitled "Drama," it is based on Greek motifs, although it is an ode to American progress in the early twentieth century, including aviation, architecture, cinema, and the dramatic arts. The mural cost $50,000 in 1929. <www.wikipedia.org/wiki/Barry_Faulkner>. Accessed March 14, 2018.

4 Faulkner, *Sketches,* p. 158, reports that Russell Pope, the architect involved gave him the commission for the murals at the National Archive in Washington, DC.

the rotunda of Washington's National Archives administration building, entitled "The Charters of Freedom" (1952). One depicts the personalities signing the Declaration of Independence (1776) and the other, the personalities at the Constitutional Convention (1787). The composition uses eighteenth century European pictorial conventions where a sovereign's image is represented with court and family before columned architecture and background landscapes. Faulkner's co-opting this model for US identity was highly successful. However, instead of static poses, his groups of figures have "movement" and convey a sense of activity and mission. The traditional sovereign's hand gestures of control are dispensed with and the composition conveys equality rather than hierarchy. Historical figures such as Benjamin Franklin stand loosely next to each other, close to the picture plane, with George Washington in the center anchoring the group. America's historical heroes are given a *gravitas*, an air of authenticity with democratic allure. Faulkner's frescoes became a national shrine.

From the outset, Smith conceived of a mosaic mural for the apse of Impruneta's chapel and made it clear that Barry Faulkner would be his artist of choice. Smith wrote: "Faulkner is particularly well qualified and experienced with mosaic technique."[5] Likewise Faulkner, a close friend, understood and appreciated Smith's work.[6]

Faulkner was no neophyte in working for military cemeteries, having created mosaics, "Angels of Victory," for *Suresnes* and *St. Mihiel*, both memorial chapels in World WarI cemeteries in France.[7] Their stylistic eclecticism prefigures his "Remembrance" at Impruneta.[8] It is to be noted that these "angels" vacillate in their message between celebrating victory,

5 Letter, Smith to Harbeson. RG 117, Entry 9, Box 72, Folder 687, (6–1–48 to 3–31–49), (December 28, 1948).

6 Faulkner, *Sketches*, p. 152. He describes spending a cosy weekend with Smith and his family in their home.

7 Ibid. p. 137. Charles Platt (with sons William and Geoffrey), architects of the US military cemetery (WWI) in Suresnes, France furthered Faulkner. (Platt became AARome director in 1930.). Platt secured the commission for Faulkner to create a mosaic for the cemetery's chapel, showing a soaring female figure holding a palm frond. (In St. Mihiel's chapel Faulkner's Romanesque "Angel of Victory" is less peaceful. She holds a palm branch (emblem of peace) and brandishes a sword (emblem of victory).

8 The style of the figure seems to have been a standard for Faulkner, informed by an eclectic mix between Romanesque, Early Renaissance and a touch of classical figures.

exacting revenge, and to a lesser degree, messaging peace and mourning for the loss of life. Generally, such "mixed" messaging remained the case in World WW II iconography.[9]

Harbeson advocated to North for a Faulkner mosaic at Impruneta. Even though an expensive feature, "… it is one that always had an impact on laymen visitors."[10] North, worried about expenses, had suggested finding someone else. Faulkner's mosaic would be about four times more expensive than comparable works at other ABMC cemeteries. Still, there was the precedence that Faulkner had created images for the chapels in *Suresnes* and *St. Michiel* for considerable sums. As would be the case with Sidney Waugh, Faulkner, with a patriotic urge, eventually accepted a large reduction in fee.[11]

At Impruneta, Faulkner's mosaic (seven meters high by eight meters wide) covers the chapel's apse. A tall female figure called "Remembrance" stands on a cloud. The ABMC booklet describes it as: "Lilies of the Resurrection in her arms, crocus strewn field of marble headstones set among trees showing the first buds of spring—symbolizing new life." The allusion to death, a helmet resting on a sword, lies below her.[12] Her beauty and aura of gentle softness deflect from morose thoughts of young men in the throes of suffering and dying. The golden mosaic background transports the viewer into a spiritual realm, free from reality's heaviness, as in early Christian and Byzantine mosaics. She floats upon a cloud and gently holds a basket of lilies. Two fine trees about to burst into bloom frame her. Her slight downward glance and kind knowing expression recall a Madonna. She is an allegory of springtime and nature's beauty. Smith saw to it that

9 Robin, "A Foothold in Europe," p. 134. Robin judges the cemeteries to be failures, in part because modifying European models to produce an American style produced unconvincing ahistorical architecture and art that he labelled as "conservative and unimaginative" and "undramatic artifacts of patriotism." Together with "uncertain symbolism" they fail to move and say anything." Furthermore, their location off the main roads, in bucolic areas, is a hindrance.

10 Letter, Harbeson to North. RG 117, Entry 9, Box 72, Folder 687, Florence (4–1–49 to 3–31–50), (February 17, 1950).

11 Letter, Faulkner to Smith. Faulkner estimated $30,000 as the cost for the mosaic. Faulkner reduced his fee to $12,000. RG 117, Entry 9, Box 71, Folder 687, (12–1–50 to 6–30–51), (May 17, 1951).

12 ABMC booklet on Impruneta.

the mosaic would be illuminated with natural light by adding a roof window, hidden from the visitor.

Fig. 37. Chapel with Faulkner's mosaic, "Remembrance."

"Remembrance" occupies a celestial sphere, above the ground level where crosses (and one Star of David) are depicted. Describing her, Harbeson emphasized her message of Resurrection: "White lilies stand for Resurrection, helmet on palm branch ...sycamore and elm are shown at first bud in spring. Flowering fruit trees are in full blossom. Grave markers at the bottom are crocus, dewdrops, and other earliest spring blossoms."[13] Renewed life, a resurrection promise of sorts, answers death. From a cultural perspective, this message, as that depicted in Manship's "Resurrection" at Anzio/Nettuno, reveals the US's denial of a soldier's death as a banal finality.

13 Letter, Harbeson to Shaw reminding him that sycamore and elms are typical trees for America and that flowering fruit trees in full blossom are emblems of life. The NYHS McKim, Mead & White, correspondence file, PR 042, Box 451, (November 16, 1953).

Sidney Waugh, Sculptor

Smith was equally adamant that Sidney Waugh be his choice of sculptor. "Sidney's style and approach are more in line with what I visualize for Florence."[14] Smith further argued that Waugh was qualified because of his "long study in Italy." When approached by Smith to collaborate, Waugh had major public works to his credit. He had mastered the creation of allegory and motion in Art Deco, and his sculpture had evolved an illustrative quality.

Success had come to Waugh early in his career. He had won the Rome Prize in 1929 with a statue entitled "Steel." US sculpture at the time often depicted the "working class" and was sometimes criticized for its socialist connotations. "Steel" showed a laborer in worker's garb with arms and huge hands drooping tiredly. Its style is realistic, its surface simplified, and its face generic and serious. Touches of classicist generalization ennoble the figure.[15] While a fellow at the AARome, Waugh developed a full-figured style. Informed by Rodin, it was highly expressive of "motion." His statue "Primitive Force" shows a muscular male in exaggerated, twisted *contrapposto*. Italy's King Umberto II purchased examples of his works.

He worked with both reliefs and full sculpture. His style came into full development in the 1930s with muscular, statuesque forms drawn from classicism.[16] In Art Deco style, he portrayed both nature's forces and human efforts. His realism was sufficiently abstracted from objective reality to avoid socialist associations. He expressed allegories of US life and aspirations with exaggerated musculature and Art Deco patterning.[17]

---

14 Letter, Smith to Harbeson. RG 117, Entry 9, Box 72, Folder 687, (8–26–47 to 5–31–48), (February 25, 1948).

15 A newspaper clipping in Eric Gugler Papers, 1889–1979, in a file on Waugh's "Steel" when it was exhibited at the Grand Central galleries, New York, The Smithsonian Institute Archives, Archive of American Art Washington, D.C. The stylistic influence of the Belgian sculptor Constantin Meunier is evident. (The latter's subjects principally were working men).

16 Article in *The Art Digest*, (February 15, 1933), (author not noted), sees Michelangelo's influence in "Primitive Force." Source as in fn.15.

17 Relief figures in brass representing primitive science as an American Indian and modern science by scientists in everyday garb. The two face each other, indicating continuity of American genius. Buhl Planetarium (1939) in Pittsburgh, Pennsylvania.

His numerous commemorative coins bear testimony that he had become a significant creator of emblems and symbols as well, representing everyday life in ways that were eminently readable by his generation of Americans.

New York's 1939 World's Fair, which gave design and exposure opportunities to many US artists, featured Waugh's work. A skillful and celebrated glassware artist as well, he created a beautiful 300-pound siren titled "Atlantica" for the Corning Glassware Company. The siren's long hair, split fishtail, and ocean waves swirl in elegant circles. At the same time, he demonstrated his plastic sculptural skills in a huge group titled "Manhattan," prominently displayed on the fairground. The blocky, massive, highly sculptural piece represented a couple with a child, allegories of Manhattan, and was unabashedly monumental. Their heads and torsos emerge from the tops of abstracted verticals not unlike skyscrapers, lending a serene humanity to the physical cityscape. Its lower surfaces show allegorical figures in half relief. Art Deco surface treatment render muscles and folds in generalized, abstracted zigzagging. Waugh's blend of force and calmness reflect the Great Depression generation's vision of energy calmed by security of purpose. He contributed to giving a pictorial voice to America's energetic rise.

Waugh's fees for public works had been substantial. His proposed fee of $50,000 for Impruneta would exceed the amount allotted for the architect(s). In support for the artists of his choice Smith wrote: "I am placing more than the usual dependence upon the painting and sculpture elements and have subordinated all other decorative elements of the design to them."[18] Securing federal funding was becoming more difficult and Harbeson, feeling he needed to remind Smith of fiscal realities, wrote that there would "… not [be] more [funds] for exceptional merit, nor more for an 'historical' or 'allegorical' subject."[19] Undeterred, Smith clung to his vision. Based on a report from the National Sculpture Society (March 29, 1949), North suggested that it might be a good solution to find different artists.[20]

---

18 Letter, Smith to Harbeson. RG 117, Entry 9, Folder 657, (4–1–50 to 11–30–50), (September 6, 1950).

19 Undated letter, Harbeson to Smith. RG 117, Entry 9, Box 71, Folder 687, (12–1–50 to 6–30–51).

20 Letter, North to Harbeson. North argues that it would not be "psychologically right" to ask Waugh to reduce his fee. RG 117, Entry 9, Box 71, Folder 687, (12–1–50 to 6–30–51), (April 3, 1951).

The question was raised as to whether fine detailing was needed for a sculpture mounted so high above the ground (atop the pylon). North referred to the memorial of *Montfaucon*, (France), a stone sculpture eight meters high mounted atop a high column, noting that Waugh's costs compared unfavorably.[21] He tried, unsuccessfully, to make Smith reconsider his insistence on Waugh.[22]

Waugh, not wishing to forgo his opportunity to contribute, finally accepted a reduced total compensation of $22,000.[23] He would have half-size plaster models delivered to "shipside, port of shipment" at his own expense. Bronze would require a full-sized model and thus be too expensive, so the final sculpture would be in stone and would be executed by local artisans. The horizontal soaring figure (Smith had stipulated horizontal) for mounting atop the pylon would dominate the grave area. Its message would be one of peace.[24] Harbeson wrote: ".... the symbolism of the statue, which surmounts the pylon and hovers over the fallen, is bearing the olive branches of peace, not the sword of victory. Flying beside her, out of cloud-like forms, is the American eagle, guardian of justice, and honor [for] the peoples of all lands and all creeds everywhere."[25] Waugh was instructed to proceed in May of 1951.[26]

21 Letter, North to Harbeson. Montfaucon cost $8,500 for the model; It was 24 feet high whereas Waugh's figure would be only 21-foot high. Detail must be broad rather than finely indicated. RG 117, Entry 9, Box 71, Folder 687, Florence (4–1–50 to 11–30–50), (October 19, 1950).

22 Letter, Harbeson to Smith telling him that contemporary financial restraint necessitated the reduction of the sculpture. He enclosed a suggestion of his own, a drawing of a flying figure, calling it "Homage to Brave Men." There is no record of this drawing. RG 117, Entry 9, Box 71, Folder 687, (12–1–50 to 6–30–51), (April 27, 1951).

23 Letter, Harbeson to North telling him about Waugh's reduction in fee and that he thought it fair. RG 117, Entry 9, Box 71, Folder 687, (12–1–50 to 6–30–51), (May 18, 1951).

24 Letter, Smith to Harbeson RG 117, Entry 9, Box 72, Folder 687, (6–1–48 to 3–31–49), (December 28, 1948). In the ABMC's booklet it "represents peace."

25 Ibid. The figure for Florence according to Harbeson could represent "peace" or an "homage to brave men." Supportive as usual, he added, "It could have a good deal of nobility."

26 Date can be deduced from correspondence around that period of time. RG 117, Entry 9, Box 71, Folder 687, (12–1–50 to 6–30–51).

Fig. 38. Waugh's "Angel of Peace", view onto tender, mournful grardian angel.

Fig. 39. Waugh's "Angel of Peace", view onto fierce American guardian eagle.

From a distance, as one enters the cemetery, the sculpture seems to ooze out over its triangular base. Only as one approaches does its aesthetics take effect. One sees a soaring female figure with bare torso clutching an armful of palm fronds. Her long hair masses with full, cumulous clouds. Her fronds form a thick bundle in front of her full-breasted torso. An arm and a leg sweep back, streamlining with her gown in flying motion. She seems sculpted by the winds. This effect is the result of several versions. Waugh experimented with the position of her arm.[27] The form of the arm reaching back is masterfully expressive of motion. The American eagle flies at her side as if to accompany her, one of his wings covering her protectively. Her sorrowful glance goes toward the graves, toward which she seems to be flying. Her position expresses sheltering and protecting. In her powerful, sculptural features, one senses motion and yet, iconic timelessness, conferring honor and gratitude to those that lie buried below. However, the eagle introduces a fierce, martial note, in contrast to the aura of peace of the "angel."

Classicism's idealized rendition of the body is combined with the decorative character witnessed in Art Deco, as in the fan-like arrangement of the clouds and the figure's clothing, but the sculptural massiveness of Art-Deco elements draws from cubism, well-known in the US at that time. North was never pleased with its artistic qualities, and wrote: "…the eagle is presumably the American emblem" and then went on to ask for an explanation from Waugh as to what the "floating figure" meant. Again in 1957, North grumbled that one could not really see it unless one stood directly below it.[28]

Two reliefs mounted above inscription panels in the south atrium testify to Waugh's ease in using ornamental aspects of Art Deco to facilitate composition. Flat forms are arranged in patterns that evoke life and movement. The American eagle's attentive stare, turned backwards, seems to quicken the inscription below: "TIME WILL NOT DIM THE GLORY OF THEIR DEEDS." The relief titled "The Spirit of American Youth" draws on Etruscan funeral figures in the gestures of the arms and in its archaic face, and on Michelangelo in the rendition of the heavy muscles. Integrated into clouds that resemble slender slabs of stone, he floats upwards, his face canted downward in quiet mourning. Stars suggest that he is moving toward the heavenly sphere. Europe's classical past provided appropriate models for such a grand theme.

27 Waugh's figure underwent several versions. The version chosen, the arm swept back, best expresses motion. The NYHS McKim, Mead & White collection, PR 043, Box 452, (Jan. 14, 1953).

28 Letter, North to Waugh barely concealing his disapproval. The NYHS McKim, Mead & White Collection, PR 042, Box 452, File 9, 1957, (May 20, 1957).

Impruneta's art articulates soothing as well as a dimension of immortality through the forms and message of its sculpture. Its emphasis is on peace. The ABMC writes: "Not like Roman emperors, or Napoleon's triumph or victory, but the spirit of peace, peace to the world, peace to the souls of the beloved dead, peace to the hearts of those who love and grieve for them."[29] In sum, Impruneta's art and architecture message neither triumph nor gratitude nor peace nor tragedy alone, but an amalgam of all, befitting the changing and uncertain times of their genesis.

29 From a commentary by Markley Stevenson. The NYHS McKim, Mead & White collection, PR 042, Box 453, (January 29, 1958). Not all sculptures of the ABMC's WW II cemeteries were allegories of peace. At Epinal, a large relief on the memorial building by Malvia Hofman is very martial showing charging infantry in lockstep with bayonets and bombs.

Inscriptions: Herolds of Ideology

> "I am afraid if the building is dedicated without inscriptions in the appropriate places, we will never get any (inscriptions) and the result (will be) that our design will be lacking human warmth and feeling."
>
> James Kellum Smith[1]

In museums, obligatory in all the ABMC's foreign cemeteries, inscriptions were part of the didactic program. How was the individual soldier's death to be given meaning when even the term "death" was to be avoided in text and in art? "Comrades in Arms," Paul Manship's bronze sculpture at Anzio/Nettuno, depicts two young men who seem very much alive. We have seen Eric Gugler's mystical cabala linking victory with the cosmos. At Impruneta, Smith's plan included inspirational texts that would confer meaning to a soldiers' death. His design provided empty panels for text on the inner wall of the chapel's atrium. But who would choose the text(s) and what would be the choice?

Fig. 40. South atrium with inscription tablets.

---

1 Letter, Smith to North telling him that the oration attributed to Pericles and recorded by Thucydides would be fitting. The NYHS McKim, Mead & White Collection, PR 042, Box 461, (June 26, 1956).

Smith, a humanist by education, obviously held that quotes from classical sources would be appropriate. As early as 1948, before the Impruneta site was secured, an English department member from Amherst College (Smith's alma mater) offered fourteen translations of the Roman "*dulce et decorum est pro patria morire*."[2] The letter containing these translations was addressed to "Paul," most likely to Paul Mitchell, head of the ABMC's historical commission that was in charge of providing historical texts. Such fervent imputation of death to service of one's nation was evidently not considered proper or sufficient, but alternative suggestions were not forthcoming.

Time passed and by mid-1956, Smith made it clear that inscriptions were needed in order to finish the project. Referring to Abraham Lincoln's Gettysburg Address, he reminded General North of the effectiveness of its text on the walls at the Lincoln Memorial in Washinton D.C.[3] Smith appealed to Waugh and Faulkner to help him choose quotes.[4] Both artists were evidently educated and well versed in the humanities. Already in 1948 Waugh had recommended quoting from Lincoln's Gettysburg Address.[5] Faulkner, probably at the same time, sent a handwritten version of Thucydides's "Pericles Funeral Speech."[6] Their suggestions began with Homer and Euripides and moved through the ages including (among many others) Shakespeare, Shelley, Psalms, and the US President during WWI, Woodrow Wilson.[7]

North reminded Smith that General Eisenhower preferred short descriptions and that "… my own feeling goes towards prayers and hopes rather than to assertions to what the souls of the dead are now doing."[8]

---

2 Letter signed /s/ GFW from the Department of English at Amherst College to Paul (Mitchell). The NYHS McKim, Mead & White collection, file "Inscription Material," 1956, (May 28, 1948).

3 The NYHS McKim, Mead & White collection, PR 042, Box 461 file 9, 1956, (June 26, 1956).

4 Letter, Smith to Faulkner and Waugh. The NYHS McKim, Mead & White collection, PR 042 Box 461 file 9, 1956, (May 10, 1956).

5 Letter from Sidney Waugh, (most likely addressed to Smith). The NYHS McKim, Mead & White collection PR 042, Box 451, (June 21, 1948).

6 Undated handwritten document by Faulkner. The NYHS McKim, Mead & White collection, PR 042, Box 453, file "Inscriptions Material," (1956).

7 The NYHS McKim, Mead & White collection, file "Inscriptions Material," 1956, (May 10, 1956).

8 Letter, North to Smith. The NYHS McKim, Mead & White collection, PR 042, Box 451, (July 6, 1956).

North added, almost wistfully, that people wouldn't read long inscriptions: "Only the military students will plow through our historical texts."[9] He did set to work finding texts, yet by November no decision had been made (except that Latin text should not be used).[10]

In December of 1956, pressed, the ABMC finally made a decision.[11] From the lists of quotes provided by Smith they chose an abridged "Pericles Funeral Oration," a quote from Cardinal Newman and a quote from the Episcopal *Book of Common Prayer*. The Pericles speech promises that "the dead shall never die" and consoles with "the glory of the dead, the love of honor." This ancient classical source, which had survived through two and a half millennia, would provide a connection to Athenian heroes from the "Golden Age of Pericles." The historical authority of Greek civilization's apogee would convey an aura of eternity and nobility to contemporary US fallen heroes. This view matched that of Gilmore Clarke (Rapuano's partner) who worked tirelessly to maintain

---

9 Letter, North to Smith. The NYHS McKim, Mead & White collection, PR 042, Box 451, (October 4, 1956).

10 Letter, North to Smith telling that nothing was decided and no Latin texts should be used. The NYHS, McKim, Mead & White collection PR 042, Box 451, (November 9, 1956).

11 Letter, North to Smith telling him that his suggestions for inscriptions have been approved (on December 27, 1956). All inscriptions are quoted in full. Pericles speech, in an A.E. Zimmern translation, is divided over three full panels:

> They faced the foe as they drew near him in the stretch of their own manhood and when the shock of battle came they, in a moment of time at the climax of their lives, were rapt away from a world filled for their dying eyes not with terror but with glory … Such were the men who lie here. They received each for his own memory praise that will never die and with it the grandest of all sepulchers, a home in the minds of men. Therefore, do not mourn with the parents of the dead who are here with us, rather comfort them, let their burden be lightened by the glory of the dead and the love of honor alone is not staled by age, and it is by honor that the end of life is cheered.
> The ABMC booklet notes that the following text is from Cardinal Newman (Sermon XX) from the Episcopal prayer book. "O Lord support us all the day long until the shadows lengthen and the evening comes and the river of life is over and our work is done, then in thy mercy grant us a safe lodging and a holy rest and peace at last." The latter is also the source of "O god who art the author of peace and lover of concord defend us thy humble servants in all assaults of our enemies that we surely trusting in thy defense many not fear the power of any adversaries."

western culture's Greek origins. Periclean consolation lay in glory and honor, per General Marshall.[12]

If such high-minded concepts could not provide consolation for mid-1950s US sensibilities, the two Biblical/Christian texts that did without concepts of heroism and honor might suffice. These were humble supplications for mercy, rest, and peace (Cardinal Newman) and God's help in defending against enemies (*Book of Common Prayer*). Peace and lasting memory were evoked in an oft-used quote from Ecclesiastes 44: "Their bodies are buried in peace, their names liveth evermore." It follows from the above that, in 1956, the ABMC was aware that the meaning ascribed to WW II deaths needed a religious/spiritual component.

Markley Stevenson (the ABMC's supervising landscape architect) wrote an article on the task of sculpture, which applies here to the inscriptions: "to reproduce works that will have a meaning to the bereaved, that will call forth a response of consolation and hope and acknowledge death with honor."[13] Notably, victory was not emphasized but rather, consolation and hope. Words from General Eisenhower's memoir *Crusade in Europe* are inscribed in large letters on the Impruneta museum's north wall: "Freedom from fear and injustice and oppression is ours only in the measure that men who value such freedoms are ready to sustain its possession, to defend it against every thrust from within or without."[14]

In some ABMC's sites the US's European WW II campaign was linked to the medieval crusade and St. Louis (the crusader):[15] On the chapel wall at the St. Avold cemetery in France, the statue of George Washington stands with King David, Constantine, and King Arthur in a sculptural group at the feet of St. Avold. Clearly, unabashed co-opting of European

12 George C. Marshall, "Our War Memorials Abroad: A Faith Kept," *National Geographic,* vol. 111, (June 1957), pp. 733–738.

13 "Our Memorials Abroad," Markley Stevenson. The NYHS McKim & White Collection, File 8, (January 1, 1955); *National Sculpture Review,* vol. 5 (winter 1955); (American Battle Monument Issue article).

14 A quote by Churchill from *The Birth of Britain* was also considered. The NYHS McKim, Mead & White collection, PR 042, Box 451, (no specific date given).

15 Linking the US campaign in Europe to the crusade of St. Louis can also be found at the Draguignan cemetery's chapel: a large fresco includes a scene where St. Louis stands on the walls of the *Aiges-Mortes* and sets sail from a port in the region. *Sainte Chapelle* is shown behind him. The ABMC's brochure reminds that the "US crusade" was symbolized by a crusader's sword that carried the emblem of Supreme Headquarters, Allied Expeditionary Force.

icons to heroicize America's victory had little appeal for Smith, as he chose Waugh and Faulkner whose works express empathy and consolation.

The importance Smith ascribed to inscriptions can be seen in his search for the fitting lettering style.[16] Lettering was to him, as it was for Gugler, a work of art in and of itself. He decided on *Romagne*.

16 Smith considered American models, such as an inscription in the Schenectady City Hall by McKim, Mead & White (1927) as well as Roman examples, among them from St. Peter's in *Vincoli* (Roman Gravestone 11th century). The NYHS McKim, Mead & White collection, PR 042, Box 451.

A Slap on the Wrist. Riparian Rights: A Lesson in Manners

"The American Battle Monuments Commission ... desires, insofar as possible, to maintain their lawns and other plantings in a green and flourishing condition throughout the entire year."
Colonel William Walker[1]

In December of 1949 the ABMC ceremoniously assumed custody from the AGRS[2] of the Impruneta site and with it, the issue of water supply for irrigation of the large green spaces and the atria's water basins. Soon they were confronting the issue of riparian rights and realized they had created an adversarial situation with Florence's government. The previous September, the ABMC had estimated its water needs at 280 cubic meters (280 metric tons) per day. Italian authorities were shocked, as the AGRS had asked for less than 25 cubic meters per day.[3] Colonel William Walker attempted to appease the Italian authorities: "I assure you that it is not the intention of the American Battle Memorial Commission to impair the interests of the present owners of riparian rights."[4]

Standard construction practice required the water supply to be secured before the construction of headstone beams and headstones, and Florence's Department of Public Works (*Genio Civile*) was taking no

1 Letter, ABMC's Colonel Walker, forwarded by *Minstero dei Lavori Pubblici Direzione Generale dell'Urbanistica delle Opere Igieniche* to *Directore Generale Commissare, Tinti.* RG 117, Entry 9, Box 72, Folder 687, Florence (4–1–49 to 3–31–50), (July 20, 1950).

2 Described in detail by American consul Waller in Florence to the Department of State: Representatives of the Catholic, Jewish, and Protestant faiths gave prayers, invocations, and benedictions. Col.Walker gave a short address. Army Col.Watson spoke on behalf of the AGRS. Enzio Aiello of the Italian Carabinieri sounded "Taps" after the benediction. RG 117, Entry 9, Box 72, Folder 687, Florence (4–1–49 to 3–31–50), (December 19, 1949).

3 Letter, Ministry of Public Works to the AGRS, Rome. RG 117, Entry 9, Box 72, Folder 687, Florence (4–1–49 to 3–31–50), (June 3, 1949).

4 Letter. ABMC's Col. Walker to the *Genio Civile,* Florence. Florence State Archive, Box *Genio Civile,* (January 19, 1950).

action.[5] Attempting to save face with the ABMC (Washington), Walker wrote: "As you know, the question of riparian rights in Italy is a very complicated and sensitive subject, dating back to the Roman Empire."[6] Walker complained to North that the *Genio Civile* was willfully obstructive and unfriendly and made clear that only it (*Genio Civile*) would decide how much water would be granted. The ABMC, taking matters in their own hands, hired an American geologist who sank test borings. The ABMC then considered drilling 15 meters deep to alluvial deposits (beneath the south-east side). The *Genio Civile* had planned to drill on the west side.[7] His report stating the ABMC's preferred drilling site had been forwarded to the *Ministero del Lavori Publici, Sezione Generale Urbanistica Opera Igenica, Roma,* through the ABMC channels that bypassed Florence's *Genio Civile*.[8] Florentine authorities viewed this not only as an arrogant breach of protocol, but also as an attempt to steal from the river, as unauthorized taking of riverbed water would constitute a "thief well."[9]

Florence's *Genio Civile* claimed water drilling to be its legal prerogative.[10] The ABMC's Florentine authorities informed Colonel Walker that the cemetery could draw no water from the watershed, whether downstream or upstream, unless they (the *Genio Civile*) gave permission and drilled the wells. Present riparian rights owners downstream could not be "offended" by concession. Drawing water from upstream would not work either, because of ancient concessions that were in place.[11] In other words,

---

5 Letter, North to the ABMC, Rome, saying that the water supply must be certain before a water distribution system, headstone beams and headstones could be installed. Florence State Archive, Box *Genio Civile,* (February 3, 1950).

6 Letter, Col. Walker to the ABMC Washington. Florence State Archive, Box *Genio Civile,* (November 8, 1949).

7 Ibid.

8 Water Supply Report from Frank C. Foley, geologist to the ABMC dated after a visit to the site on September 22 and 23. Florence State Archive, Box *Genio Civile,* (September 30, 1949).

9 Col. Walker to the AMBC, Washington, informing that proceeding independently produced a rather major turmoil with the *Genio Civile.* Florence State Archive, Box *Genio Civile,* (October 22, 1949).

10 Letter, *Ministero dei Lavori Pubblici* to the ABMC Rome informs that a Ministerial decree was sent to the *Ufficio Genio Civile* in Florence about water sounding work. Florence State Archive, Box *Genio Civile,* (November 8, 1949).

11 Letter, *Genio Civile de Firenze* to Col.Walker. Two hundred eighty cubic meters for the irrigation during the summer will require drilling for underground water. Florence State Archive, Box *Genio Civile,* (January 3, 1950).

legal action would be useless as only local authorities could make an exception to the law. Their point being made, the *Genio Civile* then conceded that they could and would drill for water to provide the daily 280 cubic meters.

The ABMC waited, but no action was forthcoming. Colonel Walker asked in a letter to North whether the ABMC shouldn't drill on their cemetery grounds without local approval.[12] Frustrated, Walker then addressed a polite letter affirming the *Genio Civile*'s authority, assuring them that the ABMC had no intention to violate riparian rights, and that it was interested only in beautifying the cemetery. Reminding them of their previous assurances, Walker was forced to ask when the *Genio Civile* might proceed with drilling. The *Genio Civile* then clarified their policy that no water could be taken from the river unless it would be returned to it.[13] The ABMC then assured the *Genio Civile* in Rome that most of the water would indeed find its way back to the Greve River, as groundwater or run-off.[14]

When would drilling begin? In the near future, it was hoped.[15] A face-saving compromise was reached, with the ABMC's request being granted and the *Genio Civile* doing the drilling, assisted by the ABMC. Some additional "back and forth" business concerned altering the river's course and flow, but the "take away" from all these dealings was that the ABMC could pursue its plans: irrigating its cemetery, straightening the river, and fortifying its banks only after deferring to the local Florentine *Genio Civile* authority.

The forgoing exchanges witness an attitude among (Tuscan) Italians that they were not subject to Rome's authority where local issues were concerned and moreover, that Italy was not a defeated and occupied country, but rather a full partner in the defeat of fascism. Furthermore, Italians insisted on being treated as such by Americans.

---

12 Letter, Col. Walker to the ABMC, Washington. Florence State Archive, Box *Genio Civile,* (January 27, 1950).

13 Letter, Col. Walker to North. Florence State Archive, Box *Genio Civile,* (January 18, 1950).

14 Letter, Shaw, the ABMC Washington to the ABMC Rome assuring that most of the water would flow back into the river. Florence State Archive, Box *Genio Civile,* (January 18, 1950).

15 Letter from Col. Walker to the *Genio Civile*. Florence State Archive, Box *Genio Civile* (January 18, 1950) and his letter to *Genio Civile*, Florence. Florence State Archive, Box *Genio Civile,* (August 18, 1950).

VII. Sic Transit Gloria

"I still await some sort of coverage in the American Press—and in the Architectural Forum or Record, the latter without much hope, as they seem to be run by narrow-minded men of small souls."
John Harbeson[1]

1941–45
In proud memory of her sons
and in humble tribute to their sacrifices
this memorial has been erected
by the United States of America

Inscription found at all United States military cemeteries

Although the ABMC saw itself as a guardian of American values, the importance of their works seemed to diminish, much to the dismay of those involved. This process and the main reasons for it are reflected in archives. In brief, the immediate post-war era saw changing fiscal realities as well as other changes in US American culture. Securing congressional funding was becoming more troublesome and the art world's taste for modernism was rising.

The Korean War marked the end of those few years of victory and peace for the US. Congress threatened to curtail the ABMC's congressional allotment and this caused alarm.[2] In Washington, resentment was voiced that so much money was already flowing to Italy through the Marshall plan.[3]

---

1 Letter, John Harbeson to J. K. Smith, expressing disappointment that major US professional journals (*Architectural Forum* and *Architectural* Review) did not note that Impruneta would be dedicated on July 25, 1960. The NYHS McKim, Mead & White collection, PR 043, Box 453, (July 27, 1960).

2 Letter, Harbeson to Smith: "Due to the Korean war and the large program of defense spending, the President has given directives to the budget committee to cut expenditures wherever possible … erection of crosses and planting of grass and trees. All expenditures will be scrutinized." RG 117, Entry 9, Box 71, Folder 687, (12–1–50 to 6–30–51), (November 28, 1950).

3 Memorandum on meeting at McKim, Mead & White office with Smith, Harbeson, Rapuano, and Vegezzi. Harbeson explained that it was uncertain that there would be

Alarmed North pleaded for a budget comparable to that of the post-WWI period, and then traveled to New York to negotiate reduced fees for sculptors and painters.[4] He asked Sidney Waugh to furnish a model of his sculpture for Impruneta in order to strengthen his position during his appeal to the Senate.[5] These efforts were successful, although it was not helpful that Senator Benton (Connecticut), together with his wife and son, visited Impruneta and upon seeing temporary wooden crosses being painted, called the project a misappropriation of funds.[6] (He also asked what the Star of David marker meant.) Learning that marble crosses would be available in five months apparently appeased him.

On April 6, 1949, in his speech at a ceremony that marked the end of the burials at Impruneta, US Consul Waller (Florence) cited the Roman dictum "*Dulce et decorum est pro patria morire*."[7] He ranked these words with Lincoln's Gettysburg Address. His speech contained words such as "precious blood," "witnesses to the truth," and the exhortation: "The living must resolve to be worthy of that sacrifice." His words reflected the US government's and his generation's unquestioning conviction that the deaths of American soldiers in WW II were honorable and heroic. The promise of the Resurrection was a very important attribute of the mindset Waller articulated that day. God would receive the fallen into the "Kingdom of Heaven."

---

money for the fiscal year of 1952. Also "… that this money would be spent in Italy where the ECA officials are busy allocating so many millions to bolster the economy of that country and making a spectacle of themselves." The NYHS McKim, Mead & White collection PR 042, Box 453, (August 9, 1951).

4 Letter, Harbeson to Smith, "… the foreign situation is uncertain and frightening. We believe it is essential that we be prepared to relate all budgets as related to precedents so far considered justifiable." RG 117, Entry 9, Box 71, Folder 687, (12–1–50 to 6–30–51), (undated).

5 Memo, White, (Smith's office) about meeting with North. The NYHS McKim, Mead & White collection, PR 042, Box 452, (May 9, 1952), (correspondence files).

6 Letter, ABMC Impruneta to North reporting the visit of Senator Benton (Connecticut) to the site on June 5, 1950. RG 117, Entry 9, Box 71, Folder 687, (4–1–50 to 11–30–50), (July 6, 1950).

7 Letter to Secretary of State Acheson. RG 117, Entry 9, Box 71, Folder 687, (4–1–50 to 11–30–50), (April 6, 1949).

Critics Pan ABMC's Projects

Domestically, the ABMC's works did not attain a hoped-for status, either as important patriotic messengers or as works of art. One reason was a decisive turn by the country's elite critics away from classicism in architecture. In November 1949, an exhibition was held at the Art Alliance in Philadelphia (Harbeson's home). Smith's model was displayed (strangely, without the pylon).[8] The exhibition was moved to Boston, then to New York. At all three venues, it failed to stimulate public consciousness and appreciation for the ABMC's work and worse, reviews in the Art Alliance bulletin's special edition on the exhibition (from November 4 through November 27, 1947) were derisive. The official text read: "... the war memorials exhibition is the absolute expression of American architecture designed to memorialize our dead on foreign ground." A reviewer of it quipped: "... but no one was likely to claim that these cemeteries had matched the sad triumph of war by any comparable triumph of art."[9] Where the US press showed any interest at all in the ABMC's projects, commentary was usually not favorable. Aline Louchheim (The New York Times), in a review of the above-mentioned exhibition, lamented that these cemeteries would "... stand forever as symbols of America, our spirit and our aesthetic." She registered surprise that the architecture was so conventional, and the planting so rigid and symmetrical. She equated "stripped-down classical" with its capital-less rectangular posts with "government building modern" and "dictatorship modern." Labeling the style "anemic," she questioned whether it was really necessary to look to the classical past to find "dignity and beauty." In addition, her article bemoaned the use of architects who had neither a profound understanding of new architecture nor the sufficient imagination to take on the challenge of symbolic, significant building(s).[10]

---

8 Exhibition of US War memorials abroad from November 4 through November 27, 1949 at the Art Alliance in Philadelphia. The NYHS McKim, Mead & White collection, PR 042, Box 452, File no. 7.

9 Ibid.

10 Aline Louchheim, "Memorials to Our War Dead Abroad," *The New York Times*, (January 15, 1950), sec. 2. p. 10.

In Italy, the mood changed for the worse. News arrived that "unpleasant anti-American riots" occurred in Rome.[11] At Nettuno, where the American architects and artists had so serendipidiously imagined a sort of homecoming, vandalism became a problem that needed to be dealt with.[12] North became increasingly disturbed when the US press showed little interest in the ABMC's projects. In 1950 he wrote to the publisher of the *American Legion Monthly* (a magazine mostly subscribed by veterans) complaining that they had failed to report on the October 1949 dedication of the *Suresnes* cemetery (near Paris, where the ABMC had an office) despite the presence of the American commander in chief in Europe (General Handy), two senators, and French troops and dignitaries.[13] The US press (except the military-oriented *Stars and Stripes*) ignored the ceremony even though there had been a general and timely notification.

In 1955, the ABMC's supervising landscape architect Markley Stevenson made himself the *de facto* spokesperson for the ABMC's message and wrote an article for the *National Sculpture Review*'s winter 1955 issue, which was dedicated to the ABMC's works.[14] As if to forestall art criticism's sting, Stevenson reminded the *Review*'s readership that the US had spent more efforts on scientific and utilitarian pursuits than on artistic creation and that the ABMC and its artists were striving to express "... the forces that make America great" in an art form that would be "... capable of lifting humanity to a more perfect understanding of what this nation really represents—the freedom of mankind." Stevenson pointed out that the ABMC wanted works that could be understood by relatives who were "... removed from the great centers of population, or of modern art movements" and that its purpose was "... not to foster the evolution of art forms." He made the proud point that no two of the memorials were alike and that all were "magnificent in aspect and prospect." He extolled the professional qualification of the artists and the fact that they were all

11 Letter, Paul Meserve (horticulturist) to Griswold, Pittsburg AIS 2001.10 Box 13 Folder 6 (January, probably 1956)

12 Letter, North to Harbeson, (November 28, 1955, RG 17, Enty 13, Folder 687 Anzio (11–1–55) through (4–5–56)

13 Letter, North to the publisher of the *American Legion Monthly*, Alan C. Macauley. The NYHS McKim, Mead & White collection, PR 042, Box 453, (February 13, 1950).

14 Markley Stevenson, "Our Memorials Abroad," American Battle Memorial Issue, *National Sculpture Review,* spring 1955, p. 5. Harbeson wrote a page using almost identical wording. It contains 25 photos.

American-born. He reminded the readership of the challenges the architects needed to deal with due to the fluctuating numbers of burials (many were repatriated) as well as those of the sculptors, whose task it was to provide consolation and hope, and that death was worthy of honor. He did not fail to mention that the cemeteries were also meant to celebrate American military accomplishments. The illustrations included a model of Smith's memorial with the pylon and "close-up" photographs of Waugh's "Peace" at Impruneta and Manship's "Comrades in Arms" at Anzio/Nettuno.

Harbeson, in the same issue of the same publication, used many of the same words, re-emphasizing that the artistic style was chosen in order to be understandable by the common man.[15] Already in 1950 Harbeson had declared that the ABMC was not willing to experiment with the taxpayer's money.[16] Lee Lawrie, the ABMC's sculpture consultant, had high hopes for the future perceptions of his and his colleagues' works. In his words: "Some of it, I do believe, will in time be classified among the great memorials as the Victory of Samothrace or the Lion of Lucerne."[17] Lawrie, comparing the ABMC's sculpture to the art of Phidias, Chartres, as well as to Francois Rude's "Departure" on Paris's *Arc de Triomphe*, unwittingly implied that the "common man" was familiar with and appreciative of such works. Such commentaries bear witness to early salvos of the "culture wars" in the US during the 1950s, where rightist political forces viewed modern trends in art as subversive and communist-inspired, especially if government funded.[18]

Again, in 1958 Markley Stevenson pointed out that the message of Waugh's sculpture was "… peace to the world, peace to the souls of the beloved dead, peace to the hearts of those who love and grieve for them."[19] Eleanor Roosevelt (a relative of E.R. and an ABMC commissioner) wrote that the chapels "… designed and built by some of our foremost architects and adorned with statues and mosaics by some of our finest sculptors and

15 Ibid. p. 9.

16 See fn. 2.

17 "I think these local people are greatly impressed with the way we honor our dead. The gallant, the beloved, the remembered." Letter by Lee Lawrie to the editor of the American Sculpture Review, vol. 3, 1955. The NYHS McKim, Mead & White collection, PR 042, Box 451, File 8.

18 Jane deHart Mathews, "Art and Politics."

19 Description of Waugh's sculpture's message. The NYHS McKim, Mead & White collection, PR 042, Box 453, (January 29, 1958).

artists are superb as works of art."[20] She emphasized that local people "flock to admire our memorials."[21]

Marshall and Eisenhower Come to the Defense

In 1957, the iconic General Marshall sought to enthuse the US public about the ABMC's works through an article in the *National Geographic*, a periodical that at the time was widely subscribed.[22] With words reflecting ideological security he reminded his countrymen that during WW II 15,000,000 had served in uniform, and 360,810 had died overseas. In his view, the fallen were heroes. He quoted from the Pericles funerary speech: "… on the battlefield their feet stood fast, and in an instant, at the height of their fortune, they passed from the scene, out of their fear, but on to their glory." The fallen were given "home, beauty, and remembrance" in the ABMC's foreign cemeteries. The chapels were representative of the (US) American people and visiting a cemetery was a "pilgrimage."[23] The cemeteries should be reminders that the freedoms Americans enjoy were owed to the fallen, whom he referred to as "sons of every free man." Marshall claimed the fallen had "bought" with their lives the right for the US to have cemeteries abroad. From a more political perspective, Marshall wrote: "I have always felt that the establishment of American war cemeteries abroad was of great international importance."Their role was also to remind those abroad of the "… sacrifices the United States made for the common cause." Marshall's article acknowledged the implications of geographical distances and travel costs unavailable to many, and mentioned that Congress might subsidize travel costs for next to kin. Indeed, Italians had invested in hotels at Anzio

20 From a letter to the editor of the National Sculpture Review, winter 1955. The NYHS McKim, Mead &White collection, PR 042 Box 45, File No. 8 (Jan. 1, 1955).

21 Ibid.

22 Marshall, "Our War Memorials Abroad: A Faith Kept" *National Geographic*, vol. 111, (June 1957), pp. 733–8.

23 Term coined by Milton Lehmann, "We Visit the Ghosts of Anzio," *Saturday Evening Post,* vol. 222, issue 13, pp. 140–8, (November 21, 1949).

expecting Americans to come to pay homage to the landing site and battlefields, and their hopes were dashed.[24] Few came.

A long article by Howard Walker followed Marshall's in the *National Geographic* titled: "Here Rest in Honored Glory."[25] It recorded with words and photos the dedication of six memorial cemeteries in that year, and told the story of the campaigns in the name of liberty.[26] The article blended personal encounters with *reportage* in order to entice its readership. Several pages were devoted to the dedication ceremony at Anzio/Nettuno and included its two maps. A caption under an image of a military honor guard in front of Gugler's temple reads: "Sicily-Rome Cemetery is Hallowed." Its style and lavish illustrations bespeak an attempt to raise awareness of the existence of the cemeteries abroad that few people at that time would ever be able to see in person. In the journal *Sunday News* (May 1959), an article appeared with an aerial view of Impruneta, titled "They're Still Remembered." The title belies the fact that public memory was fading, almost chidingly reminding readers of Eisenhower's words of a decade earlier: "They will never be forgotten."[27]

Established art critics were less kind. The American Institute of Architects (AIA) journal's May 1960 issue contained an article by its executive director, Edmund Purves that was pejorative of the ABMC's art. Writing from a critical standpoint Purves observed that the cemeteries reflected an "unhealthy sentimentality." North wrote in rebuttal complaining of the "flagellation of the erection of battle monuments."[28] He reminded Purves that all architects involved were members of the AIA and defended their work as "superb" and "awe-inspiring." He pointed out the long historical tradition of honoring military achievements. He quoted ancient Egypt and defended the ABMC's monuments as "pure structural design and execution" and therefore more appropriate than any erected by private initiative or

24 Marshall, "Our War Memorials", p. 733, tells the story of an exemplary Gold Star Mother who had saved every penny to be able to visit her son's grave, with the obvious intention to incite visits by families.

25 Walker, magazine staff member, (*National Geographic*, vol. 111, (June 1957), pp. 9, 739–769.

26 Ibid. Italy's Prime Minister Antonio Segni is quoted: "It was in the logic of things that the men of America should fight together with us, in the name of liberty and independence."

27 Eisenhower's words sent to the dedication ceremony in *Coleville-sur-Mer.*

28 North's letter to the editor of the *A.I.A Journal* (June 18, 1960). The NYHS McKim, Mead & White Collection, correspondence file PR 042, Box 453.

separate combat units.[29] The ABMC had "… durably recorded the achievements and sacrifices of our Armed Services with the cooperation of members of the American Institute of Architects." North closed his letter with: "… the kinfolk of the dead appreciate the memorials."

Critiques of the ABMC's work did not improve. To the contrary, in 1965 (the year of the US commitment of massive forces to South Vietnam) the highly influential *New York Times* published an absolutely scathing review of the ABMC's cemeteries' sculpture titled "Our National Pride: The World's Worst Sculpture."[30] Presaging deep changes in the direction of US cultural history, the author judged it "just plain bad," observing that the sculpture did not address the "agony of the men who died." The article asserted that glory and beauty were incompatible with war's brutality and that war was the result of the power struggle and had little to do with the notion of sacrifice. Canady (the author) ascribed the reason this sculpture even existed (referring to marble replicas exhibited in the US) to political connections between Washington's power elites. Canady labeled The National Sculpture Society (founded in 1898) as the "… most sterile, esthetically most tightly hidebound and reactionary stone-hackers effective lobby in Washington." He went on to describe the sculptural works as "… blocky, antiseptic wastes of marble suggesting the forms so dear to Hitler and Mussolini but without the sheer, brutal weightiness that made those monuments perversely impressive."

The ABMC's cemeteries did have enthusiasts and remained in the consciousness of groups such as the Veterans of Foreign Wars (VFW). The November 1966 issue of their periodical (The VFW) featured a photo of Manship's "Comrades in Arms" on its cover.

29 The ABMC's was mandated to control all elements. E.G., two statues, one bronze, the other marble were made as tributes to fallen comrades, by the 361st and 363rd regiments of the 91st division. These were moved in agreement with the regiments "to display them in perpetuity." (The ABMC Florence brochure.)

30 John Canady, "Our National Pride: The World's Worst Sculpture," *The New York Times*, (July 25, 1965.) He reviews an article in an issue of the quarterly *National Sculpture Review*. The article is on pp. 6 and 25. Illustrations are on pp. 2–24.

Such was symptomatic of the growing disconnect between groups in the US at the time. Fault lines had long since begun to appear in US culture, as reflected in Alan Ginsberg's highly influential *Howl* (1955). The "beat generation" came into being, and racial injustice became a national issue. As the Vietnam morass escalated, many Americans of the WW II generation experienced their secure and static worldview, as enshrined in cemeteries abroad, suddenly changing. Their pride in American rectitude was challenged as the younger generation increasingly saw the Vietnam War as a perverse abuse of cherished notions of loyalty, honor, glory, and patriotism. One result was the term "generation gap" coming into being. America's hubristic belief in its "exceptionalism" was shaken.

In 1965, the year of a massive US military commitment in Vietnam, North published an article titled "In Proud Remembrance" in the *National Sculptural Review.*[31] It cited words of (General and former President) Eisenhower, who spoke in terms of "America's great artists" and the "beauty" and "dignity" of the monuments. North reiterated the ABMC's mission, explained the Star of David on tombstones and lauded sculpture's symbolic power. He praised the "unprecedented cooperation between the arts." North, near the end of his life and amid the Vietnam era's turmoil, was first and foremost an old soldier, praising rows of gravestones: "… their patterns standing in military array upon the broad green lawns never fail to impress the visitor with the tremendous sacrifice in human life made by our fellow countrymen."[32] In short, he represented one side of the generation gap, recalling "the good war." This represented the sad conclusion to North's tireless efforts to create public reverence for works of beauty in the tradition of Paul Cret that, in his time, had seemed so luminous.

"They will never be forgotten," but sadly they had been, at least by many in a younger generation, as though they had been elegantly gift-wrapped and put into storage. In an article "Gone, and Mostly Forgotten: A

31 *National Sculpture Review*, World War II memorials issue, vol. 1, Spring (New York: 1965).

32 Ibid. p. 24.

Veteran's Lament" a Mr. Gaile who had fought in Anzio lamented his generation's lost message. He was quoted: "Today they don't even know what Anzio was. Most people aren't even interested."[33]

33 Christopher Wren, *The New York Times*, (June 1, 1999).

VIII. Robert Tischler's Continued Hegemony in Italy: Continuity and Change

"It is difficult to recount what mental efforts were needed to throw off the chains that the political pathos had created."
Rudolf Hillebrecht, architect[1]

This work uses style and message expressed in soldiers' cemeteries to illuminate aspects of respective national cultures. In Germany's case, rebirth followed catastrophe, but this was a process that needed more than a generation's time, a process in which Germany's VDK played its part. The VDK's well-documented, laborious quest for new forms and a new message evolved gradually, as did Germany's culture at large. Its chief architect Robert Tischler's design hegemony in Italy provides both a point-of-departure and a thread of continuity we can follow. For this reason, this work examines Germany's Italian cemeteries in "stepping stone" fashion, beginning during the NS period and ending with Germany's final Italian cemetery.

Shortly after WW II ended, the VDK grappled with finding new footing while beginning its forensic work in Italy (1947).[2] Robert Tischler continued his cemetery work, both in Germany and abroad (North Africa).

1 Werner Durth, *Deutsche Architekten: Biographische Verflechtungen 1900–1970,* Friedr. Vieweg & Sohn, (Braunschweig/Wiesbaden, 1986) (*Es ist schwer zu schildern, welcher geistigen Anstrengungen es bedurfte, um sich jener Fesseln zu entledigen, die das politische Pathos erzeugt hatte.)*, p. 374. As with Tischler, Hillebrecht was more apolitical than hypocritical. His projects ranged from designing workers' housing with Gropius in 1934 to projects using forced labor during the NS period in Hamburg, to Hannover's leading city-planer in the post war era (1948). See <https://de.wikipedia.org/wiki/Rudolf_Hillebrecht> (May 2, 2015). Accessed March 15, 2018. His career is an example of the continuity of work of Germany's functionary elite from the NS period to post-war Germany. This continuity brings up the issue of suppression "*Verdrängung*" and managing the past "*Vergangenheitsbewältigung*" that complicates evaluation of Tischler's post-war work for its remarkable aesthetic properties.

2 Italy, Germany's ally until 1943 was very accommodating toward the efforts of Germany's VDK.

Tischler belonged to the so-called "functionary elite" whose expertise was considered indispensable for Germany's destroyed society and infrastructure. Before his death in 1959, he designed four large cemeteries in Italy. What little we know of him suggests a secretive and private individual, temperamental and impulsive, and his dealings with the VDK were not always harmonious.[3] Yet, he remained steadfastly in charge of Italian projects. After his sudden death, two of his cemeteries were completed under the supervision of another architect. The last to be built (at the Futa Pass) was awarded to still another architect in 1959, shortly before his death. Any evidence of Tischler's reaction to this obvious slight perished with his death, when a fire, rumored to have been deliberately set, destroyed his office and its contents in Munich.

Prior to the VDK's cemetery construction, Germany's ca. 107,000 fallen soldiers in Italy remained where they had been interred, at about 2,000 different locations. Following the Geneva Convention, their remains were *in curati* and further action would be at Germany's expense (*proprie spese*).[4] The VDK began discussing "collection cemeteries" in Italy as early as 1950.[5] Tischler had already traveled the Italian countryside, which he evidently considered to be his domain, in search of cemetery sites, and over the years kept on urging the VDK to acquire lands he selected, where possible.[6] Tischler imagined organizing Italy into four

3 Cross-reference to a furious letter under chapter IX, fn.7.

4 Italian Cabinet to other government agencies: Re: Disinterment of foreign soldiers on Italian soil. Archive of the Ministero degli Affari Esteri, *Affari Esteri* 1950–1957, busta 123, (July 6, 1949).

5 *Richtlinien für die Gestaltung der Kriegsgräberstätten des 2, Weltkrieges im Ausland,* (1950). The AA Archive, Bestand 92, 602–85.20, 271. The VDK's collection cemeteries were located in Sicily, south Italy and in the regions near Anzio/Nettuno, Bologna, Pinzano, Pordoi, and Sardinia.

6 Letters, Tischler to the VDK leadership. Tischler's advised purchasing land for a large cemetery (later at the Futa-Pass) and prior thereto: "… to purchase suitable parcels of land, that after construction to be returned without cost to Italy. In return, one could receive the assurance of eternal rest and eternal usage, as was the case in former agreements." ("… *bereits jetzt geeignete Grundstücke von uns gekauft würden, die später dann nach vorgenommenem Ausbau allerdings kostenlos zurückübereignet sein würden. Als Äquivalent könne man dafür aber die Zusicherung des ewigen Ruherechts und der ewigen Nutzung fordern, wie es auch bei früheren Vereinbarungen der Fall gewesen sei.*") By which means Germany would acquire the lands in the future after the accord was not clear. The VDK Archive, file *Geländefrage* A.100 884, (March 22, 1957).

zones. (In 1954 the VDK decided in favor of five regional "collection cemeteries.") These would include stone structures.[7] Formal cemetery projects, however, were not allowed before a formal diplomatic accord was in effect (December 1955).

Archives reveal that Italy's post-war government and its population, in general, were respectful toward the remains of German soldiers. A 1958 Italian Ministry of Defense publication on the subject, in courtly Italian language, reads:

> "Our country, mistress of civility and finesse, is inclined to honor those who fell on our soil while doing their duty, and to recognize them as sons, of both a fatherland and of a distraught mother."[8]

In the Salerno region, an elderly widow (Signora Apicella), a stereotypical Italian old woman, thin and dressed in black, wandered about searching for remains of German soldiers, and reburied them. She used only her hands and simple tools. She wrote to Germany's President Heuss (hand-written letter, December 2, 1959) that she felt sympathy for bereaved German mothers and was following God's voice. She added that these young soldiers had only done their duty.

That notion, so effortlessly expressed by both the Italian Ministry of Defense publication and Signora Apicella, betrays a moral security that official Germany and the VDK could not enjoy. Any reference to "duty" or "fatherland" was effectively stricken from their vernacular, simply unutterable by German officialdom, until a generation had passed. The gradual process of Germany's finding its voice, and articulating what the war experience had taught is revealed through art in the chapters that follow.

---

7 The AA Archive, VDK "Aufgaben", (March 3, 1958) n.7.

8 ("*La nostra patria, maestra di civilità e di gentili sentimenti si è curvata sui Caduti che giavevano in ogni sua zolla e lo ha riconscuito figli di una Patria e di una trepida mamma, morti nell'adepimendo dei loro dovere*.") The first paragraph from a publication by the Ministery of Defense, (*Commissariato De Cura e Oneranze salme cadudi in Guerra. Statistica e Sintesi delle Opere in Italia e all'Estero).* (undated, likely around 1958 judging from images, because models of Monte Cassino are shown, but not the Futa Pass model.) The Signora Appela's image and letter can be found in the AA Archive, Bestand 92, 602–85.20. 271.

Pomezia: Italy's Gift to Germany

"Admonition to the Present and Future. In death's presence there shall be no pomp, but rather tasteful decency down to the smallest detail…it goes without saying that not every site that was built in the past is still exemplary today."
Robert Tischler[1]

Fig. 41. Pomezia. View toward modest memorial baldachin. Note pathway too narrow for parades. The feel of a park pervades.

Architect Robert Tischler's earliest Italian cemetery in the post-WW II era is located at Pomezia (near Rome) where 27,433 soldiers are interred

1 *("…Mahnung für Gegenwart und Zukunft. In der Nachbarschaft des Todes dürfe kein Pomp herrschen und müsse man auch in der Einzelheit Takt zeigen [es] sei nicht jede Anlage, die früher einmal gebaut wurde, heute noch vorbildlich.")* The VDK Archive, A.100–1030, (1949).

on five hectares. Its design dates from 1957 and it was dedicated on May 6, 1960.[2] The site and its layout were not the VDK's choice, but were a gift from the Italian government to Germany in 1947, as marked on a tablet in today's cemetery in order to provide a burial place for the ca. 2,000 German remains exhumed at Anzio/Nettuno. Thus, the Pomezia cemetery was initially established as a consequence of the US's (the AGRS's) initial Anzio/Nettuno project. The VDK (and Tischler himself) preferred hill sites, but Pomezia lies on a plane. A central path with a cross halfway divides the grave areas that are laid out in a grid-system, an American norm for land use.[3] Simple wooden crosses, painted white in straight lines, served as grave markers.

The town of Pomezia is situated at the edge of the Pontian marshes and is one of the many towns on swampland recovered by drainage that was founded under Mussolini's leadership.[4] Such drainage projects provided arable acreage and the poor from different parts of Italy, especially the north-eastern part, settled these towns. It is likely that the cemetery site belonged to the state and no expropriation was necessary, especially since it was made available in short time. The Allies bombed Pomezia in the winter of 1944. The local population welcomed the erection of a German cemetery in its vicinity.[5]

The Tischler re-design retained the cemetery's central axis and the graves were organized within a grid of rectangles.[6] He enclosed the site through a low peripheral wall of sedimentary stones of different sizes, which lends a calming horizontality as it subtly defines the cemetery's presence within the surrounding expanse. Hills are seen in the distance. Artfully located trees, now mature, further suggest a hushed sanctity and modesty befitting the surrounding landscape.

---

2 A session of the Review Committee (*Prüfungsauschuß*). At the VDK Archive on June 7, 1957. The VDK Archive, "Die Jahre 51–58", A. 100–182.

3 The importance of the grid system in the American landscape is treated in the works of Paul Groth (UC Berkeley) and J.B. Jackson. It is not clear whether the Italian contractor that re-buried the German remains consciously copied the American grid system.

4 On Mussolini's new towns see *Città di Fondazione Italiane, 1928–1942* (Latina, Novecento, 2011). The catalogue of the exhibition edited by Giorgio Pellegrini. On Pomezia see p. 171 ff. text with many illustrations.

5 Based on an interview with the resident on-site custodian (2013).

6 For an illustration of this provisory cemetery see *Sacrari Militari della Seconda Guerra Mondiale,* Rome, 1974, provided by the *Commissariato Generale Straordinario per le Onoranze ai caduti in guerre.*

Fig. 42. Tischler's lay-out plan for Pomezia. Does not include the reception building. Note use of trees to create *gravitas* for modest memorial.

Although Tischler would have preferred a more "architectural" monument design at Pomezia, he was only able to add a garnishing of architectural elements with some tree and shrub planting to its pre-existing "garden" design type (*gärtnerische Anlage*). One such element is the entrance building in the form of a low, solid, very simple, calmingly well-proportioned structure with a gabled roof. One enters from the side through another small rectangular building and with a turn, stands in the wide and transparent atrium.

Light enters a triad of openings at the building's facade, illuminating wrought iron grillwork.[7] Square pillars without bases or capitals set widely apart separate wide openings on the building's cemetery side. Natural light flows freely through the building's space. Their sober clarity, modest dimensions, and exquisite spacing render an impression unlike the stony mass that had characterized Tischler's pre-war designs. Here, the visitor pauses and beholds an immediate and unencumbered view onto the cemetery. To the side is a modest, sober, and thoughtful reception area with books, a few brochures, and lists of those buried, a modest reception/presentation such as is included at all the VDK's WW II cemeteries. There is no mystery, no ritual of entry, no darkness, but rather an immediate transparency. Tischler's recognizable mastery in color, texture, dimension, and landscaping elicits an experience diametrically the opposite of that produced by his earlier works. An era had come and gone. Their ponderous reddish stone is

7 Artful, wrought-iron works such as grills at windows and doors and high free standing crosses can be found in all German WW II cemeteries in Italy. Free from ideological burden, they serve to embellish the otherwise sober architecture.

replaced here by light-colored limestone and travertine. Slabs of varying dimension favor the horizontal and surfaces are hewn to produce shadows in the Italian sun.

The eye follows the straight and modestly narrow central path leading to a monument. The path's center strip is of stone slabs flanked by borders of smaller square paving stones that seem to invite private walking and individual emotions. Tischler maintained the central way, laid out in 1947 when it was likely chosen for convenient access to the graves, and designed its pavement so as to be too narrow for parades. When the question was raised as to whether a German military band might play at the cemetery dedication, the VDK, mindful of Italian sensibilities, consulted the Italian Foreign Ministry and the Ministry of Defense. The band was allowed to perform, without incident or pejorative comment.[8]

Fig. 43. Baldachin with Schmoll genannt Eisenwerth's sculpture.

8 Note, *Commissariato,* Rome to the VDK to be forwarded to the German Ministry of Defense. German Foreign Ministry, (April 27, 1960). The AA Archive, Bestand 92, 602-85.20.

From the entrance portico, the monument appears at the path's end. It takes the form of an open baldachin appearing small and unimposing. However, its white travertine beneath foliage of cypress and stone pine exerts a magnetic pull enhanced by a central sculptural group, seen but not recognizable in detail. Approaching, the visitor begins to notice the monument's "squareness," presented within its own square paved piazza upon its three-stepped square socle. The visitor sees a large sculpture in the center of the baldachin that reaches from its base to its ceiling. It beckons to be seen. Masterful use of modest dimensions renders monumental power, a recognizable Tischler specialty.

Figures emerge from a central massive pillar. The sculptural group (by sculptor Fritz Schmoll Genannt Eisenwerth) recalls a Romanesque figure pillar. It represents a grieving family of two men, a mother figure, a young woman, and a child. Four soldiers hold their helmets in their hands in sad tribute. The young woman with the child, presumably a widow, stands beside one man, her hand softly on his shoulder. The child looks upward in search of the fallen, missing father.

Fig. 44. Detail of sculpture. Note the fine stone mosaic work overhead.

The impact of Schmoll genannt Eisenwerth's sculptural group is all the more salient when one recalls the sculpture he created, which no longer exists, "Germany, Awake," at the interior of the Tischler's pre-war Annaberg, a truly monumental male of green porphyry with massive visage. As if rising from sleep, he brings his mighty torso to a sitting position with eyes yet closed and one arm angled, as if stretching from sleep. Its Munich style (*Münchner Bildhauerschule*) blended archaic Greek sculpture with modernist abstraction and surface simplification.[9] The message, national strength awakening, was unmistakable. At Poemzia, the sculptor continues to work in his style of rendering figures with abstracting surfaces caught in the moment between stasis and movement, but here, with eyes closed, they exude painful and modest meditative introspection, responding to the new reality that loved ones are vanished forever for a lost cause. The tall brother/husband/father figure's wide open eyes express distress, but also disbelief. The triumphal energy expressed in "Germany Awake" has been replaced by a mourning that is neither quiet nor resigned, a mourning that will not pass.

From Dante's *Divina Commedia Paradiso* 3.85 is engraved: "Our peace lies in his will," (*Unser Friede liegt in seinem Willen*) inscribed in both German and in Italian. One reads it only when looking back over the grave fields toward the cemetery's entrance/exit. (Similar messages are found in other German WW II cemeteries.) An American researcher has advanced the theory that this choice of Dante attests to Germany's desire to rejoin the civilized world.[10]

A "comrade's grave" lies beneath the baldachin. The names of those buried are engraved on its base, and these include two women. Thus, although modest in scale and message, Pomezia's light-bathed baldachin might be considered as a crypt monument, a minimalized *Totenburg* without walls. In such altered form, Tischler was able to realize his conviction that the most appropriate way to inter soldiers was the "comrade's grave."[11] The square formed by rows of cypresses surrounding the memorial might

9 Waldermar Grzimek, *Deutsche Bildhauer des zwanzigsten Jahrhunderts: Leben, Schulen, Wirkungen,* (München; Moos, 1969). The author discusses the Munich school of sculpture.

10 Virginia Jewiss, "Picarda's Peace in a German War Cemetery" *MLN,* vol. 127, no. 1, pp. 119–129, John Hopkins University press.

11 Letter, Tischler to the *Vorstand*. The VDK Archive, A. 10–182 (January 23, 1957). He lamented the fact that others disagreed with this "soldierly note."

be seen as an echo of a "*Heldenhain*" (a hero's grove), that was such a prominete feature in Tischler's designs for WWI cemeteries, but without the latter's ideological message.[12]

Tischler was a trained landscape architect and Pomezia is park-like. Pine and oak soar to frame every view and answer straight lines and right angles. A sense of symmetry demonstrates clarity of design and message. Cypresses planted in four very straight rows are used to effect, emphasizing the square of the baldachin. Burial areas are organized into squares with shrubbery toward the edges. Today, the pines are tall, and oak crowns soar above wide-spaced crosses to impart a beautiful calm that softens the original plan's geometry. (The stone pine's beauty, so emblematic of Rome, reminds one of Ralph Griswold's Anzio/Nettuno design.[13])

At Pomezia, (as at *La Cambe* in Normandy) Tischler prevailed with his controversial symbolic crosses. An early photo, a "bird's eye view," shows these used in combination with small stones bearing names.[14] This arrangement was replaced in the 1970s with smaller erect crosses bearing multiple names on each side. Crosses were thus arranged in order to simplify ground keeping.[15] Some crosses stand closer together, because those already buried were not moved as new bodies continued to arrive.[16] Thus, the burial areas were never expanded and a large section toward the monument remains grassland beneath tall trees, enhancing the cemetery's peaceful "feel." The monument thus claims its own space, distanced from the grave area. Today the grounds are level, uniformly grass-covered, and shaded from the bright southern sunlight. Some years after the cemetery was dedicated the VDK was able to purchase an adjacent olive orchard.

Pomezia's aesthetics and message reflect Germany's beginning to process its hard and bitter recent past, and a new beginning of Tischler's personal artistic journey as well.

---

12 See Lurz, "Ein Stück Heimat," p. 66. Nature promising eternity in the commemoration of soldiers' death is an ancient German tradition. Even though its ideological cargo was downplayed after WW II, it cast a long shadow.

13 Tischler's plan seemed to have fewer trees than the mature cemetery. See the VDK Archive, kunsthandwerk-pomezia pdf.

14 Willi Kammerer, "Erinnern und Gedenken: Kunst auf Kriegsgräberstätten," *Volksbund Deutsche Kriegsgräberfürsorge,* kunsthandwerk-pomezia.pdf, 2002. The VDK Archive.

15 This arrangement created grass lanes between rows of graves wide enough for lawn movers to pass. Interview with cemetery caretaker on August 2013.

16 Ibid.

Costermano: Modesty and Mourning

"Even the loftiest of cemeteries from both World Wars shall and must, with time, decay and become places of admonition where their focus, which is the memorial to death, shall remain."
Robert Tischler[1]

Fig. 45. Costermano. Note the natural feel of the site. Meadows border grave-fields dotted by Tischler's symbolic group crosses.

In accordance with the accord of 1955 (cited previously) the Italian state acquired sites for German cemetery projects. The Costermano site lies above the Garda lake. Its natural setting reflects the VDK's post-war preference.

1 *("Auch die großen Friedhöfe der letzten beiden Weltkriege würden und müßten im Laufe der Zeit zu Mahnstätten werden, auf denen nur der Mittelpunkt, also das Totenmal, die Zeiten überdauere")* The VDK Archive, A.100–1030, (1949).

Costermano was completed after Tischler's death and his successor, architect Gerd Offenberg, should also be credited for its style emphasizing the lightness of walls, wide openings, and transparency. Its modest grave fields hold the remains of 21,920. Costermano was dedicated in 1967, the year Offenberg became the VDK's chief architect.

The entrance building is approached via footpath that leads behind a hill and then straightens before a broad stone building with flat roof.[2] The visitor encounters a consistent flow of light and openness. Two high and unadorned wall panels of fine vertical stonework, connected by a lintel, jut forward to frame the entrance. Between them, broad and inviting steps spill forth and lead upwards. While the perpendicular wall panels recall the entrance to Pinzano's stadium, which was designed to emphasize NS notions of grandeur, at Costermano these are used to frame openness. Unlike Tischler's pre-war tightly controlled entrance dramas, Costermano's entrance is not shrouded in mystery. There are no angular turns as still found in Pomezia. The visitor's sightline is unimpeded and reaches through the entrance building to the cemetery beyond, to rest on the green of trees. Such transparency and directness is likely due to Offenberg's modernist aesthetics.[3]

Grave areas are reached via an opening as high as the entrance building. Fine filigree in wrought iron fills this opening. Gates are integrated into it and their denser pattern in fine metalwork is clearly delineated against the light. At the front of the building a rectangle of square window openings repeats the grillwork's squares. Stone's mass contrasts with the light's transparency as shapes and differing textures work artfully with one another.

---

2 More research would be needed to see whether these details come from Tischler or from his successor Gerd Offenberg, who finished Monte Cassino's entrance building in a 1960's modernist style.

3 Post-WW II German Modernism in architecture is characterized in line and plane, right angle, and lightness of mass and color, width of window and openings, and dearth of ornamentation. Plans are simple and clear. See <http://germanpostwarmodern.tumblr.com> for examples. Accessed March 19, 2018. Such unadorned planar aesthetics has its roots in German architecture of the 1920s (where stone's massive character was emphasized).

Fig. 46. Costermano, entrance. Note modernist, open transparency rendering unhindered view.

Having crossed a modest terrace and into the cemetery proper, the visitor finds a mosaic of a dove holding an olive branch in the floor (by Maria-Theresia Mecke-Steger, 1967).[4] This theme, displayed at this important spot, symbolizes "Admonition to Peace" (*Mahnung zum Frieden*), by then the VDK's official slogan.[5] From there one's view is free to roam over the site's landscape, over the hillside, over meadow-like open spaces where wild flowers abound around trees, and up to the chapel on the right. The natural topography was left basically untouched. The terrain slopes gently downward on the left and on the right mounts toward the knob of a hill. This free panorama without managed, restricted views reveals a comfort with openness. During this same period, a recovering West Germany (the FRG) was prospering and its citizens were making first forays as private citizens exploring the wider world.

4 See fig. 7.

5 Ahlhorn's letter to Adenauer in which he cites "*Mahnung zum Frieden*" (admonition to peace) as the VDK's central slogan. The AA Archive, Group B 92, B and 271.

One finds one's self in a landscaped environment (*gärtnerische Anlage*) composed of several parts. The first and main impression is one of naturalness. Footpaths meander in curved lines among graves in rows arranged along the site's natural contours. Low earthen berms (dams) accompany. Calm prevails, engendered by the even lowness of these rows and enhanced by uniform planting of heather. Mono planting was to make the grave areas blend in with nature and render an impression of modesty. The rows' ends are rounded, not angular. Grass-covered swaths, wide enough for lawn mowers to pass, separate the heather-covered rows.

Individual graves have flush markers in the form of the VDK Maltese–type crosses (by then standard), bulky and short and symbolic (and used elsewhere, at times standing erect as in Bourdon, France). Each bears two names. The heather growing higher than the ground level makes the markers appear as denticles in a Greek freeze in rows with undulating edges. The grass strips of light green make patterns of landscape art. In the fall, the heather turns into a blanket of flaming magenta.[6] Tischler's groupings of three symbolic crosses are to be seen on the hillside grave areas to the right, set in wide yet calculated intervals. Their verticality and massiveness assure their impact.

Tischler's architectural "lasting elements" do not dominate. A free, modestly wide staircase branching off a main narrow cobbled pathway leads upward to the higher region with two more grave rows and to the monument in the form of a Romanesque chapel. This simple building with gabled roof displays Tischler's qualities of abstraction. Its walls of brownish natural stone tightly and horizontally layered appear stark, lacking adornment or compositional divisions. Aesthetic distinction comes from the entrance design. In typical Tischler manner, the entrance is given emphasis. Two broad strips of larger paving stones lead to the door that is framed through the use of heavy monolithic stones jutting from the lintel. The stones cut deeply through the wall into the interior. Their reddish color contrasts with the darker stone of the chapel's wall thus adding gravitas. We see a continuation of entrance emphasis typical of Tischler designs during the late Weimar Republic era, but here, there is no intent to create

6 Monoculture, specially heather was used throughout many WWI sites and particularily, for temporary sites. It carried connotations of "Germanness." In WW II cemeteries "monoculture" had practical and financial reasons.

mystery.[7] Two mighty cypresses at the chapel's narrow end anchor it's meaning into a funerary context and romanticize the structure's starkness.[8]

Fig. 47. Tischler's cubist chapel with main entrance off-center, on blank side wall. Entrance framed by signature monolithic stones. Note flush grave markers within monoplanting of heather that turns bright magenta in the fall.

Before passing under the awe-inspiring monolithic lintel, the visitor may tarry on a small terrace and gaze over a low wall onto the grave fields below. The terrace itself is a conflation of stone. Where the chapel's stark wall meets this terrace, stone meets stone at right angle without softening transition. A long stone bench is set against the chapel's side, but it does not seem to invite one to sit comfortably and be part of this symphony of stone.

One descends three steps into a simple, open hall-like space. Its walls are of light stone in the shape of bricks. Sparseness is enlivened only by a

7 Chapel entrance at Lissey near Verdun. Kuberek in *Unglücklich das Land*, p. 79.

8 Tischler favored cypress trees and used them to create a funerary context, as seen at Pinzano.

band of vertically set stones close to the roofline, which contrasts with the stonework's otherwise pervasive horizontality. Modest though the building may seem, its interior has a quasi-sacred effect. Daylight pours in through single round windows high above, on both ends of the space, to create a subdued effect as it falls upon tombstones on its floor. These bear the names of those buried in a comrades' grave beneath.[9]

Fig. 48. Chapel, interior, with Wimmer's "Kneeling Youth."

A bronze sculpture of a nude youth kneeling on his left knee, head slightly turned down and away from the knee, looks upon the comrades' grave. His right hand rests on the upright knee rendering the impression of action, as if he had just paused to kneel in order to quietly mourn.[10] The sculpture quotes from classicism's *contrapposto*. Its proportions are intact, but its somewhat

9 A large inscription is a dedication to all victims of war. "*Den Opfern von Krieg und Gewaltherrschaft*. 1939/1945." (To the victims of war and violence).

10 The VDK's *_richtlinien* of 1958 (cited) prioritizes connection to the families of the dead.

androgynous limbs and head are elongated, lacking detailed definition of muscles and surface, face and hair. Generic rather than individuated, it represents humanity. Its sculptor, Prof. Hans Wimmer of Munich, had first-hand near death experiences at the Russian front.[11] In its blend of classicism and abstraction the sculpture follows the "Munich school," but the posture and his elongated slender limbs remind one of Wilhelm Lehmbruck's work (1881–1919).[12] This tragic artist's work, (he committed suicide in 1919), deemed "degenerate art" (*entartete Kunst)* by Hitler's "experts," comported ideological authority in Germany's post-WW II period, when interest in Lehmbruck's work became revived.[13] Two nude male sculptures, both elongated and thin, titled "Rising Youth" (*Emporsteigender Jüngling*, 1913) and "The Fallen" (*Der Gestürzte,* 1915–1916) portray young life rising, followed by the tragedy of suffering and loss.

Thus, Costermano's sculpture also incorporates Lehmbruck's horror of war. He refused to give meaning to death in war but instead, focused on the despair that follows. Wimmer's Youth, through mourning, moves past despair and onto hope, and this sense is comparable to the quotation from Dante at Pomezia, messaging humanist order.

In its guidelines, the VDK specified how to refer to the war's battles. At Costermano a large map of Italy, made of colored tiles, stands in an open shelter outside the chapel and is surrounded by tall trees. Symbolic crosses indicate where on the Italian peninsula of those interred at the cemetery lost their lives. (The bodies of 21,920 are buried at Costermano, killed at 1,654 different locations in 37 provinces, often exhumed from communal cemeteries.) The setting for the map's record of death is effective. In stark contrast to the US cemeteries that provided highly designed maps illustrating their campaigns that led to victory, specifics of campaigns are not re-told.

---

11 Hans Konrad Röthel, "*Der Bildhauer Hans Wimmer. 1907–1992.*" Mit einer Einleitung von Max Huggler, München 1964. Also see Uta Kuhl (hrsg. vom Schleswig-Holsteinischen Landesmuseum Schloss Gottorf), *Hans Wimmer, Das plastische Werk*, Göttingen 1999, ISBN 3-88452-789-4[3]; and Vincent Mayr, "Ein "unverbesserlicher Gegenständlicher" – Der Bildhauer Hans Wimmer in München, Bayerischer Landesverein für Heimatpflege," *Schönere Heimat*, 2009/Heft 4, S. 237.

12 Christa Lichtenstern, *Plastik als Kraft, Beobachtungen zur Lehmbruck-Rezeption in der Bildhauereri im 20. Jahrh.,* (Georg Kolbe Museum, Berlin, 2002). There is no major work on Lehmbruck. Most materials treat specific examples such as Renée Price, *New Worlds: German and Austrian Art 1890–1940*. (New York; Neue Galerie, 2001).

13 Suggestion, Dr. C. Fuhrmeister in conversation. (November 2013).

A meadow bordered by varied plants ranging from shrubs to high trees overlooks beautiful Lake Garda and its richly cultivated lands. This marks the site's zenith.[14] Sharp tops of cypress trees quote from funerary tradition, as does a heavy basalt altar, a stone monolith upon three stubby pillars. An eight-meter-high artfully shaped wrought iron cross (1967) stands alone, isolated.[15] Although it was not required by the guidelines (from 1958 and 1965) the VDK erected iron crosses at all post-WW II cemeteries in Italy, crowning their funerary iconography with the Christian symbol.[16]

14 Werner Lindner, in richtlinien (1954) emphasizes the beauty of the surrounding landscape p. 5. It became a standard criterion for all German cemeteries in Italy. All have beautiful views over the Italian landscape.

15 Willi Kammerer; *Volksbund Deutsche Kriegsgräberfürsorge Erinnern und Gedenken: Kunst auf Kriegsgräberstätten*, VDK Archive, kunsthandwerk-costermano pdf. The artist is Manfred Bergmeister, he also did the stunning iron work at the entrance. Klaus Manfred Bergmeister, Schmiedearbeiten-Smithery-Les Oeuvres Forgées. Tübingen; (Berlin: Wasmuth, 1997), p. 183. The book provides many illustrations.

16 At Quero, originally without explicit Christian symbolism, a tall cross was erected in 1979. I wish to thank Peter Paessler, Archivist of the VDK Archiv for this information.

Cassino Crypt Design: Tischler's Fallback

> "Emperors, kings, princes, popes, cardinals, and great nobles did bury and continue to this day to bury their dead not in graveyards or greatest monasteries, but in family or church crypts."
>
> Robert Tischler[1]

This chapter explores a project that was abandoned, but is included because its story reveals the VDK formulating new tenets in the midst of warming Italo/German relations. The result was the VDK's re-thinking its style and message in response to a spectrum of public opinion.

Seven thousand German soldiers died around the Monte Cassino monastery. In the spring of 1954 Cassino's mayor (*Senatore)*, Pier Carlo Restagno, requested that Germany build a central memorial cemetery (*Zentralehrenstätte*)[2] and thus, even before the German-Italian accord was in force (December 1955), a cemetery project was under discussion. Robert Tischler created a detailed design and two models and strongly advocated in favor of this project referred to herein as "Cassino I." He had envisioned a "follow-on" to his imposing *Totenburgen* at Tobruk in Libya (dedicated on November 25, 1955) and one under construction at El Alamein (Egypt, dedicated in October 1959). In the summer of 1957, Cassino I was abandoned.

It was imagined that a hill site with a crypt could accommodate the tens of thousands of additional remains of those who had fallen in battles along the nearby "Hitler" and "Gustav" lines. The site lay at a switchback in the only roadway leading up to the famous monastery, Monte Cassino,

1 Letter, Tischler to the Vorstand (Presidential Council) *("Kaiser, Könige und Fürsten, Päpste und Kirchenfürsten, die grossen Klöster und Standesherren betteten und betten heute noch ihre Toten nicht auf Friedhöfe, sondern in Familien oder Kirchengrüfte.")* The VKD Archive, A.100–1031_Tischler-Symbolkreuzgruppen pdf, (January 23, 1957).

2 Note on meeting in the German foreign ministry that informs that Cassino's mayor had approached the German government several times asking that a memorial at Cassino be built. The VDK Archive, A.100–857, (April 12, 1954).

and seemed unmatchable for veterans as well as for the VDK.[3] A medieval castle ruin on a neighboring hill would be restored and not only would anchor the design, as a photograph of the land shows, but provide a coveted pedigree of the medieval past. Tischler's design would be an equivalent in size and significance.[4]

The site was comprised of two land parcels, and thus there were two owners. One was immediately willing to sell and Cassino's mayor Restagno believed the other owner to be the monastery, which had expressed support for the project. The actual owner, as it was discovered, was not the monastery but the Vatican, which took the matter under advisement.[5] Restagno, fully in favor of the project, promised to try to obtain Vatican agreement to sell, but by 1956 he began wavering, restraining Tischler's enthusiasm for the site with reasons such as "lack of parking space" and "potential congestion on the road to the monastery."[6] He added that the Vatican might need ten years to come to a decision. Soon, the monastery itself became a major stumbling block. The castle ruin was the *Rocca Janula* built by the monastery's monks in the 9th century to defend against attacks by the Moslem Saracens, and was considered of historical significance. The ruin should remain unmolested. When the monastery's abbot expressed opposition, the mayor immediately followed suit. Such yielding to ecclesiastical authority astonished the VDK's general secretary, Margraf.[7] In masterly Italian diplomatic

3 Dr. F. Debus relates a short history. German soldiers held their positions until May 10, 1944. Debus emphasizes that comrades would be disappointed were the place of actual fighting not be the burial ground. *Die Kriegsgräberstätte Cassino,* VDK Archive, A.100–857, (November 30, 1956).

4 Ink drawing by Tischler showing his design. Report on meeting between the VDK's general secretary Otto Margraf, Restagno, and the Abbot of Monte Cassino. The VDK Archive, A.100–857, (March 20, 1957), p. 4.

5 When it became known that the Vatican was the owner, the process of acquisition became more complex. At first it seemed that the Vatican favored the project. It seems likely that Vatican recalcitrance was one of the reasons the project did no go forward. Note from the VDK Rome to the VKD Germany, VDK Archive, A.100–857, (August 2, 1956).

6 Restagno drove Tischler to the site in order to make his point. See report from the session at the VDK's presidium on August 7, 1956. The VDK Archive, A.100–857, (May 24, 1957).

7 See fn. 32. Seifert, from the VDK's re-interments commission, in his report about the meeting, expressed his surprise at the sudden capitulation of Restagno to the Abbot after the former had, for a good year, expressed his enthusiastic support for the VDK's plan. The VDK Archive, A.100–857, (May 13, 1957).

manner, the mayor retorted that the German honor would not be sufficiently represented if the project were built in the vicinity of already existing important monuments.[8]

Time was of the essence for the VDK as the remains of ca. 20,000 German soldiers had already been disinterred, put into sarcophagi, and storage had become a problem. Restagno offered the use of a military barracks for their temporary storage. Alternative cemetery sites were suggested, one being not too steep and which had a "nice panorama." With regret that two years had been lost, the VDK officially abandoned the project in Rome, at the Commissariat (headed by General Ricagno and Captain Uberti), and by mid-1957 Tischler had chosen a hill site at nearby Caira.[9]

Tischler had invested his personal *cache* in the Cassino I project and submitted full plans to the VDK with sketches depicting his vision of the entire site, including the castle ruin.[10] A huge crypt building was to be located on a hill site opposite the ruin. Thus, the ruin was to become part of a theatrical stage set with romantic pathways and trees growing at their edges. He sketched a dead tree in the foreground and a lone crooked oak at the access road, representing life and death. The crypt would stand on a platform of leveled ground with a broad staircase leading to its entrance. A hefty wall would span a small valley between the two hills.[11] Both crypt and ruin would bespeak valor and strength. Tischler even imagined that the castle might be reconstructed.[12] His final design showed it so, reborn, towering over the site, without scars of the past. It would serve as an aesthetic

8 Report on a meeting between Margraf, Restagno, and the Abbot of the monastery on October 3, 1957. The VDK Archive, A.100–847, (October 3, 1957).

9 This may be deduced from a site plan of the Caira hill from May 15, 1957. A site plan from September 20, 1957, four months later, made for the *Ministero Della Difesa* already shows the outline of the enclosing wall for the cemetery. The VDK Archive, Cassino_ Altplaene (old plans), 00007075, (May 15, 1957) and 00007076, (September 20, 1957).

10 Tischler wrote to the Vorstand that his sketches were ready. The VDK Archive, A 100 _ Tischler, (October 22, 1956).

11 The image ressembles a woodcut.The only VDK WW II crypt building in Italy is in Motta St. Anastasia near Catania, Sicily, built by Diez Brandi, who built many churches in Germany. VDK brochure *Deutsche Kriegsgräberstätten*. Issue on Italy, entry; Motta St. Anastasia.

12 This plan shows more definition of the landscape, but the dead tree in the foreground is still there. The VDK Archive, Cassino _Altplaene, 00007078. Undated, but it can be safely assumed that it is from November 1956. Other drawings pertaining to this project are dated within November 1956.

and historic counterpoint to his edifice. The wall connecting the two seems to allude to the several fortified defense lines that straddled the entire Italian Peninsula, which were built during WW II. Located near the Gustav line, the Monte Cassino monastery was considered a fortress by the Allies and completely destroyed by bombing in mid February 1944. (Following the destruction, German soldiers actually did use the monastery ruins as a fortification.[13])

Fig. 49. Tischler's model for grandiose, massive crypt at Cassino (not executed). Note existing the castle ruin is shown restored.

For the crypt, Tischler proposed a squat tower-like entrance and a long single story lateral wing that would serve as the crypt proper. The tower set the tone, projecting forward and having canted sides suggestive of Egyptian/Assyrian architecture as in the design for Pinzano. The entrance was to be a huge recessed rectangle that contained an enormous door and a round window above. The long crypt had narrow window openings, recalling a castle's crenels. A wall curled around the end of the crypt enclosing an open court recalls Quero's *Laufgraben* and Pinzano's terrace. Large iron-grilled openings at the rear of the structure were to admit light.[14] Thus, even though it did not emphasize typical *Totenburg* "centrality" its ambitious structure and its purpose locate the design in the *Totenburg* orbit.

The design of its interior confirms vicinity to NS concepts of grandeur. A sectional drawing of the crypt provides details on its interior.[15] The

13 My thanks for information about the fall of Monte Cassino go to Prof. Piero Polodoro, Rome.

14 The VDK Archive, Cassino_Altplaene, 00007081, (November 6, 1956).

15 The VDK Archive, Cassino_Altplaene, 00007096, (September 15, 1956).

entrance over a flight of stairs would lead to the court of honor (*Ehrenraum*) where a list of the dead was to be provided. A votive stone (*Gedenkstein*) was to rest on a platform framed by two pillars, all set upon on a three-stepped stage that spanned the width of the space. The long lateral tract, the actual crypt, was to be accessible from the entrance hall via a corridor that ran along its entire length, with light admitted through small window slits.

Again, the repertory of pillars appears, repeating the entrance hall's pillars, to separate the corridor from an empty paved hall and accompany the visitor while walking through empty space. "Emptiness" would echo death. (The architectural elements employed to create this impression were similar to those envisioned for Pinzano.) Bodies were to be interred beneath, safe at rest, yet accessible through a staircase. The VDK by that time insisted on sarcophagi with each person identified.[16]

Tischler experimented with light, delivering different versions of this crypt, different in detail rather than in basic concept. He would use either smooth or rough stone. The backside would be windowless, the front window openings resembling the arrow slits keeping the interior dark. Alternatively, grilled openings or semi-circles recalling classical Roman baths would be employed such that the building's end would be bathed in light pouring in through a large semi-circular Roman window.[17]

Sketches show distant mountains, uninviting, looming over the site as if questioning man's dialogue with nature. As at Pordoi, a soldier's death would bear the stamp of eternity. The castle ruin would serve as a *memento mori* of bygone worldly power, indifferent to its decay, in this way bestowing an aura of eternity. Resurrection symbols, and even crosses symbolizing suffering were to be absent. The design and its message would seem to serve not the dead, but death itself, framed as sacrifice with pretensions of grandeur echoing the nationalist mindset of the 1930s.

Not surprisingly, Tischler's vision immediately triggered a passionate discussion.[18] Different ideologies of appropriate forms of burial revolved around crypt versus earth burial. In December 1956, the year in which Tischler finished his design, he was given to understand that not all members of the VDK supported it and he was not pleased. The VDK decided definitively that there would be an earth burial with singular

16 The VDK Archive, Cassino_Altplaene, 00007080, (November 6, 1956).
17 The VDK Archive, Cassino_Altplaene, 00007019, (November 10, 1956).
18 The VDK Archive, A.100–857, in January 1957.

graves (August 7 1956).[19] Tischler argued that this particular site's topography would not permit an earth burial and in a letter to the VDK, he defended the crypt building as the traditional form of burial for kings, emperors, popes, and prince electors, etc. He referred to crypts in Egypt, India, and China as the most imposing buildings in the history of the world, and "… in this monumental building a specially dignified form of the crypt is possible and is planned."[20]

Early in 1957 the VDK Presidium established a commission comprised of persons from different walks of life to assist in their decision-making process.[21] Among its members was Dr. Georg Banasch, a Catholic prelate active in Berlin. Banasch became the commission's chairman and also Tischler's most consistent and vociferous adversary. Equating crypt burial with "mass grave," he took the position that it was in conflict with the Catholic Church's mandate for an earth burial.[22] He remained adamant in his position despite the VDK's president, Gustav Ahlhorn's, attempt to intercede.[23] Ahlhorn argued that the "mass grave" label was inappropriate since each soldier would have his own sarcophagus, that WWI cemeteries in Italy were all ossuaries, and thus, crypt burial would not diminish the status of the buried. The prelate remained intransigent. Crypts, he maintained, were nothing more than ornamented mass graves and the VDK should build honorary cemeteries, not mass graves.[24] He allowed that the crypt burials in Tobruk and El Alamein were a necessity because of desert

19 Letter, VDK to Tischler, (October 31, 1956), informs that crypt burial will not find unanimous support on the side of the VDK. In a long letter, prelate Dr. Banasch reminds that the VDK wishes to remain firm in its decision to use only earth burial. For both letters see the VDK Archive A.100–857, (February 1, 1957).

20 Letter, Tischler to VDK Vorstand. *("…wohingegen in dem monumentalen Bauwerk eine besondere würdige Form dieser Gruft möglich und geplant ist.")* However, at that time the controversy about the crypt design for Monte Cassino had increased and a go-ahead for Tischler was thus postponed. The VDK Archive, A. 100–1031, (January 23, 1957).

21 The Commission consisting of Domkapitular Berlin-Charlotttenburg Prälat Dr. Banasch, Senator A.D. Emil Theil, Amtsanwalt Schmidt Hahnenfeld, Kreis Werder-Aller and General Friedrich Hosbach was already in place by August 7, 1956, as may be deduced from the directorate's *(Presidium's)* report on meeting of that date. The VDK Archive, A.100–857, (August 7, 1956).

22 Letter, Dr. Banasch to the VDK's President Ahlhorn. The VDK Archive, A.100–857, (February 14, 1957).

23 Letter, Ahlhorn to Dr. Banash. The VDK Archive, A.100–857, (March 1, 1957).

24 Declaration from Banasch. The VDK Archive, A.100–857, (January 2, 1957).

conditions, but it was hard for families as crypts prevented any personal relationship between the families and the dead. Furthermore, he maintained that the VDK's accomplishment of identifying 90 percent to 95 percent of those fallen in Italy would be nullified through mass burial at Cassino. The VDK, he proposed, might as well cremate the bodies and have urn burials, as that would make the whole enterprise less costly.[25] And why, he lamented, could Germany not build cemeteries in flat valleys, as do other nations (a question later to be raised in Rome with regard to the persistent pursuit of the Futa-pass site).

Another member of the commission (H.A. Hahnenfeld, a veteran) advocated in favor of both *Totenburgen* and crypt burial.[26] For him, proximity to the war's battles was of paramount importance and the fact that the site's topography did not allow for an earth burial was a decisive factor. He paid admiring homage to Bitolji, Quero, Tobruk, and El Alamein as striking and memorable monuments.[27] He stated that people did indeed like the *Totenburg* and did not resent crypt burial. He found it "consoling" and more suitable to fulfilling the VDK's task to guarantee the right to eternal rest (*ewiges Ruherecht*) than did individual graves. Within a generation, he argued, families would no longer visit, and he therefore advocated in favor of *Totenburgen,* as they would provide for longevity. "Experience has shown that the named *'Totenburgen'* have lost none of their sublime beauty in comparison to individual graves." He used his status as a veteran to remind the VDK that they were to act as custodians of all of the people, including the veterans. As one, he would prefer crypt burial and the VDK should erect more *Totenburgen.* Yet another voice was heard, that of a veteran who had fought at Monte Cassino. In a letter, he explained that he had heard of this controversy and wished to comment as a veteran.[28] He considered the term "mass grave" to be pure propaganda, and that religious confessions should not be an issue. A soldier's experience took him from

25 Ibid.

26 Letter, Friedrich Albrecht Hahnenfeld, Amtsanwalt, to the VDK Vorstand. A member of the commission, head of his region's VDK (Ortsgruppe), he emphasizes that he writes as a veteran and Lutheran Christian. The VDK Archive, A.100–857, (July 3, 1957).

27 See Kai Kappel, "Die Totenburgen von Tobruk und El Alamein- strategische Memorialarchitektur für die Bundesrepublik", *RIHA Journal* 2017, "War Graves."

28 *("Erfahrungen ... haben gezeigt, dass die aufgezählten "Totenburgen" im Gegensatz zu den Einzelgräbern von ihrer erhabenen Schönheit nichts eingebüsst haben.")* Letter, Otto Schlicht, Oberlandesgerichtsrat to (retired) Senator Emil Theil, the VDK Landesverband Bremen. The VDK Archive, A.100–857, (May 15. 1957).

his family and put him in union with his comrades and hence burial among comrades was not only appropriate, but also desirable. The veteran further maintained that a monumental crypt would be a lasting and powerful message for peace. The site near Monte Cassino would be perfect, as many would visit it.

In the midst of the controversy prelate Dr. Banasch argued that adding the remains of ca. 15,000 soldiers from other areas to the ca. 7,000 that actually had fallen around Monte Cassino would amount to falsification of history. In addition he asked whether building a monument next to castle ruins would not be an attempt at stylistic triumph over other nations, which erected more modest (less pretentious) cemeteries: "We, the Germans should become more modest after the events of the Nazi time."[29] The VDK's president Ahlhorn responded politely, defending crypt burial and challenged the prelates' notion of modesty.[30] In consideration of all arguments, crypt burial in a prominent building on this site was deemed acceptable by the VDK.

The VDK and Tischler had been optimistic about the prospect of success.[31] Tischler worked on the project for a good two years and was well paid, but then at a meeting in the spring of 1957 in Monte Cassino "the last nail was driven into its coffin."[32] The monastery's abbot had the last word. The site would not be made available. The VDK was stupefied because the abbot had been a supporter until then. The VDK had learned a painful lesson, that designing prior to firm possession of the chosen site could be a

29 Letter, Banasch to Ahlhorn. *("Wir Deutschen sollten nach den Geschehnissen der Nazizeit bescheidener werden.")* The VDK Archive, A.100–857, (February 1, 1959).

30 Letter, Ahlhorn to Banasch. He politely defends crypt burial and addresses the prelate's demand for modesty by pointing out that the Polish cemetery in Monte Cassino, overdone with marble, was certainly not a model of modesty. At least the Italian people did not think so. The VDK Archive, A.100–857, (March 1, 1957).

31 Letter, Tischler to the Vorstand shows that he presumes his design will be carried through. He would travel to Rome. The VDK admonishes him to show restraint in his talks. The VDK Archive, (October 31, 1956). Tischler took this admonition as an insult and reacted furiously. Why not send someone else, he asked. Tischler nevertheless did travel to Rome. The VDK Archive, (October 22, 1956); The VDK Archive, A.100–857, (November 14, 1956).

32 Report on the session of March 10, 1957 at Monte Cassino, signed by Seifert. Present were Senator Restagno, the Abbot of Monte Cassino and the VDK general secretary Margraf. The report informs that the VDK had been already shown alternative sites. The VDK Archive, A.100–857, (May 13, 1957).

demoralizing and expensive detour.[33] (This experience is similar to that of the US's ABMC. Designs for Impruneta and Anzio/Nettuno were underway well before the actual sites were officially made available. The US's moral authority (and financial power) prevailed while Germany's did not).

The veterans' points were likely widely held and evidently contributed to the VDK's realization that it needed to formulate new guidelines.[34] One wonders what might have been brought into being had the VDK's members been military men, as was predominantly the case with the ABMC. One further wonders whether the VDK would actually have allowed Tischler's grand crypt design to go forward had the decision to terminate it not been made for them by the Italians. In any case, heated discussion about burial practices, designs, and veterans' points of view prompted the VDK to formulate new burial guidelines, which were finalized and published in 1958.[35]

33 Plan of May 15, 1957. The Caira site was secured by 1957. The VDK Archive, Caira_Altplaene, 00007075.

34 Response from Eulen to Hahnenfeld's position *(Stellungsnahme)* in favor of crypt burial. (See fn. 26). The VDK Archive, A.100–857, (March 15, 1957).

35 Ibid. Eulen reassures Hahnefeld that his letter will be cited in preparatory meeting for formulation of new guidelines of May 24, 1957. The VDK Archive, A.100–857, (March 15, 1957).

IX. Cassino/Caira: Between Tradition and Innovation

"Foreign critics are already talking about it, that building such memorials represents today's greatest cultural challenge."
Robert Tischler[1]

Fig. 50. Tischler's Cemetery at Caira. Note only half of the hill is occupied by cemetery.

For the VDK, the "Cassino I" project was clearly a debacle. Even those Italian entities that had proposed it in the first place had "turned their backs" on it. The monastery had been in favor of using the site for a crypt prior to the Vatican's becoming involved, and even the Vatican had been favorably disposed early on. After this debacle, the VDK resolved not to

1 *("So sprechen ausländische Kritiker heute schon davon, dass die Schaffung derartiger Ehrenstätten zu den grössten Kulturaufgaben der Gegenwart zählt.")* Letter, Tischler to the VDK Vorstand. Tischler defends crypt burial. The VDK Archive A.100–1031_Tischler–Symbolkreuzgruppen, (January 23, 1957). Werner Lindner in *Richtlinien* also saw the curating of graves as a cultural task (*"Kulturaufgabe)."*

continue with crypt burials (to have only earth burials)[2] and to proceed more prudently. In the future, the VDK made sure it had approval from Germany's Foreign Ministry, and its president would secure Italian government blessing (working with the Ministry of Defense) before a project would go forward. (By contrast, the US's ABMC did not need consent or approval for their designs from any person or entity in Italy).

The VDK found itself in a difficult situation. Their people had exhumed thousands of bodies and temporarily stored them in parish churches, railway stations, and even an army barrack, offered by Cassino's mayor Restagno. More were arriving and the issue was pressing. A full two years of Robert Tischler's work, his many automobile trips over the Alps and his dreams of a grand eternal monument had been for naught. His documented reaction is not to be found but one cannot help but imagine it.

Cassino's city government wanted a major construction project and the need to re-bury was urgent. Restagno suggested additional sites and showed them to Tischler, and to other VDK dignitaries. None were found satisfactory, including the Colle Marino-Caria.[3] Tischler then undertook to travel ca. 4,000 km in 10 days, revisited four potential sites and had another close look at the Colle Marino-Caira site.[4] He then recommended acquiring it.[5] The VDK dispatched a team to re-evaluate it.[6] When Tischler learned of this, he was furious. In a torrid letter, he expressed his indignation and even

2 The cemetery at Motta St. Anastasia at Catania (Sicily) was dedicated in 1965 and is an exception. 4,561 remains, in individual sarcophagi, are contained in two crypts. See the VDK brochure, Italy. Entries are in alphabetical order.

3 Report from a sightseeing trip. General Secretary (of the *Vorstand*) Margraf rejects the site, "… out of the question for the German honor site.", *("…kommt für die deutsche Ehrenstätte nicht in Frage.")* There was no adequate access road, he noted. The VDK Archive, A.100–857, (March 20, 1957). Later, the VDK built a new road.

4 An earlier Italian WWI cemetery at Monte Marino-Caira (1915–18) had provided provisory graves for over 1000. The VDK Archive, Caira_Altplaene, 00007105, (March 29, 1957).

5 Letter, Tischler to the VDK directorate. The VDK Archive, A.1008–57, (March 29, 1957).

6 Session, directorate (*Vorstandssitzung*). The commission consisted of General Hosssbach, member of the VDK directorate and Dr. Lindner, the author of several guidelines and member of the VDK's Committee of Experts, Stiebeling, vice chair of the administration (*Bundesgeschäftstelle*), and von Münchhausen, head of the VDK Rome. They also consulted with local construction professionals and inspected the Futa-Pass. The VDK Archive, A.100–857, (May 11, 1957).

threatened to abandon the VDK altogether.[7] He saw the VDK's action as a direct challenge to his professionalism. The VDK's directorate (*Vorstand*) obviously, soothed Tischler and made amends. Tischler wrote that he believed he could transform the Colle Marino-Caira site into a splendidly tasteful cemetery, especially as it had a flat hilltop perfect for a monument that would be visible from afar, from the main road to Naples, which was to be improved in the near future.[8] The VDK decided in favor of Tischler's vision for a cemetery, herein called simply "Caira." Architect Gerd Offenberg completed the project after Tischler's death and in his dedication speech (April 23 1965), he gave Tischler full credit for having found the site as well as for having furnished the design.[9]

In the spring of 1957 the VDK formally accepted the local government's offer of the Caira site[10] and Cassino's mayor and Frosinone's provincial prefect initiated the site acquisition process, forwarding matters to Rome.[11] The prospect of a German soldiers' cemetery in Caira was most welcome since the region had high unemployment.[12] An article in the weekly *Il Gazzettino del Lazio* expressed "great joy" over the plans for erecting a "magnificent"

7 Letter, Tischler to *Bauleitung*, "If you value my professional views so little that you have to send a commission after me, then the more purposeful way would be to spare me from the task of choosing sites altogether and have a commission search for them in the future." *("Wenn Sie meinen fachlichen Feststellungen so geringen Wert beimessen, das Sie mir zur Nachprüfung eine Kommission nachschicken müssen, dann wäre der zweckmässigere Weg doch der, mich mit Platzauswahlen überhaupt zu verschonen und die Plätze künftig von Kommissionen suchen zu lassen.")* The VDK Archive A. 100–857, (June 13, 1957).

8 The VDK Archive, Caira_Altplaene, 00007105, (March 29, 1957).

9 Gerd Offenberg became the official chief architect of the VDK after 1967. He had finished Costermano. For the full text of his speech see the VDK Archive, A100–857_offenberg. His speech reveals that new, more democratic times had begun. He mentioned and thanked everyone involved in the construction, from artistic talent, construction specialist, gardeners, the German youth (for having collected funds), to the simplest worker. Praise for the VDK in Kasssel and in Rome ended his speech. A team spirit seems evident from his paying special tribute to the Italian contractors. Friendships were forged. The VDK Archive, A 100–852.

10 Report on meeting with Monte Cassino's city government. The VDK Archive, A.100–857, (March 3, 1957).

11 There is no archival indication that the VDK, supported by the Aussenamt, contributed financially to the purchase. Most likely the Italian government purchased the site in agreement per the accord of December 15, 1955. The VDK Archive, A.100–857, (November 4, 1957).

12 Mayor's letter to the VDK. The VDK Archive, A.100–857, (July 18, 1975).

and "monumental" German cemetery.[13] The VDK's directorate sent its project plan to Germany's Foreign Ministry (*Auswärtiges Amt*) for approval.[14] It needed about a year's time and on April 22, 1958 the project was approved and ready to move forward. Tischler received his formal commission for the cemetery design on May 27, 1958 with a "slap on the wrist" from the VDK, reminding him that he must adhere to their new guidelines specifying only earth burial. A detailed burial plan for 22,577 fallen soldiers was soon underway.[15]

Tischler's Design

The VDK decided on a solution based primarily on landscaping (*gärtnerische Lösung*), Tischler's least favorite genre.[16] Yet, Tischler went to work within the VDK's prescribed parameters. He would deliver a plan by October along with cost estimate (DM 3,944,575), a design,[17] and two models.[18] Some changes were made and the cost estimate revised.[19] Changes were approved for submittal to various permitting agencies on

13 Article from September 19, 1957. The VDK Archive, A.100–857, (September 19, 1957).

14 Note (November 4, 1957) expresses the intention to send the design to the AA (*Auswärtige Amt*). The VDK Archive, A.100–857. By February 28, 1958 there was no response from the German Foreign Ministry (AA). The VDK Archive, A.100–857, (February 28, 1958).

15 Report on meeting and conversation with VDK's General Secretary, Margraf. The VDK Archive, A.100–857, (April 22, 1958).

16 *Vorstandssitzung* session protocol. Also stipulated was that walls were limited to one meter in height, except the outside wall in the south would be 2.5 meters. The VDK Archive, A.100–857, (October 30, 1958).

17 Letter, Tischler to Vorstand. The VDK Archive, A.100–857, (August 18, 1958).

18 Letter, Tischler to the Vorstand including two models. The VDK Archive, A.100–857, (October 29, 1958).

19 Protocol, experts' committee's session, (*Sachverständungsbeiratssitzung)*. Acceptance of design and costs was forwarded to the Secretary General (Vorstand) for approval and and then later forwarded for approval by the Auswärtige Amt (AA). Costs had risen to DM 4,319,310 from DM 3,944,575. The AA (German Foreign Ministry) would pay DM 2,054,205 and the VDK DM 2,265,105. The VDK Archive, A.100–857, (December 5, 1958).

December 12, 1958.[20] About six months were required for approval by German agencies (the VDK's Presidium and the German *Auswärtiges Amt*),[21] and also by Italian governmental entities. Before the year's end the VDK's general secretary, Margraf, hurried off to Rome to facilitate the Italian governmental approval process, submitting plans in person to General Uberti at the Defense Ministry's *Commissariato De Cura e Onoranze dei Caduti in Guerra*. Uberti was enthusiastic and particularly liked the chapel-like entrance building with its three large rectangular openings toward the south, a gabled roof and a tower with a belfry.[22] Uberti's *Commissariato* apparently granted permission for the VDK to begin preparations while "official final" approval by the Italian government remained in the offing.

A VDK document bears testimony to impressive expertise in project estimating, planning, and scheduling, as well as to the pressing need to rebury so many stored bodies.[23] Thorough project planning would coordinate the work of all persons and parties involved. The VDK's plan even took into account the time needed to receive permissions from the various Italian government ministries.[24] With all permissions in place, the groundbreaking ceremony (*Spatenstich*) took place on April 20, 1959, about two years after the choice of the site.[25] Construction itself was seen as a great logistical and

20 The first chapel version came with an explanation: There were large windows toward the south to admit light. The chapel would be accessed via a few steps under a sheltering roof. See fn. 18.

21 Final approval from the AA dates from February 16, 1959. The VDK Archive, A.100–857, (June 10, 1959).

22 Note (*Aktenvermerk*) from Otto Margraf. The VDK Archive, A.100–857, (September 18, 1959).

23 In his speech for the dedication ceremony, Offenberg gives particular tribute to the VDK's logistic accomplishments, having mastered the task of reburying 20,000 dead from 25 different locations. The VDK Archive, A.100–857, (April 23, 1965).

24 Document from the VDK's grave registration (*Gräbernachweis*) demonstrates the tremendous logistical task of the VDK to unbury, identify, and rebury large numbers. The document outlines the sequence of construction phases: preparation of grave areas as the first objective, followed by planting, erection for the custodian's building, and lastly, embellishment and inscriptions. It also addresses the need for a clear chain of command, necessary for decision-making. The VDK Archive, A.100–857, (December 12, 1958).

25 Permission was granted on January 23, 1959 by the VDK's Presidium. Note from the VDK's departmental manager (*Abteilungsleiter*) to the *Bauleitung*, The VDK Archive A.100–857, (February 2, 1958); The VDK Archive, A.100–857, (February 16, 1959).

engineering achievement.[26] At the 1965 dedication ceremony, Offenberg praised the Italian contractors for their accomplishments.[27]

Fig. 51. Tischler's early model with entrance through chapel. Note tall cross atop massive base.

Tischler's two 1958 models show that his concept of placing grave sites on ellipsoid terraces around the hilltop was in place from the beginning. His mastery of topography is evident in that he was able to disguise the fact that the site included only half of a hill. Seen from a distance, terraces and the planting invite the visitor to imagine the whole hill transformed into an Etruscan *tumulus.*[28] Thanks to skillful planting the totality of the site can

---

26 The first phase of construction began with earth moving, retaining walls, and grave fields. 4,890 M^3 of stones from quarries near Rome and Siena were transported in 800 truckloads and set and worked by hand. Together with the cement foundations, the total amounted to 8,700 M^3. Cut and fill work was extensive. 5,000 truckloads delivered 30,000 M^3 of soil. Once at the site, transportation up the hill proceeded over existing mule paths. The VDK Archive, A.100–857, (June 25, 1959).

27 Ibid. Italian involvement in the construction is fully acknowledged: architects Munaretto and Pirani; a Dr. Tausch from Feltre and Engineer Longo from Cassino who created the grave fields in the shortest time; the Henraux firm (Quereta), its engineer Dr. Casini assisted by Mr. Palombi; the C. Ge. Company (Rome) and its engineer Dr. Casini for the aboveground construction (*Hochbauarbeiten*). Note that all contruction firms were Italian.

28 Piero Polidoro offers an interesting interpretation of the tumulus, theorizing that its shape derives from Dante's paradiso: <http://www.lumsa.it/sites/default/files/ateneo/rivista/atlumsa8.pdf>. accessed March 15, 2018.

be seen only from the air or from a nearby mountaintop. The models also showed Tischler's vision for the hilltop, which included a huge cross on a mighty protruding bastion. The VDK subsequently revisited the design, determined the bastion to be too bombastic and monument-like, and eliminated it.[29] There would be neither a monument on the hill's top nor a prominent cross, which must have been a major disappointment for Tischler.

Although the project's approval was likely facilitated by General Uberti's partiality to its chapel-like entrance building, by early 1959 new tastes had emerged. Tischler chose to realize a more modernist entrance structure instead of a chapel.[30] For this, the term "sanctification space" (*Weiheraum*) had been renamed the "hall of honor" (*Ehrenraum*), thus stripping the space of its sacredness.

Fig. 52. Caira's entry building in clear, cubic form. Entrance framed by monolithic stones. Note mature cypresses working with design.

The entrance building testifies to Tischler's preference for a cubistic structure with a strong entrance accent, created by massive monolithic stones that frame the door. The lintel seeming markedly ponderous for the narrow entrance it roofs.[31] An airy window opening surmounts the entrance and

29 Note of experts'committee meeting, recommending rethinking cross's monumental socle. The VDK Archive A.100–857, (November 20, 1958).

30 This model also shows planting concentrated at the outer terraces and in regular intervals on the hill. Atop, a dense grove is envisioned. The VDK Archive, A.100–857.

31 Tischler was able to maintain his aesthetic vision of an entrance, as at Costermano's chapel.

works in subtle counterpoint to the lintel below. A few steps lead through the doorway. Artful wrought iron work is seen. The first view falls on the lettering on a wall in Bauhaus script of the 1920s. Its message reads: "20,057 German soldiers lie at rest in this cemetery."[32] Graves are yet to be seen. The message, standing alone, vitalizes the visitor's imagination.

Tischler died suddenly in August of 1959. Offenberg preserved Tischler's concept, that the visitor must pass through the entrance building in order to gain access to the cemetery, and that this would be an orchestrated experience. Such orchestrated entry involves a sequence of spaces. First, the visitor crosses a bridge over a ravine where a path leads along a low stone retaining wall executed in local fashion. After several sections of stone steps on a slightly curved path the visitor of today arrives at the travertine entrance building framed by thick and tall cypress trees.[33] The visitor has yet to see anything of the cemetery proper.

Tischler's constricted entrance between its *patina* of old travertine darkened by cypress trees continues to set the somber funerary mood. Offenberg proposed a re-design for the entry building's façade (1963) with a broad staircase spanning its width and leading to a wider entryway. This would have done away with Tischler's shadow, his mysterious play of light and dark, narrow and wide. Entering would be a more "wide-open" experience.[34] Offenberg's re-design was never executed.

Germany's rise to prosperity was accompanied by artistic and ideological development, spawned in part by Italians who shortly after 1955 had migrated northwards in great numbers to find work. Reciprocally, by the mid-1950s Germans were traveling in mass and Italy was a popular tourist destination. During the short period between Cassino I's cancellation and Caira's inception, the VDK adjusted to Germany's remarkable

---

32 *(20,057 Deutsche Soldaten sind auf diesem Friedhof zur letzten Ruhe gebettet.)* This lettering and the VDK emblem are done by Peter Offenberg (the architect's son). The VDK Archive, A100–857_Offenberg. No date of work given.

33 The VDK drawing shows the entrance design with an access road. The VDK Archive, Caira_ Altplaene, 00006868. No date given. Similar drawings dated June 1963.

34 Offenberg imagined a façade that would have admitted light. The VDK Archive, Caira_ Altplaene, 00006743, (September 11, 1963).

cultural development toward modernist aesthetics. The VDK's new vision is articulated in its 1958 "guidelines." Yet, the hesitancy to include a monumental aspect in its commemorative repertory for cemeteries in Europe remained.

The Caira project as completed bears testimony to such development. The visitor turns to climb a long staircase between high walls of travertine and emerges into Offenberg's wide interior space that is strikingly airy and modernist. The mystery and sacredness of darkened interiors with a traditional guided shaft of light is nowhere to be seen. This "room of honor" is a clear cube where the roof seems to have been slid aside and opened to the sky.[35] It is bathed in light. Its smooth and bright travertine forms clean geometric shapes with straight lines. Looking up, the visitor beholds tall cypress columns and wide stone-pine umbrellas, traditional funerary emblems.[36] On its floor, a mosaic pattern set in small light stone is subtly antithetical to the past use of dark stone for interior floors.[37] Without heaviness, a large rectangular floor-to-ceiling window above the entryway stairs frames a view onto nature. Light streams in and sets into relief a Latin cross of transparent glass crystal in a gilded frame. The lean lines of a travertine bench complete the architectural stage for a bronze sculpture group of a woman and a man, the space's spiritual focus, set amid architectural brightness and openness, as if to beckon hope and renewal.[38]

35 The building's proportions possess geometric harmony: 1 unit roofed, 2 units open, another unit roofed. The square building is halved, one half for entrance and stairs and the other for the open roof section. The terrace outside corresponds to the inside (walled) width. The VDK Archive, Caira_ Altplaene, 00006830, (February 6, 1961).

36 The entrance hall with its rectangular opening in the ceiling was fully designed by 1961. The VDK Archive, Caira_Altplaene, 00006861, (February 6, 1961).

37 Similar fine abstract stonework is used at Pomezia for the ceiling of the monument's cross vault.

38 The space's lightness within expresses German society's sense of hope. It contrasts and at the same time strengthens the sculpture group's message of mourning and consolation.

Fig. 53. Caira, Offenberg's modernist entrance hall with open roof and Müller-Diefenbach's "Mourning and Consolation."

Sculpture and its Message

The VDK had shifted its focus from the fate of the soldiers to the pain and loss felt by their families. This presented new architectural and sculptural opportunities, but an ideological challenge as well. Tischler had denoted his "chapel's" interior as a "sanctification space" and reserved in it a place for a sculpture with the notation: "generic character."[39] Just a few years later, the Suse Müller-Diefenbach (Tübingen) sculpture "Mourning and Consolation" (*Trauern und Trost*)[40] was delivered and set in place. It sits aside from the sightline onto a small door, which marks the grave area's

39 Tischler's models indicate "an unspecifying inscription." ("*eine allgemein gehaltene Inschrift.*") The VDK Archive, A.100–857, (October 29, 1958).

40 The VDK information sheet on Monte Cassino/Caira *Informationsblatt.*

entrance. Its place near a wall, beneath the open roof, is bathed in natural daylight. Its dark bronze against creamy light travertine catches the visitor's attention immediately. The sculpture's two figures sit upon a low, light-colored stone stage. Intensely intimate and relaxed body language seems to invite the visitor to share their space. The male figure (quite a bit larger than the female) has one leg bent, the other framing her as she kneels and nestles into his sphere, as if into a zone of tender protection as he lays a hand softly on her shoulder. Lowered faces with downcast glances bespeak the tranquility that follows quiet resignation. The two figures are neither parent and child, nor man and wife, but simply male and female. The intimacy they share shows a Germany honoring personal strength and tenderness. The sculpture states powerfully that tenderness was and remains mourning's consolation (*Trost*). The privacy of their message and their generalized style effectively beckon to all grieving families. The sculpture group gives mourning a presence, following the VDK's guideline that families as well as the living generation relate to the dead not only symbolically, but also experientially.[41]

The sculpture's forms are neither classical nor realistic. Their mass and breadth (particularly the male's) deviate from realistic anatomy. A generalizing abstracted surface treatment indicates neither clothing nor nudity. Its roughness suggests catastrophe and scarring. Their faces are out-sized and generalized, elongated and with generic short hair. The style moves toward abstraction and expressionism, as seen in other examples of contemporaneous German art.[42] Müller-Diefenbach's "Mourning and Consolation" presents a humility that is antithetical to Nazi paroles and beliefs of the pre-war period, which had so distorted the German psyche. Its whispered answer to past hubristic pride works because it is so artistically genuine. Such sculpture as part of the entry experience effectively designates this space as the cemetery's ideological focus.

41 *("Sie müssen die persönliche Beziehung der Hinterbliebenen und der lebenden Generation zu den Toten überzeugend zum Ausdruck bringen, und sollen nicht nur symbolischen Charakter haben.")* The VDK guidelines, (March 19, 1965). VDK Archive.

42 For a general overview of German sculpture of this time, see Werner Hofmann, *Die Plastik des 20. Jahrhunderts*, (Frankfurt A.M.: Fischer, 1958). See also Peter H. Feist, *Figur & Objekt, Plastik im 20. Jahrh.,* (Leipzig 1996); Christa Lichtenstern, *Voraussetzungen und Entwicklungen naturästhetischer Perspektiven in der Skulptur und Plastik nach 1945 in Marburger Jahrbuch für Kunstwissenschaft*, (Marburg: 1993).

The informed visitor recalls Käthe Kollwitz's sculpture "The Parents" at the cemetery in Vladso (Belgium) depicting herself and her husband heartbroken, kneeling before the grave of their son Peter, who was killed in WWI at 18 years of age. Kollwitz's prestige rose in post-WW II Germany.[43] She refused to dignify their son's death with meaning.[44] "The Parents" shows father and mother separated, each kneeling on its own socle, each isolated in its own inconsolable grief that has literally turned to stone. Neither of the two sculptural groups offers facile meaning.[45] Both abjure politicization, instead setting the soldier's death into an intimate private sphere. In contrast to "The Parents," however, "Mourning and Consolation" offers hope. It might signify a father's tender touch creating union between two parents. They share their grief and find consolation through it, urging life to go on.

The shift in ideology is also noticeable in a bronze shrine by Maria Luise Wilckens in the custodian's house.[46] It contains a list of the dead. Of special interest is the depiction of the Old Testament story, "The Youths in the Oven," complete with the song's text. Offenberg (again, Tischler's successor) chose the theme, saying that the text had inspired his prayers and consoled him during his own wartime experience at the front. "This prayer is Jewish poetry, but all poetry is humanity's patrimony" (*Dieses Gebet ist jüdische Poesie, aber alle Poesie ist ja Gemeingut der Menschheit*).[47] One

43 Reinhart Kosselleck, "Kriegerdenkmale als Identiätsstiftung der Überlebenden." *Identität*, (München, Werner Fink Verlag, 1979), pp. 255–276. The author points out that Käthe Kollwitz, who worked on this sculpture for twenty years, created a singular monument because rather than focusing on dying, she expressed that death was the death of an individual, and what his death meant for the surviving parents. P. 267.

44 Kosseleck ranks the work among a few others, such as Lehmbrucks "Gestürzter" (Falling) in that it questions the search for meaning. See "Introduction," p. 19 in "*Der politische Totenkult. Kriegerdenkmale der Moderne.*" Reinhart Kosseleck and Michael Jeismann eds. (München, Fink, 1994).

45 Rudy Koshar theorizes "... that Käthe Kollwitz's sculpture was also moved in part by the sense that adults had sent innocent youth to a profoundly destructive slaughter." Therefore, a sense of guilt went into its making. *From Monuments to Traces: Artifacts of German Memory 1870–1990*, (University of California Press, 2000), p. 99.

46 Little information is available about the artist. She created pieces for churches in Germany. Thomas Bindle, Arch-Abbey Sanct Ottilien, Bavaria treats her in an anthology for an exhibition in Altötting, Bavaria, 2000. EOS-Verlag (Bavaria), 2000/01.

47 Letter, Offenberg to Dekan W. Trepte, Lutheran theologian, then president of the VDK. The VDK Archive, A.100–857_offenberg, (April 14. 1965).

might impute an inspiration parallel to the placing of the Dante quote at Pomezia, a blessing for Germany's striving to return to the realm of the civilized.

The Cemetery's Grave Area

The visitor leaves the "hall of honor's" soothing travertine and enters a realm of heavy grey stonework, coming into a court where all the cemetery's design energies seem to focus. This transition resonates in counterpoint. The visitor pauses to glance back. The sculpture appears, now surrounded within ponderous monolithic posts and lintel, Tischler's signature. Two bells (from 1962) crest the open roof (another remnant of Tischler's original chapel). The visitor senses the divide between openness and tight control, a struggle between countervailing forces. In a mosaic on the court's floor, a dove bearing the olive branch intermediates between two realms as it looks back toward the sculpture. The dove, the VDK's new emblem (by Theresia Mecke-Steger of Munich), clearly declares modesty and peace.[48]

Tischler's terraced retaining walls appear as a horizontal composition in heavy stone.[49] They speak in Tischler's design language, where mass connotes eternity. A wall on the right side of the court invites the visitor to look over it. Below are two terraces full of graves.[50] Walls, pavements, and stairs are solid and powerful stone entities that meet at right angles. The visitor cannot help but sense beauty in Tischler's geometric harmony. A marker (executed in stone), at the edge of the staircase that leads to the graves on the hillside, presents the plan of the cemetery (34 grave fields) in metal relief.

48 The VDK pamphlet (*Informationsblatt*) offers no information on the artist.

49 The VDK Archive, Caira_ Altplaene, 00006866, (June 18, 1963).

50 The two lower terraces' width spans a quarter of the hill.

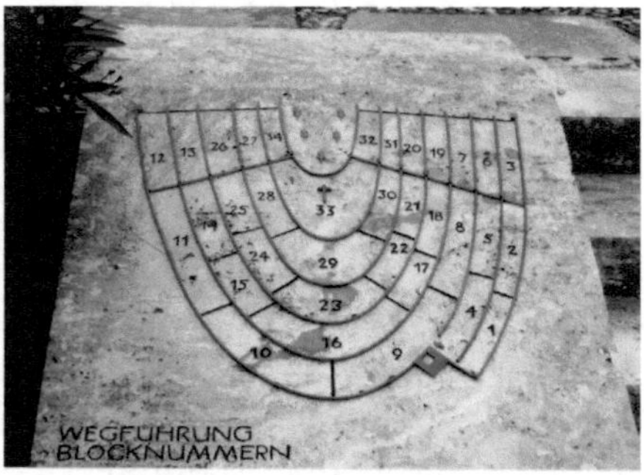

Fig. 54. Lay-out plan engraved on stone at entrance to the grave fields.

Tischler's designs display his ability to create uniqueness from each hill site. All such sites are different, and none ideal for such a large-scale burial. Tischler's solution at Caira was eight ellipsoid terraces behind eight supporting walls. Massive stonework determines its aesthetic character (as at Futa Pass, but Caira's 2,620 meters of wall is actually longer than Futa Pass's ca 2,000-meter wall).

The cemetery half of Caira's hill was transformed into six stacked half-ellipsoids and two quarter-ellipsoid terraces extending outward from the site's straight rear border, such that wider burial areas exist further from the rear border. Thus, the visitor mounting from the entrance first confronts the widest strips of graves, and climbing toward the rear border encounters progressively narrower ellipsoid terraces. The arrangement of graves along these curved terraces presented a major design challenge of which a visitor most likely will not become aware.[51] Detailed layout plans shows six double rows of graves, narrowing to five, and so on all the way down to one row near the rear border.[52] Due to curvature, rectangles could not be uniformly arranged in straight rows, so individual graves were rotated in order to fit the available space. The sizes of graves were also varied for this purpose. Individual grave space narrows but never shortens.[53] (By contrast, the US was able to obtain more ideal sites, cultivated flatlands, and therefore could follow a fixed scheme of burial patterns and erect

51 All plans for grave location from January 1959 onwards demonstrate the enormous effort that went into fitting the graves into the available space. The VDK Archive, Caira_Altplaene, (ca 1959–1965).

52 The VDK Archive holds hundreds of detailed layout plans, Altplaene.

53 The VDK Archive, Caira_ Altplaene, 00006949, (date not given).

monuments). Grassy areas and paths equal out incongruities to the eye and a uniform planting of St. John's wort on the grave rows covers these adjustments.[54] Cypresses and stone pines dot the grave fields in irregular intervals, often with dramatic effect, such as when two cypress trees frame a view over the lovely surrounding cultivated landscape.

Tischler envisioned a system similar to that first deployed at Costermano. Symbolic crosses would work in combination with flush grave markers, each bearing four names. Larger single gravestones, he argued, would turn the cemetery into one single paved area and there would be no place for plantings.[55] His plans show terraces sparsely planted in order to set his symbolic crosses into full view, distributed in regular intervals. Some grave fields were given over exclusively to planting in order to avoid monotony, and trees and shrubs were planted thickly at the peripheries. The ever-controversial issue of crosses and markers, in the end, was solved with a system where in the middle of six rectangular graves (where possible) a single Maltese Cross would be set, bearing up to six names, three on each side (the highest number permitted by the VDK).

Horizontal grave fields on a terraced hillside demanded a logical system of pathways, which formed an important part of the aesthetic character of the whole. Pathways were laid out so as not to give the impression of "packaging" that might diminish somberness. Paths called "*Weinsteigen*" (vineyard pathways) at intervals, lead across the grave area to the next terrace's wall, where stairs along its side provide access to the next higher (or lower) terrace level. Grave fields appear contiguous and the pathways have a uniform look: pebbled-cement aggregate, with wider pathways having larger stones set in the aggregate. The views from the outer pathways overlook olive trees, the surrounding landscape, and farmhouses.

54 A strip's shape changes most drastically around the turns of the half-ellipsoid. The VDK Archive, Caira_ Altplaene, 00007011 and 00007008; The VDK Archive, Caira_ Altplaene, 00007003. Dates not given. Some layout plans for grave fields are dated February 1960.

55 Letter, Tischler to the Directorate (*Vorstand*) imploring the VDK to provide names as soon as possible as to enable finishing the cemetery plans. Construction would take two years, and grave marking an additional four years. The VDK Archive, A.100–857, (January 30, 1959).

No Hilltop Monument

The VDK placed great importance on a dignified comrades' grave (*Kameradengrab*) that was a mass grave or "common grave." In Tischler's WWI cemeteries the mass grave- was the ideological and spatial center. Soldiers fougth together and died tother. At Pomezia, this lies under the monument's floor and at Costermano, under the chapel's floor. Caira's common grave as originally envisioned (in the 1958 plan) was to be located beneath a hilltop bastion-like platform surrounded by a grove, upon which a large cross would be mounted.[56] The VDK's committee of experts eliminated this bastion-like platform.[57] Tischler evidently felt the site's hilltop should be special in some sense, as he then suggested that the hilltop be an apt setting for the graves of the unidentified.[58] The VDK argued that many would be identified during reburial and should some remain unidentified, disintering them individually and reburying them as a group would be inappropriate. In principal, mass graves at prominent areas were to be avoided.[59] Tischler countered that he did not envision a mass grave but singular graves, but to no avail. His suggestion was rejected.

After Tischler's death other architects called for a monumental design on the hilltop as it seemed to "cry out" for one. A plan was produced showing a round-paved plaza accessible laterally via pathways, with a round structure at its center. "Comrades" (those whose remains could not be separated) would be buried beneath, and steles commemorating military units would surround it.[60] Nothing came of this. Even the idea of randomly placed trees was deemed undesirable, as it might be associated with a heroes' grove (*Heldenhain*). As the VDK's Christel Eulen put it: "German soldiers' cemeteries in foreign countries are messengers of

---

56 A plan of the whole design shows a projecting bastion on the hilltop. There are several versions that show Tischler's attempt to design the hilltop properly. The VDK Archive, Caira_ Altplaene, 00006955, (December 12, 1958); The VDK Archive, Caira_Altplaene, 0000681.

57 The VDK Archive, Caira_ Altplaene, 0000681, (November 20, 1958).

58 Note, Tischler to the VDK Vorstand. The VDK Archive indicates that the VDK wanted to revisit the issue of cross's base, A.100–847, (February 24, 1959).

59 The VDK Archive, A.100–857, (November 12, 1958).

60 The VDK Archive, Caira_ Altplaene, 00007084, (November 2, 1962).

our thoughts and feelings."[61] The VDK held fast to its decision that there would be nothing on the hilltop that might suggest a grove, a comrades' grave, or a monument.

The solution to the location of the comrades' grave was to install it on the terrace below the hilltop where it would share place with the unidentified, which thus did receive the special honor that Tischler had wished for them. This unassuming design became the spiritual center of the grave fields. This terrace's style is modest, consisting of a low encircling wall, flush graves, a few erect grave stones with the comrades' names, and a VDK high cross standing on an unassuming platform at the outer end of this terrace's curve.

Fig. 55. Cemetery in harmony with the surrounding landscape.

61 Speech given at the convention and exhibition of 50 years of the VDK in Berlin, 1959. Sonderdruck-Juni. VDK, 1959, p. 7: "*Deutsche Soldatenfriedhöfe im frenden Lande sind Botschafter unseres Denkens und Fühlens.*"

The planting design, modest and rhythmic, invites the visitor to embrace and experience the comforting curves of Tischler's ellipsoidal terraces.[62] St John's Worth uniformly covering the grave-fields has a calming effect. (There are no flowering shrubs). Cypresses, umbrella pines, oaks, and junipers frame the path at irregular intervals in order to to appear random, indicating the VDK's de-emphasizing strict order (after 1960). This seemingly natural order facilitates blending the cemetery into the surrounding landscape. This concept is contrary to the US vision, which isolated the cemetery through thick planting so as to protect it, as if a piece of America were inserted into foreign soil: "... American visitors immediately sense that they are on an isolated bit of American ground amidst the land of another country. Most of these national cemeteries are completely enclosed by walls of trees and exclude any potential outside visual disturbance."[63] Through irrigation, light green Saint John's wort with abundant yellow blooms supports the calm order of the design and contrasts with dark cypresses. The rear border was densely planted with natural flora. This existing condition was conveniently employed to disguise the fact that only half of the hill was made available to the VDK. Near this border all the retaining walls make an inward turn and come to an end while pathways continue and finally merge into one that meanders between the cemetery and the property line, an untended buffer zone left to wilderness.[64]

At Caira, Tischler's urge to monumentalize had been consistently thwarted.[65] His followers tried as well to emphasize Caira's hilltop, but nothing came of their efforts either. Perhaps, due to the Monte Cassino I experience, the VDK found the monument building to be a thorny issue and preferred to avoid such controversies. This reveals the VDK's insecurity with identifying or inventing appropriate memorial language, and makes it more interesting to explore the processes that lead to the mighty monument that would later crown the final cemetery at the Futa Pass.

---

62 Planting plan, the VDK Archive, Caira_ Altplaene, 00006689, date not included in copy.

63 James M. Mayo, *War Memorials as Political Landscapes: The American Experience and Beyond*, (New York: Praeger, 1976) p. 101.

64 The plant list, compared to the one of the Futa Pass, is modest. The VDK Archive, Caira_ Altplaene, 00007100, (March 25, 1965).

65 Protocol, meeting of Committee of Experts (*Sachverständigungsbeiratsitzung)*. The VDK Archive, A.100–857, (November 12, 1958).

X. The Futa Pass: New Form Makes an Appropriate Statement

> "We like to lead our dead toward the light, and bed them on heights for their last rest."
> Christel Eulen[1]

Site Acquisition

The Allies mounted a major assault against the Axis's fortified "gothic line" in the rugged Apennine mountains (1944–1945) between Florence and Bologna. Requirements for a VDK cemetery site included useable landscape, accessibility, and as with US cemeteries, a site should be located in the vicinity of actual battlegrounds. A large "consolidation cemetery" in the Apennines would accommodate the remains of the many thousands, which lay scattered. Robert Tischler traveled to Italy in December of 1956 to inspect sites the Italian government suggested, evaluating each's location and condition (*Ort und Stelle*).[2] He discovered the Futa Pass site during this trip while inspecting a site that lay below, and was attracted to the land dominating it.[3] Several sites had been under consideration, some suggested by

1 Christel Eulen, Recording Secretary (*Schriftführerin*), widow of the VDK founder Dr. Emmo Eulen (1919) and highly influential personage at the VDK, reports on her visit with the Ministry of Defense Commissariat (*Commissariate Generale*) in Rome. The VDK Archive, A.100–885, (November 11, 1957). For full quote and context see fn. 16.

2 Tischler's letter to the directorate reports that he found the Futa Pass hillside better than the land below (that had been suggested from the Italian side). His report called it "grazing land." The VDK Archive, A.100–884, (March 29, 1957).

3 Session of the directorate: The document evidences that Tischler had chosen the Futa Pass hill. The VDK Archive, A100–884, (November 3, 1957).

the VDK's office in Rome (*Dienststelle*),[4] but the Futa Pass, according to Tischler, was the only possibility in that region.[5]

A letter hints that for "historical reasons," he doubted that the land would be made available by its owner(s), whereas his Italian counterparts did not express reservations. His letter mentions that the land was used only for grazing.[6] (Unlike the US, the VDK knew that they would not receive any land that had agricultural value.)[7] He estimated it could accommodate 26,000 bodies and its only problem was the existence of a spring. One owner, he speculated, might be satisfied if he could be supplied with water by means of a water main.[8]

Several VDK members inspected and approved Tischler's recommendation.[9] However, response from the defense ministry's *Commissariato Generale*, the decision maker, was not forthcoming. Instead, other sites were offered, such as Pietramala (Tuscany). Tischler disliked it. Another site, Cesena near Rimini, was suggested and Tischler was asked to inspect it on his way north.[10] He rejected the Cesena site because it was valuable land and would be needed for housing developments in the near future and

---

4 A letter from the VDK *Vorstand* to Tischler *(Bauleitung, München)* informing him that suggestions for other sites had come from Rome on December 5, 1956. The VDK Archive, A100–884, (February 29, 1957).

5 A document bears evidence that the *Bauleitung* in *München* (under Tischler) had endorsed the Futa Pass hillsite on March 29, 1957. The VDK Archive, A.100–883, (March 29, 1957).

6 "Site does justice to the war-historical requirements (route of withdrawal of Wehrmacht over the Apennine). The site is excellent and its conditions are most suitable for the establishment of a design." *("Platz wird den kriegshistorischen Erfordernissen gerecht (Rückzugsstrasse über den Apenninen). Die Lage ist hervorragend und die Geländeverhältnisse sind für die Errichtung einer Anlage bestens geeignet.")* Photos of Futa Mountain by Tischler under Traversa/Italian B 7743; The VDK Archive, A.100–884; A 100–182, (March 29, 1957).

7 Tischler emphasizes that Italy would not give any cultivated land. The VDK Archive, A.100–884, (May 29, 1958).

8 See fn. 17 for details on owners.

9 General Hossbach reports on his viewing of the Futa Pass (and Cassino) sites with Dr. Kinder, Stiebeling, von Münchausen. He expressly mentions that Tischler had found the site. (The inspection tours of the experts' commission were a usual procedure). The VDK Archive, A.100–884, (June 21, 1957).

10 Baron von Münchhausen had looked at the site with Ricagno, made photos and maps, and had given them to Tischler in Rome, asking him to look at the site on his way back. The VDK Archive, A.100–884, (November 12, 1957).

it would likely be very costly.[11] He traversed, again, the whole region and again, found the Futa Pass the best solution.[12]

The bilateral accord stipulated that Italy would purchase the land and make it available to Germany in perpetuity. (This had been the case for sites at Pomezia, Costermano, and Caira.) The Italian site acquisition process was, as before, Byzantine and Tischler suspected that a "sticking-point" might involve money. He suspected that the Italian government might not offer a fair price and/or might not be trusted to pay as agreed.[13] Indeed, he cited the example of an aggrieved landowner at the Costermano site (near Lake Garda) who would not permit entry onto his property until he had been paid. Delays in burial proceedings caused by this incident were costly. Tischler did not let the opportunity pass to reprimand the VDK: Had he been kept apprised of the situation in Italy, such delays would never have happened.

In Kassel, with no word coming from the *Commissariato Generale*, Tischler urged the purchase of alternately suitable sites that could later be liquidated should they not be needed. Archives reveal Tischler's stressing the need to assert the "right of eternal rest" as the legal basis for Germany to erect new cemeteries.[14] Christel Eulen urged the directorate (*Vorstand*) to write to Rome and inquire forcefully about the delay. Eulen's late husband had been one of the VDK's founders, and she herself had been in its service since 1957. She soon took matters into her own hands and traveled to Rome herself to firmly demand an answer from Ricagno, the head of the *Commissariate*.[15] His excellency reminded her that he had offered a different site, Cesena near Rimini. Facing an impatient, strong-willed, and determined woman, he asked: "Why do you [Germans] always choose sites on heights, and do not accept the sites being offered?" Eulen answered: "We like to lead our dead toward the light, and bed them on heights for their last rest."[16]

11 Tischler's report. The VDK Archive, A.100–884, (November 7, 1957).

12 Letter, Tischler to Margraf, reports the results of his inspection tour in the summer of 1957. The VDK Archive, A.100–884, (August 28, 1957).

13 Ibid.

14 Session, Committee of Experts. The VDK Archive, A.100–883, (May 1, 1957).

15 Protocol by Christel Eulen of the meeting with Tischler. He emphasizes that Italy would not give any cultivated land. The VDK Archive, A.100–884, File Geländefrage, (May 29, 1958).

16 ("… *warum wir immer Plätze in der Höhe aussuchten und nicht mit uns angebotenen Plätzen vorlieb nähmen*?" Christel Eulen responded, "… *dass wir unsere Toten gerne dem Lichte zuführten und sie auf Höhen zur letzten Ruhe betteten.*") See fn. 1.

Her fortitude was apparently effective. Ricagno resolved to try (once again) to negotiate with the principal owner of the site at the Futa Pass, a widow named Contessa Gilberta Minganti.[17] Von Münchhausen had reported to the VDK's general secretary, Margraf, two years earlier that Ricagno had attempted to negotiate with Minganti, but the young Contessa was wealthy, very energetic, and a "German-hater" (*Deutschhasserin*).[18] Tischler had been correct: supplemental funds from Germany would be needed.

Frau Eulen remained firm on acquiring the Futa Pass property, and also insisted that the cemetery be of the "landscaped" type with some architectonic focus (*gärtnerische Gestaltung mit baulichem Schwerpunkt*). The VDK's design center in Munich (directed by Tischler) did not favor a mountain cemetery with main emphasis on landscaping. Tischler began to waiver. Frau Eulen reminded him that he himself had discovered and advocated for this particular site in the first place,[19] that the *Bauleitung* had previously reviewed and approved the landscaping issue and that this had driven the decision to pursue the acquisition of this site.[20] In addition, to abandon efforts now would be to lose face before the general commissariat, which by then was in renewed negotiations with Contessa Minganti.

Contessa Minganti wanted to know what made her piece of land so desirable and was informed that the battles having been fought over it would make it appropriate as an "honor site" (*Ehrenstätte*). Further, its openness would make it accessible to visitors. (Interestingly, a US party had viewed sites in the vicinity years before, and rejected them because the curvy roads would not facilitate easy access.) The VDK administration office in Rome

17 There were two other owners: Count Pontello who owned a parcel at the northern end and a Mr. Achille Cetica at the southern end. Count Pontello demanded the erection of a fence between his estate and the cemetery. Cetica wanted his cartway replaced. A cartway would interfere with the design. Letter, v. Münchhausen to Oesterlen about ways to offer alternatives to satisfy Cetica's demand. AdK Archive, Dieter Oesterlen Archive, Büro Ordner AR-Nr. 66, (June 25, 1960).

18 Note from Margraf. The VDK Archive, A.100–884, Mappe Geländefrage, (March 12, 1957).

19 Letter, Eulen to Tischler. She refers to his travel report from May 25, 1956 confirmed by him on March 29, 1957, and his positive view of the site, as well as that he had agreed that a "gärtnerische Anlage" would work. The VDK Archive, A.100–884, (February 19, 1958).

20 Ibid. See also a note that confirms that the Vorstand had planned a "gärtnerische Gestaltung" for the Bologna-Florence area.

had described the land as infertile and stony, with thin topsoil supporting very little plant growth, and hence having no agricultural value. Available photos support this argument, showing nothing but empty stretches of land covered with wild grass, an occasional shrub, and pocked with bomb craters filled with stagnant water.[21] Minganti then claimed that the site was the "most beautiful" section of her land. As negotiations dragged on, the VDK urged its office in Rome to resume its search for alternative sites, but not to let this be known.

Frau Eulen, by now the Futa Pass's persistent champion, realized that a design proposal would be needed for negotiations. On March 4, 1958 Tischler sent a design proposal to Kassel.[22] He calculated that the site could accommodate remains of 20,000 to 25,000. He wrote that there would be a "dominant structure" on the hilltop surrounded by terraces. It can be deduced that his design included a castle-like monument on the hilltop. The Directorate (*Vorstand*) had not anticipated any unusual problems with this design[23] and asked their office in Rome to try to acquire the site by all means. Tischler's design was forwarded, with the note that it was not binding.[24] One can assume that Tischler thought that by now a grandiose piece of architecture would be accepted since relations between Italy and Germany had improved. (The design proposal unfortunately cannot be found.[25]) He was proven wrong.

The silence emanating from Rome alarmed the VDK as well as all Tischler's critics. The VDK sent Seifert (interment section) to Rome to expedite matters[26] and von Münchhausen passed on to him the rumor that

21 Letter from the VDK general secretary's office to the VDK Rome office. The VDK Archive, A.100–884, (March 20, 1958).

22 Letter, Tischler to the VDK directorate. The VDK Archive, A.100–884, (March 4, 1958).

23 Report on the meeting of the directorate. The VDK Archive, A.100–884, (January 12, 1958).

24 Airmailed letter from the VDK general secretary's office to the VDK Rome. The VDK Archive, A.100–884, (March 20, 1958).

25 Italy's Department of Defense claims that it does not hold any Tischler drawings. (telephone conversation).

26 Report, VDK's Seifert on trip to Italy from April 22 to April 27, 1958. *("… Stimmen gehört von Münchhausen und Pirani aus dem Generalkommisariat, die gegen den gegenwärtigen Entwurf [Trutzburg] sprechen. Ergänzend sei geäußert worden, dass ein religiöser Anklang (etwa in Kapellenform) geeigneter sei zumal die Bischöfe von Bologna und Florenz nicht deutschfreundlich eingestellt seien.")*

Tischler's "*Trutzburg*" (fortification-like) design was not pleasing. Reportedly, Contessa Minganti would not allow such a structure to be built on her land. The Contessa had become adamant in resisting expropriation, arguing that giving up her land was not a matter of "life or death" and therefore she had a legal right to resist.[27] In addition, von Münchhausen passed on that he had learned that since the bishops of Bologna and Florence were not well disposed toward Germany, a building with religious connotations such as a chapel would be more suitable.[28]

Tischler was summoned to Kassel and informed that the cemetery would have to accommodate 34,500.[29] He again suggested asking Italy to go forward with an alternative site but Frau Eulen protested and challenged Tischler, once again, to rethink his design. Two months later he submitted a concept to accommodate 34,000 and, frustrated by the silence in Rome, repeating his request that the VDK ask the commissariat for alternative sites.[30] Eulen realized that a different site might have to be found but wanted the VDK to conduct the search in secret. In June of that year (1958) the VDK Directorate journeyed to inspect the Futa Pass site in person. It was discovered that the Italian government had offered Contessa Minganti only the equivalent of DM 3,000 (*Deutschmarks*) and that she had retained a lawyer (G.L. Gualani of Bologna) who prepared a document requesting DM 24,000 and five conditions, including the retention of hunting rights. Negotiations went forward and the VDK made an offer of DM 20,000 with a rejection of all five conditions.[31] Minganti accepted the offer.[32] On March 25, 1959 the Futa Pass site was secured.[33]

---

27 Ibid.

28 Ibid. Rumors of a castle-like Trutzburg design spread. It might have been a Totenburg form ressembling Annaberg.

29 Note on conference between Tischler and the directorate. The VDK Archive, A.100–885, Mappe Geländefrage, (April 12, 1958).

30 The VDK's Stiebeling's note to *Bauleitung* in reference to the note from the directorate. The VDK Archive, A.100–885, Mappe Geländefrage, (September 9, 1958).

31 VDK's v. Lutzau rejected all conditions. The VDK Archive, A.100–885, Mappe Geländefrage, (January 12, 1958).

32 March 25, 1959 a file note that on March 4, 1959, DM 12,000 was paid for 8,5 ha. Then on September 3, 1959, an additional 5,160 QM were purchased for DM 2.000. (The remaining 6,000 DM was paid by the *Auswärtige Amt*). The VDK Archive, A.100–885, Mappe Geländefrage.

33 Note confirming that the site was secured. The VDK Archive, A.100–885, Mappe Geländefrage, (March 25, 1959).

Final details took some time to resolve. In December 1960, von Münchausen reminded the VDK that the contract with Minganti and Count Pontello called for fencing.[34] Architect Oesterlen would design a fine fence of stone steles (*Tessiner Steinplatten*) along the site's north side. The called-for cartway on the southern border was never built as the owner (Mr. Achille Cetica), desired a street that would connect his property to the *Via del Sole,* the motorway being planned to connect Bologna to Rome.[35] Finally in 1966, Cetica was happy to accept a new water supply connection as a substitute.[36]

34 Letter, v. Münchhausen to the *Bundesgeschäftsstelle*. The Akademie der Künste, Archiv für Bauwesen (herafter referred to as AdK Archive), the Dieter Oesterlen Archive, Büro Ordner AR-Nr. 66, (December 21 1960).

35 Letter, v. Münchhausen to Oesterlen. AdK Archive, the Dieter Oesterlen Archive, Büro Ordner AR-Nr. 66, (June 25, 1960).

36 Letter, the VDK's Ehrke to v. Münchhausen. AdK Archive, the Dieter Oesterlen Archive, Büro Ordner AR-Nr. 67, (January 28, 1966).

The VDK's Search for New Talent

"A period searches continuously for its expression…where there are new spiritual forces, new forms come into being…. The Third Reich is no more…. A new democracy struggles to find its new form."
Friedrich Tamm, architect[1]

Robert Tischler was both the VDK's chief architect and architect-of-choice for Italian soldiers' cemeteries during the early post-war years. This experienced, reliable, strong-willed, secretive, and occasionally troublesome architect delivered impressive, practical designs for the cemeteries at Pomezia, Costermano, and Caira, designs that were realized using available resources, on-time and on-budget. His designs also reveal an evolution toward a new language of style and message in accordance with the VDK's developing "guidelines" (previously discussed). The VDK was evolving, as was Germany at large, although sometimes haltingly, toward a democratic society where the value of the indivual was paramount.

Following Tischler's abortive crypt design for Cassino, the VDK admonished him to reign in his fantasies. The organization was neither comfortable with his attempt toward pompous impact, nor with his attempt to provide mysticism for contextualizing soldiers' deaths. Tischler's second, more acceptable design followed, to be built at Caira, but the issue of monumental form proved vexing. For the VDK, anything bespeaking monumentality was anathematic. The VDK had been paying attention to the newer movements in design sprouting up in Germany in the 1950s, and undertook to expand its design resources. At a meeting of its "committee of experts" in 1955 in Hannover, additional architects were discussed,

1 Full quote: "A period searches continuously for its expression. The clever, the intelligent, and above all the creative ones strive to find new expression. The search is ongoing, and new spiritual forces are bringing new forms into being. The Third Reich is no more. A new democracy struggles to find its new form. It is manyfolded." (*"Eine Zeit sucht ständig nach ihrem Ausdruck. Die Klugen, die Intelligenten, vor allem die Schöpferischen sind immer auf dem Wege, diese Form zu finden. Sie suchen sie stets von neuem und so entstehen da, wo geistig-lebendige Kräfte am Werk sind, immer neue Formen. Auch jetzt. Das Dritte Reich ist vorüber. Eine neue Demokratie ringt um ihre Form. Sie ist vielgestaltig.")* Durth, *Deutsche Architekten,* p. 327.

among them Dieter Oesterlen, professor of architecture at Braunschweig's Technical University.[2]

Germany's intelligentsia, searching for new beginnings, began discourses on cultural and moral issues soon after the war's end. For example, the literary *"Gruppe 47"* stands out as a forum in which new energies and views were passionatley discussed. In the field of design, the "Darmstadt seminars" (*Darmstädter Gespräche*)[3] provided such a forum. This was founded, curated, and led by the professor of art history Hans Gerhard Evers of Darmstadt's Technical University. Prominent persons attended such as Theodore Adorno, Alexander Mitcherlich, Max Horkheimer, and Ortega y Gasset. The sessions were open to the public and were intended to reach as many people as possible. Its first session (1950) was titled "The Image/Ideal of Man in Our Time" (*Das Menschenbild unserer Zeit*) and figural versus abstract art was a subject. (Prominent art historian Hans Sedlmayr condemned abstract imagery, while Willi Baumeister, an equally prominent modernist artist, defended it.)[4] The VDK welcomed the new aesthetic styles, as evinced by Costernmano's nude male (reminiscent of Lehmbruck) and Motta St. Anastasia's, Catania (Sicily), dead nude male, the latter a splendid figural work of art.[5] Caira's "*Trauer und Trost*," figural but dismissive of the classical canon, further demonstrates the VDK's interest in contemporary movements. The 1951 session titled "Man and His Place" (*Mensch und Raum*), however, was especially relevant for the VDK's architectural re-orientation. While non-modernist traditions continued and there was no full-hearted re-adoption of "international style"

2 The VDK meeting at the Kreuzklappe Restaurant, Hannover, about using other architects. H. Thulesius (Braunschweig) and Baurat Heinemann (Hannover), both in the VDK's decision-making ranks, appreciated Oesterlen's work. At a Committee of Experts session, Heinemann recommended him. The VDK Archive, A.100–883; A 10–182, (January 14, 1955); also see the VDK Archive, Futa Pass, Gelände Ordner, A.100–884, (November 11, 1958).

3 For a list of participants see Wikipedia under Darmstädter_Gespräche: accessed June 13, 2014. For a critical treatment of these meetings see Durth, *Deutsche Architekten*, p. 359 ff. The author describes heated controversies about Germany's renewal in art and architecture and the little impact of these meetings.

4 See H.G. Evers. *Das Menschenbild in unserer Zeit,* commissioned by the city of Darmstadt and by the committee, *Darmstädter Gespräch 1950*; Neue Darmstädter Verlagsanstalt, Darmstadt 1950. Of further interest is Otto Bartning, *Darmstädter Gespräch, Mensch und Raum.* Neue Darmstädter Verlagsanstalt, Darmstadt. 1952.

5 The choice of Hans Wimmer for Costermano is an example. His figural work went toward abstraction as his public statuary in Munich shows.

or of its German equivalent, "*Neues Bauen*," newness of design was of great interest to prominent architects responsible for rebuilding Germany's cityscapes, such as Otto Bartning, Paul Bonatz, Hans Scharoun, and Egon Eiermann, who attended.[6] Thus, the time was ripe for lively discussion of future styles in the VDK's architecture. It is unlikely that Robert Tischler attended these sessions or involved himself in this movement.

It is, however, very likely that Dieter Oesterlen attended.[7] A student of Hans Poelzig, Oesterlen's formative years exposed him to Berlin's rich creative scene of the 1920s.[8] He belonged to the generation of architects who came into their own after the war. He established his reputation in the early post-war period through the reconstruction of destroyed churches and public buildings in and around Hannover using contemporary construction methods while preserving the traditional form of the building, such as the gothic *Marktkirche* in Hannover, in the heart of the city. A co-founder of the so-called Braunschweig School in Architecture (*Braunschweiger Schule*), he practiced a rational, modernist style that was antithetical to the classicism that Speer had tainted.[9] Germany's post-WW II refutation of classicism was in complete contrast to the US preference, where for governmental buildings, classical styles ruled. Oesterlen's style resonated with Germany's appetite for modernist aesthetics. Retrieval of

---

6 Some architectect who returned from exile did not find work in Germany. Neither did all new construction follow the tenets of modernity as spelled out in the movement of the "Neues Bauen" in the 1920s. For example: In a letter to President Theodor Heuss Hugo Häring, member of the 1920s "*Ring*," an association of progressive architecture to which Gropius, Mies van der Rohe, and Hans Scharoun had also belonged, expresses bitter disappointment: "Meanwhile, the coming generation will have no chance to design in the "*neues Bauen*" style. Isn't that a shameful and nightmarish situation that leaves Germany with no hope for development of a fresh, truly anti-fascist renewal?" Häring laments the splendid career offers of architect Paul Bonatz and Paul Smitthenner who had thrived during NS times. (The latter built the VDK cemetery in Bourdon, France in 1960–66). Werner Durth, *Deutsche Architekten*, p. 345.

7 According to conversation with Petra Lander, daughter of Helmut Lander, in Darmstadt, April, 2013.

8 Oesterlen worked in Berlin during the NS period and was able to support himself and his family. See Anne Schmedding, *Dieter Oesterlen, Tradition und zeitgemäßer Raum,* (Wasmuth: Tübingen-Berlin 2012), pp. 26–36. She points out that after 1945 experience in building was in demand and having been involved with designing for the armament industry and its laborers was overlooked. P. 36. Schmedding's book is a comprehensive critical treatment of Oesterlen's work with many illustrations.

9 Ibid.

the Weimar Republic's rich design heritage is an integral part of post-war German architecture, and Oesterlen's Futa Pass design is firmly rooted in 1920's expressionism (to be explored later herein).

It seems likely that Oesterlen met the sculptor Helmut Lander at the "Darmstadt seminars." Lander, in 1951, had just relocated from Weimar to Darmstadt, and according to his daughter (Petra Lander), attended the 1951 session.[10] Lander was a mosaicist whose work was based on color and abstract shape. The Oesterlen/Lander cooperation turned out to be harmonious and prodigious. Their work reflects the "zeitgeist" of that period, which included a revival of Caspar David Friedrich in painting and the revival of *Lehmbruck* in sculpture. The two collaborated on several church projects in which Lander's abstractions in stained glass boldly cover entire walls, and form an essential and integral part of the design. Oesterlen invited Lander to design a mosaic for the Futa Pass's court of honor. Lander's work was to be of central importance to the Futa Pass cemetery's aesthetics and message. Its mosaic followed the style of his powerful stained glass works, where abstract compositions push diagonally upward. Large strands interlock. Blue dominates white. Fields overlap, suggestive of overlapping ice sheets as in Friedrich's 1824 painting "The Sea of Ice" (*Das Eismeer*), an iconic painting that was to become a spiritual and aesthetic icon for the generation of Lander and Oesterlen.[11]

10 See fn.7. Petra Lander: "They all knew each other!"

11 See Joseph Beuys' three-dimensional reconstruction of the "*Eismeer*." (Sea of Ice). Regine Prange, *Das Kristallene als Kunstsymbol; Bruno Taut und Paul Klee: zur Reflexion des Abstrakten in Kunst und Kunsttheorie der Moderne.* (Olms, Hildesheim, 1991, pp. 386 and 15–16. The sharp ice-sheets pushed over each other offer an abstraction of forms that was unusal for its time. It would be a source of inspiration for the "crystalline" in the German avant-garde art of the 1920s. Lander seemed influenced by the "*Eismeer*" in his work for the *Jesus Christuskirche in der Sennestadt*, 1960.

A Radical Departure?

"…it is the most expensive cemetery that we have ever built and as far as we can see, that we will ever build. On the other hand, it is one of the largest, in any case the largest that is built through our measures."
v. Lutzau, VDK directorate[1]

"My goals were … to offer no false pathos in the usual form. My vision was to depict the grimness of the fates they suffered, and to give war no glorification."
Dieter Oesterlen[2]

Fig. 56. Futa Pass cemetery's monument.

1 For German original and context of quote see fn.21.

2 *(".. kein falsches Pathos in üblicher Gestalt. Meine Ansicht war, das Unerbittliche des erlittenen Schicksals darzustellen und keine Glorifizierung des Krieges zu geben….")* The AdK Archive, the Dieter-Oesterlen-Archive, Oest.-Ord-300, (May 26, 1969). Norbert Fischer, *Geschichte des Todes,* p. 78. "*Das national gefärbte Pathos der Zwischenkriegszeit finded hier keine Fortsetzung.*" (Nationally tinted pathos of the interwar period finds no continuation).

Commission and Construction

The cemetery at the Futa Pass is Germany's largest in Italy, holding the remains of 30,853 fallen soldiers.[3] Begun in the spring of 1961 and dedicated on May 17, 1969, it was the last built in Italy, and the most expensive.[4] Oesterlen's design belongs to those mysterious leaps onward to something new, a stroke of genius where the "Gordian knot" of conflicting views, concepts, and requirements that had consternated the VDK was solved in a single stroke. Its design addressed the VKD's landscaped-versus-architectonic controversy and the monumentality-versus-modesty issue, fulfilling the long-sought objective of creating an impressive yet appropriate monumental form with a powerful message.

The VDK's choice of Oesterlen over Tischler clearly signaled its decision to use contemporary talent of a younger generation. The cemetery appears, at first view, as a radical departure from Robert Tischler's designs. Tischler's era had passed, but in fairness, his contribution to the Futa Pass cemetery may well be his works being the challenging counterpoint that propelled Oesterlen toward newness. When seen as such, the design represents both an ending and a beginning, mirroring Germany at large.

The VDK considered asking Oesterlen to design the Futa Pass in 1958[5] but an opportunity to design at Abbeville, near Arras in France presented

3 The cemetery is reached via the *Strada Statale* Nr. 65, between Florence and Bologna. It has seventy-two large grave fields. Its maximum elevation is 952 meters above mean sea level. The wall is two kilometers long and rises at the zenith to ca. 18 meters above grade. Of 30,853 soldiers buried, more than half had fallen in the area between the 9th and the 21st of April 1945, ca. two weeks before war's end.

4 For information on the Futa Pass see: the VDK periodicals *Kriegsgräberfürsorge* Nr. 12, 1964, 140–141; *Kriegsgräberfürsorge* Nr. 2, 1965 pp. 20–21; Alexander Koch, ed., *Dieter Oesterlen, Bauten, und Planungen 1946–63*, vol. 2 (Stuttgart: A. Koch, 1964), pp. 190–197.; Dieter Oesterlen, *Bauten und Texte 1946–1991*, (Wasmuth, Berlin: 1992); pp. 84–93; Dieter Oesterlen, "Der Totenberg am Futa Pass", *Bauwelt*, Nr. 43, (1982), pp. 1770–1777; Anne Schmedding in "Die Moderne als Tradition" in *Architecture in the Age of Empire: Die Architektur der neuen Weltordnung.* Tagungsband 11. Bauhaus-Kolloquium Bauhaus-Universität, (Weimar: 2004). pp. 515–5181 Schmedding, *Dieter Oesterlen p. 113–118.*

5 Prof. Bartmann's proposal was for 34,000 dead and DM 2,600,000 in costs. Oesterlen's was for 22,300 with total costs DM 2,250,000. The VDK Archive, Die Jahre 51–58, Sachverständigen Beirat, A.10–182, (November 20, 1958).

itself.[6] When this site became unavailable the VDK offered him the Futa Pass project and specified: "… it should not be a field of crypts but rather a place of real peace with greenery, housing for groundskeepers, a reception center at the entrance, a memorial hall, and surrounding walls."[7] The offer reached him while he was on a cruise near Greece.[8] He immediately interrupted his vacation, traveled to Florence, and together with Baron von Münchhausen, visited the site. The lonely, windswept, bare mountain struck him as "sacred."[9] He accepted his commission on June 3, 1959 and presented several sketches of his design and a model to the VDK on February 17, 1960. The VDK's decision makers were immediately captivated. "The whole site is led by a clear artistic thought…The original form that Prof. Oesterlen has chosen, finds full acceptance, because it evolves completely from the basic idea."[10]

The VDK believed that monuments build on flat land were ineffective[11] and a "*Trutzburg*" had no appeal, but Oesterlen offered an appropriate monumental form that was an organic element of the whole. His ingenious plan would result in a monumentalized mountaintop, organically springing from the single essential constructional element. His sketches were

6 Letter, VDK's Heinemann inviting him to produce a design for a cemetery in Abbeville, west of Arras. The AdK Archive, the Dieter-Oesterlen Archive, Büro Ordner-AR-Nr. 77, (January 26, 1959).

7 Letter, Heinemann to Oesterlen: *("… es soll keine Gruftanlage werden, sondern eine Friedhofsanlage (Grünanlage) mit Wärterhaus, Eingangsgebäude und Ehrenmal (Halle) und Umfassungsmauern.")* The AdK Archive, the Dieter-Oesterlen Archive, Büro Ordner-AR-Nr. 77, (April 6, 1959).

8 From Oesterlen's notes. The AdK Archive, the Dieter-Oesterlen Archive, Oest-Ord-300, (April 14, 1969).

9 See fn. 1 for source.

10 Directorate's session report. The VDK Archive, A.100–885, (February 26, 1960). For view of the model see fig. 66. ("*Die gesamte Gestaltung der Anlage steht unter einem klaren künstlerischen Gedanken, der die ungeteilte Zustimmung des Gutachter-Kreises gefunden hat.… Die eigenwillige Form, die Herr Prof. Oesterlen für das zentrale Mal gewählt hat, findet volle Zustimmung, weil sie sich so ganz aus dem Grundgedanken entwickelt.*")

11 Stiebeling, "… problem of the central monument dominating the entire site was never solved well (in Lommel), flatness was difficult to master, the architect strives toward heights." *("Problem des zentralen, die Gesamtfläche beherrschenden Males, ein Problem, das in Lommel nie richtig gelöst worden sei … Fläche immer schwer zu meistern, Architekt strebe Höhenlage zu.")* Protocol, Vorstand session. The VDK Archive, Futa Pass A.100–885, (February 1 and 2, 1962).

forwarded to Rome,[12] and it seems there was no doubt that Rome would approve, as planning commenced before formal approval.

Major difficulties and setbacks arose very soon after work began (in the spring of 1961). Nature proved intractable. Flooding from torrential rains and springs cut through earthwork, and there were several landslides. These would plague the project until effective subterranean engineering has been established in mid-1966. Rocks and boulders needed to be removed. Unforeseen faults in the bedrock were discovered which necessitated footings to different depths. Local weather conditions were so harsh that work could only proceed between April and October. Making matters worse, excavation had to be interrupted whenever unexploded munitions were found.[13]

The VDK's steering committee (*Präsidium*) under Walter Trepte became anxious about problems with the site and consequent rising cost estimates. In June of 1960, costs had been estimated to be in the DM 2 million range. By July they had climbed to DM 3.75 million.[14] In August of that same year, the official estimate was DM 5.34 million.[15] (In the end the costs reached the seven million mark). Some within the ranks of the VDK expressed regrets about the choice of the Futa Pass site and questioned whether soil conditions could ever be rendered suitable.

Ancillary preparatory work included exhuming remains and storing them at various venues, including churches and train stations, in preparation

---

12 Permission from Rome came on April 12, 1961. The AdK Archive, the Dieter-Oesterlen Archive, Büro-Ordner-AR-Nr. 77, (April 12, 1961).

13 Note that says mines were to be cleared by a company named Ferrocemento, Milano. The AdK Archive, the Dieter-Oesterlen Archive, Büro-Ordner-AR-Nr. 77, (October 1962).

14 See the session of the directorate. The AdK Archive, the Dieter-Oesterlen Archive, Büro-Ordner-AR-Nr. 66, (February 17, 1960).

15 In the spring of 1960 the sum was estimated at DM 3.60 million. Letter, Heinemann to Oesterlen. The AdK Archive, the Dieter Oesterlen Archive, Büro-0rdner AR-Nr. 66. In the fall of that year the cost estimate was DM 5.34 million. Oesterlen to engineer Prof. Jelinek (soil specialist) (September 16, 1960), the AdK Archive, the Dieter-Oesterlen Archive, Büro-Ordner-AR-Nr. 77. However, unofficially it was known already then that the budget would be DM 6.6 million (because of technical difficulties). Note from Oeterlen (March 30, 1966). The AdK Archive, the Dieter Oesterlen Archive, Büro-Ordner AR-Nr. 67. Costs finally reached the 7 million DM by January 1946. (In the VDK Archive, cost spiral can be traced under A.100–885); see fn. 28 below.

for reburial. This was a pressing logistical problem.[16] The VDK began considering the idea of a simple crypt, a solution that could be finished by 1962 and would keep expenses in check.[17]

The VDK's *Präsidium* "went into crisis mode" when they learned from the on-site Philip Holzmann company that costly, complicated underground buttressing would be needed. In meetings on January 1 and 2 of 1962 the *Präsidium* questioned the wisdom of a mountainous site for the individual reburial of 35,000 remains. Had there been compelling reasons for selecting this site, other than the late Robert Tischler's leadership and enthusiasm?[18] President Trepte admitted that he'd had reservations from the outset, but that Oesterlen's design had seduced him. He bemoaned the facts that the cost estimate had not been furnished for the first half of the year and that the need for preliminary soil research had not been recognized earlier. Again, while some doubted the wisdom of proceeding, other VDK members remained passionately in favor, reminding others that it would have been impossible to find cultivated land. The VDK would need to "make do."[19] Again, a crypt building was reconsidered and dismissed because crypt burial remained controversial.[20] Trepte acquiesced, the *Präsidium* reluctantly acceded to this reasoning, and work proceeded.[21] The VDK retained an independent soil specialist who reported that the terrain could be made suitable for the project

16 Exhumations were begun on October 2, 1961 and remains were stored temporarily in Traversa, about 1.5 km away. The VDK Archive, Sitzungsberichte, Futa Pass, A.100–885, (October 2, 1961).

17 Stiebeling's remark. The VDK Archive, Sitzungsberichte, A.100–885, (May 23, 1960).

18 Protocols of the directorate's meeting reveal that initially a plan existed to use only half the mountain as at Monte Cassino for 18,000, but then it was decided that it the whole mountain would be used for a cemetery for 35,000. Directorate's *Vorstand*) session. (The VDK Archive, Sitzungsberichte, A. 100–885, (February 1 and 2, 1962).

19 Protocols of the directorate's session. The VDK Archive, A.100–885, (August 8, 1960).

20 For the passionate controversy about crypt versus earth burial, see chapter IX under: Cassino (Unrealized): Tischler's Fallback.

21 The session of the presidency. Directorate chair (v. Lutzau) advocating for the project counseled "…it is the most expensive cemetery that we have ever built and as far as we can see, that we will ever build. On the other hand, it is one of the largest, in any case the largest that is built through our measures." ("*…es sei der teuerste Friedhof, den wir jemals gebaut haben und soweit es zu übersehen ist, den wir jemals bauen werden. Andererseits sei er einer der größten, jedenfalls der größte, der durch unsere Maßnahmen geschaffen wurde*.") The VDK Archive, A.100–885. Neue Gestaltungsvorschläge, Futa Pass, (August 12, 1960).

with proper engineering.[22] Ernst Cramer and his partner Doré, prominent Zürich-based landscape architects, and their Italian representative Ermes Crescon were retained to supervise construction.[23] To control costs, it was decided to use only indigenous plants, grass, and shrubs.[24]

The cemetery's official beginning of construction (*Richtfest*)[25] was held on November 23, 1963. By spring of 1964 the estimated costs had risen to DM 6.6 million and were expected to rise further.[26]

Landslides continued to plague construction. At one point, Oesterlen changed the course of a wall section in order to ground it on bedrock.[27] Costs kept increasing[28] and by May of 1965 funds were exhausted.[29] The Holzmann company was put in charge of on-site project management.

---

22 Futa Pass Friedhofsbau report from Jelinek (saying that soil conditions would allow construction). The VDK Archive, A.100–885, (January 30, 1962).

23 In a hand-written letter from Walter Rossow to Oesterlen, Ernst Cramer and Creston were given the construction supervision authority. The AdK Archive, the Dieter-Oesterlen Archive, Büro-Ordner–AR-Nr. 77, (August 8, 1961).

24 This was not to be the case, as there were extensive plant lists. See below under Planting Design, pp. 358–362.

25 Following German tradition, the completion of a building's structure to the point that allows covering the roof is celebrated. Oesterlen sent a speech to be read aloud. Food and drink were served to all. The AdK Archive, the Dieter-Oesterlen Archive, Büro-Ordner-AR-Nr. 77, (November 23, 1962).

26 See Oesterlen's budget. The AdK Archive, the Dieter-Oesterlen Archive-Futa Pass-Kosten, Büro-Ordner-AR-Nr. 67, (March 25, 1964). The presidium had agreed to a projected total cost of 7.079.00 already in January of 1964. See fn. 28.

27 Letter from Bournot (landscape architect and plant specialist) to Oesterlen about the latter's change of the course of the wall. The AdK Archive, the Dieter-Oesterlen Archive, Büro-Ordner-AR-Nr. 74, (May 2, 1962).

28 By 1960 estimated costs were in the 4 million range, by mid 1960 they had risen to DM 5,237,000. The AdK Archive, the Dieter-Oesterlen -Archive, Büro-Ordner-AR-Nr. 67, (15.6.60). Final cost estimate of DM 7, 079,000 (which would prove correct) was approved by the VDK's presidency on January 30, 1964. Protocol, February 13, 1964, VDK Archive A.100–885. This document notes rising labor costs in Italy having risen 44 percent between 1961–63. The most expensive items would be the tombstones, at DM 1,246,677. The AdK Archive, the Dieter-Oesterlen Archive, Büro-Ordner-AR.-Nr. 74, (March 3, 1964).

29 See note at the AdK Archive, the Dieter-Oesterlen Archive, Büro-Ordner-AR-Nr. 74, (May 1, 1965).

Such evident dedication testifies to (the Federal Republic of) Germany's renewed and growing technical and financial resources, as well as to its energies at the beginning of the 1960s.[30] At Caira, Italian companies had been used, but at Futa Pass the major share of the costs were paid to German companies. All construction materials (except stones) were transported to the site from Germany, exempt from the customs tax, per the 1955 accord.

Form is the Message

Although the VDK specified a "landscaped site" cemetery, it is Oesterlen's architecture that strikes the visitor immediately. From the *Strada del Sole*, its bold triangular form is seen reaching toward the sky and holding fast against the winds, and often, low-hanging clouds. Once having arrived at the site, the visitor finds natural growth, with the cemetery itself hidden inconspicuously. Its parking lot may host travellers running the road between Florence and Bologna, or perhaps a group of motorcyclists, but few of these road tourists actually visit the cemetery itself. There is only a small sign next to a modest staircase to inform one that a German soldiers' cemetery lies beyond.

The visitor passes a small court and enters the cemetery proper through an inconspicuous entrance squeezed in between a wall and the reception building so as to allow only two abreast to enter. Its wrought iron gate (by Fritz Kühn) is severe by design. One treads the footpath of gravel framed initially by large stones slabs. A modest wall rises on the right-hand side to keep company. The sound of gravel under foot is heard. The visitor walks onward, gaining height. The path makes a slight right

30 The Philip Holzmann company did the construction of the spiral wall, and the Mannesmann company established an extensive drainage system. Payment to Holzmann of DM 2,007,000 was approved. The VDK Archive, Futa Pass A.100–885, February 1 and 2, 1962. In the end, the Holzmann company received DM 3,207,999,98 and Mannesmann DM 247,031,86. The AdK Archive, the Dieter-Oesterlen Archive, Büro-Ordner–AR-Nr. 67, (June 10,1965). This document lists all the different companies and personalities. involved and the compensation for their work. Oesterlen received DM 165.080.02. The ca. 10,000 granite tombstones made by the Belgian company Renier cost DM 1.246,677,39.

turn, and then another. Below, terraces packed with marker stones begin to reveal themselves. The wall accompanies on the right, strengthening its presence. Another right turn lies ahead and the wall beyond has disappeared around the corner. The visitor proceeds. The vista broadens into grassy fields of flush grave markers, trees at the cemetery's far away edges, and the distant, severely picturesque Apennines. Past the turn, the path and its companion wall resume their course.

Fig. 57. Pathway accompanying the spiral wall, looking west.

Oesterlen insisted on using gray-green *pietra serena* quarried at nearby Firenzuola.[31] Stones are of unequal size and their surfaces are purposely left rough-hewn. Their interstices are wide, an aesthetic feature enabled

31 "The VDK suggested the use of *Nagelfluh* stones from the unfinished Pinzano project but Oesterlen refused. "I hold it absolutely necessary to use indigenous stones." (*… ich halte ich es für unbedingt notwendig, dass ein heimisches Material zur Verwendung kommt.)* The AdK Archive, the Dieter-Oesterlen Archive, Oest.-Ord.- 300, (October 1, 1962). Oesterlen undertook several trips to choose the stone himself. Author's interview with Oesterlen's son Friedrich.

by extensive subterranean engineering. Some jut forward to cast irregular expressive shadows on the wall.

The visitor notices angularity, a signature Oesterlen aesthetic principle, in both small and grand scale. His wall is formed in linear segments, shorter and longer, joined at unpredictable angles. With angular changes of direction, it wraps itself around the mountain, taking on height as it climbs, ever leaning into the land it supports, forcing and holding masses of earth that harbor massive fields of graves.[32] Toward the zenith, it morphs into a towering monument, a high stylus-like form atop the mountain. Oesterlen described it as an "endless spiral" (*unendliche Spirale*).

Fig. 58. Spiral wall climbing aroun the mountain.

At the entrance, the visitor first notices massive, stacked walls, none of them horizontal. Treading the path, the visitor has a sense of being on a pilgrimage. The sequencing of forms and spaces amidst voids and solids

32 In a letter to Jelinek he refers to his wall as the "unending spiral" "*unendliche Spirale*." The AdK Archive, the Dieter-Oesterlen Archive, Büro-Ordner-AR-Nr. 77, (February 27, 1960).

presents a dynamic fragmentation, which triggers a sense of anticipation. Gazing up at the zenith, the visitor notices the monument's triangular wall magically altering its shape. A slender tower appears as a sharply pointed cutting stylus, or a broad-based triangle rising to dominate the view. This constant metamorphosis of aspect and sharp angularity works in counterpoint to the grave fields' gently sloping calm and creates a restless dynamic. Its seeming ability to change shape from different vantage points reminds one of history in uproar, and matters not finding rest.

Fig. 59. Stacking of walls, the aspect changing with each step.

Near the zenith, the wall becomes much taller as it continues its steep sweep upward in straight diagonal toward the sky (reaching 18 meters). It turns upon itself in a convulsive lurch and then, as if by surprise, drops in total verticality leaving an open gap in the court it has created. This gap, closed with heavy steel steles stubbed to low height that suggest defensive fortifications, is the only visible element with martial overtones. It accentuates the wall's sudden vertical drop.[33]

33 See fig. 62. The sculptor was Fritz Kühn from East-Berlin.

The large court, open to the sky, is the cemetery's ideological center. It is eerily empty except for a powerful mosaic covering its floor and continuing, scaling the wall. The visitor mounts via broad steps, having completed a two-kilometer pilgrimage along the spiral path.

Oesterlen also provided three direct paths that ascend directly to the zenith. The one most taken begins near the cemetery's entrance, where the zenith is seen as a sharp stylus. As the zenith is approached by this more direct route, constricting and expanding spaces and changes of direction manage the visitor's experience in a way that recalls Tischler's legacy.

Fig. 60. Shortcut, pathway access direct to memorial. The image brings out cubistic abstraction and sculptural quality of Oesterlen's design.

A perpendicular change of direction reveals narrow stone stairs encased tightly by walls and leading to a forecourt, above which the huge triangular wall looms high.

Then, broad steps lead to a space that widens into another forecourt, offering relief from constriction and magnificent views over the surrounding landscape. Only after yet another turn and another climbing of broad steps does the visitor reach the court of honor.

Fig. 61. Forecourt before towering wall, view showing abrupt drop-off, and metal railing above crypt entrance.

No matter which route the visitor takes, the court's openness communicates the emptiness these thirty thousand left behind. The abrupt perpendicular drop of the wall, according to Oesterlen, was to give tangible form to the sudden ending of young lives. "In the court of honor, enclosed by writhing walls and clouds passing above, the visitor is very much alone."[34]

34 *("Im Ehrenhof angelangt, ist der Besucher allein, umschlossen von den immer höher ansteigenden Mauerscheiben and den darüberziehenden Wolken.")* The AdK Archive, the Dieter-Oesterlen Archive, Oest-Ord.-300, (October 1, 1982).

Fig. 62. The court and its emptiness.

The human zone lies between the shy's vastness and the stone's weightiness. The visitor encounters the cemetery's spiritual message most intensely when standing alone in the empty court of honor. Enclosed by Oesterlen's wall, unsettled by the power of Lander's mosaic, gazing under the restless open sky, the visitor finds no comfort, no celebratory polished marble, no stone reserved for noble themes, no figure that bespeaks consolation. The visceral message is that death is a void that cannot be filled.

Fig. 63. Court and Lander's mosaic, detail.

Helmut Lander's Mosaic. The Power of Abstraction

Helmut Lander's powerful mosaic (1962–67) covers the floor and parts of the wall.[35] Stones of different kinds and colors form two bolts that run across the court's floor.[36] Treading on them is unavoidable. Irregular, angular abstract shapes echoing the architecture join in various configurations. Some stones are assembled into larger fields and form zigzagging coherences. Like white arrows, they shoot across the floor, pulling the visitor forward, and then leap up the high wall in dynamic upward diagonals, wrapping over its narrow edge, as if clinging on over the deep. In

35 Note acknowledging completion of Lander's mosaic. The AdK Archive, the Dieter-Oesterlen Archive, Büro Ordner-AR-Nr. 80, (August 31, 1965).

36 A detailed list of the twenty different stone types from different quarries was needed. Sizes and costs are given. Total costs were DM 103,889.61. The AdK Archive, the Dieter-Oesterlen Archive, Büro Ordner-AR-Nr. 80, (October 21, 1965).

expressionist fashion their diagonal push challenges the wall's perpendicularity. Those mosaic stones on the wall are more sculptural, revealing that Lander's work is rooted in cubism.[37] Some bulge out from the surface as if driven by an invisible force.[38] This thickening of the stones increases where one bolt leaps over the narrow edge and into the abyss. Oesterlen supported Lander in his insistence on perfection and would give him full credit for his decisive contribution to the artistic and expressive effect of the court of honor.[39]

Lander named his mosaic "Bolts of Lighting."[40] This fits into Oesterlen's expressive intentions: "A flame-like ornament from black, white, and grey natural stones is like a last rearing, working with his own architectural expression … through the sudden rise and fall of the spiral wall, the sudden end of a soldier's life in war is given form."[41] (One critic interpreted its lightning forms as SS [*Schutzstaffel*] symbols.)[42] The mosaic is much more than a decorative addition. Working with the architecture, it is an outcry of

---

37 Lander to Firma Holzmann, letter with a plan: "The mosaic needs a suitable relief surface through the tilting of the tiles." ("*Das Mosaik soll durch Kippen der Platten eine reliefartige Oberfläche bekommen*" *Im Frühjahr 64 soll es gebaut werden*). The AdK Archive, the Dieter-Oesterlen Archive, Büro Ordner–AR-Nr. 80, (August 23, 1963).

38 Letter from Lander to Firma Holzmann. "… the surface of the slate must, in its basic structure, be more 'rocky.' Some stones should be taken out and replaced with more 'moving' ones. The others should have uneven edges und be partially polished." *("die Schieferfläche muss in der Gesamtstruktur 'felsiger' werden. Dabei sind bei dem bereits angesetzten Material einige Platten herauszunehmen und durch bewegtere zu ersetzen. Die übrigen Platten müssen durch Beschlagen der Ränder und teilweises Abbossen dem Gesamtcharakter angepasst sein.")* The AdK Archive, the Dieter-Oesterlen Archive, Büro Ordner –AR-Nr. 80, (September 28, 1964).

39 Oesterlen's letter to Firma Holzmann telling that Lander was not content with their work, another kind of cement was needed, since "the entire effect and the perception of the site depends on the mosaic." *("Da das Mosaik die gesamte Wirkung [ausmacht] … [und] auch Beurteilung der Anlage [davon] abhängt.")* The AdK Archive, the Dieter-Oesterlen Archive, Büro Ordner–AR-Nr. 80, (May 12, 1965).

40 *„Blitzstrahle."*

41 *("Ein flammenartiges Ornament aus schwarzen, weißen, grauen Natursteinplatten ist – einem letzten Aufbäumen gleich – über den Ehrenhof und seine steilen Mauerscheiben hinweggezogen und unterstützt den Versuch, mit dem jähen Aufstieg und Abfall der Spiralmauer dem unvermittelten Ende des Soldatenlebens im Kriege Gestalt zu geben.")* The AdK Archive, the Dieter-Oesterlen Archive, Oest. Ord–300, (October 1, 1982).

42 Dr. Martin Huzel inquired, in a letter to the VDK, whether the two high bolts were SS (*Schutzstaffel*) emblems. The AdK Archive, the Dieter-Oesterlen Archive, Büro Ordner–AR-Nr. 80, (May 23, 1965).

distress and outrage over the fate of so many lost. The mosaic's passion and intensity speak of Lander's experience with war and imprisonment.[43] But of more important impact on his work was his inconsolable distress: "… the feeling of guilt my generation will bear until their end of our lives."[44] Until his death in 2013, he lived with the horror of NS crimes, and his other works show his unrelenting search for artistic language to honor suffering.[45] In this most daring and personal artistic expression in a German soldiers' cemetery, Lander's passionate forms far exceed calculated patterning.[46] They flow forth from his frightful war experience. They testify to his generation's terror resulting from Hitler's maniacal visions. They unleash uproar, outcry, and then revolt. In his case, one certainly cannot apply the label "suppression" of historical truth. Lander donated a sizeable portion of the honorarium he received for his mosaic at Futa Pass back to the VDK.[47]

There is no figural sculpture at Futa Pass. From the middle forecourt below the court of honor, steps lead downward to a crypt. Here, the remains that could not be separated are entombed behind a wall that carries the inscribed names and dates of death. Mutilated remains are contained in specially designed VDK sarcophagi, stackable ten high.[48] Three hundred

---

43 Lander experienced war for four years as a young soldier, then was a prisoner of war.

44 (*"Dieses Schuldgefühl wird meine Generation wohl bis zum Lebensende mit sich herumtragen müssen."*) Quote is the end of a longer statement: "Only toward the end of the war the consciousness for the guilt and responsibility for the death of millions of people, undescribable suffering and injustice grew." (*"Erst zu Ende des Krieges wuchs das Bewusstsein für die Schuld und die Verantwortung für den Tod von Millionen Menschen, unbeschreiblichen Leid und Unrecht."*) from Exhibition catalogue: Helmut Lander, Malerie, Zeichnung, Plastik, Kunst und Architektur 1944–1994, Kunsthalle Darmstadt, 1994; <www.Helmut-Lander.de>, accessed March 15, 2018.

45 Lander's war experience determined the themes of his considerably large sculptural work. A sculpture from the 1960s called "Imprisoned" shows a nude male with feet locked into a heavy block. Others figures have their hands tied or are hung as on gallows. There are destroyed faces. The feeling of helplessness prevails. Seen while visiting Helmut Lander's studio in Darmstadt, made possible by courtesy of Lander's daughter, Petra Lander, April 2012.

46 The glass mosaic that is concurrent with the Futa Pass is in the *Zwölf Apostelkirche* in Hildesheim, built in 1964–67. There, Lander's thrusts become more perpendicular and simplified.

47 The honorarium for Lander was DM 27,000. Lander donated 3000 DM to the VDK. The AdK Archive, the Dieter-Oesterlen-Archive, Büro Ordner-AR-Nr. 66, (April 21, 1964).

48 Letter, VDK's Ehrke to Oesterlen. Proposal for the stacking of sarcophagi in the Kameradengrab. The plans date from February 7, 1960. The plans have a black border

ninety-six are distributed into 270 sarcophagi, an unintended "mass burial" as there was space for only that number of sarcophagi.[49]

In a separate crypt, accessible from the court, a few tombstones variously decorated are placed. These were sculpted by German prisoners of war and were salvaged when the temporary cemetery at Cervia was dissolved and the remains removed.[50] Futa Pass's main crypt is light, in the form of a rectangular horizontal with a flat ceiling. Its wide entrance admits light freely through a fine ornamented wrought iron gate (by Kühn). Although one descends into it, its elevation is still above ground. Oesterlen made it a point to admit light through three windows. This would prevent a mystical atmosphere in the tradition of Tischler. Set at eye level, the windows offer a sweeping view toward the west across the Apennines. The landscape remains an integral part.[51] There is solemnity, but the sense of immuration the visitor experiences, for example, at Quero, is absent. Remains rest not beneath a paved floor, but behind a wall that late afternoon sunlight shines upon, illuminating their names inscribed on two of the crypt's walls.

Fritz Kühn's Crown of Thorns. Suffering in Steel and Iron

Oesterlen and Lander managed to collaborate with Fritz Kühn (of Grünau, East Berlin) on other projects. E.g. in Oesterlen's iconic *Christuskirche*

---

like a letter announcing a death. Sarcophagi are 28 cm high, 68 cm long, and designed such that they could be stacked. They would contain the remains of soldiers that could not be separated. Ehrke specifies that the crypt should be dry. This crypt would be completely closed. The AdK Archive, Pläne, Arch.-Nr. 66, Images 8268/697, (August 24.1964).

49 Letter, VDK's Ehrke to Oesterlen. The AdK Archive, the Dieter-Oesterlen Archive, Büro Ordner-AR-Nr. 66, (September 18, 1964).

50 Letter, v. Münchhausen to Oesterlen. The VDK had promised the families of the Cervia dead that the monuments done by prisoners of war that had artistic and handcraft value would be used at the Futa Pass, but not the grave markers because on grave field markers had to be uniform. The AdK Archive, the Dieter-Oesterlen Archive, Büro Ordner-AR-Nr. 66, (May 13, 1960).

51 To conceive of landscape and building as one was at the center of Oesterlen's conception. *("Bau und Landschaft sind im Laufe der Zeit miteinander verwachsen.")* The AdK Archive, the Dieter- Oesterlen Archive, Oest.-Ord.-300.

in Bochum both Lander and Kühn made a cross. The *Stadtkirche* in Jever 1960 has a large cross by Kühn. Kühn went through many variations for the Futa Pass's crown of thorns, and there was a lively exchange in correspondence, which included the VDK.[52] Light, painting rectangles on the crypt floor, illuminates Kühn's large steel abstracted crown of thorns, which has a rough surface (*gesandelt)* and had been galvanized (*verzinkt*). Changing light brings this sole sculpture into focus in varying ways.[53] "It does not symbolize just Jesus's suffering and death, but the suffering brought about by the whole war."[54] The cemetery offers no promise of resurrection. Its sole biblical passage, "Blessed are they that mourn, for they shall be comforted" (*Selig sind, die da Leid tragen, denn sie sollen getröstet werden*) from Jesus's "Sermon on the Mount" is inscribed on the crypt's far wall. Christel Eulen and the VDK's president Trepte (a retired Lutheran theologian) selected this biblical passage.[55]

Oesterlen insisted on displaying Kühn's work despite restrictions and on its exportation to the West, which were intensified following the erection of the Berlin Wall in 1961. The two artists had to use ruse to circumvent cross-border restrictions.[56] It was so heavy that it had to be divided

52 Preparatory drawings and correspondence can be seen also in a private archive of Ernst-Kühn Gesellschaft, Grünau, Berlin. Files show that the GDR government kept a large part of the accounts received for his work. (Courtesy of Helgard Kühn.)

53 Tischler had used this motif in Merseburg after WW II according to a statement by Christian Fuhrmeister.

54 Note on the meeting where Oesterlen showed Kühn's design for the crown of thorns (December 15, 1964). It found universal approval (*"...diese Dornenkrone aus Stahl beinhaltet nicht nur das Leiden und den Tod Christi, sondern, wie Prof. Oesterlen sich ausdrückte den ganzen Krieg!"*) The VDK Archive, A. 100–883, (February 16, 1964).

55 Note from the VDK's Ehrke. The AdK Archive, the Dieter-Oesterlen Archive, Büro Ordner-AR-Nr. 67, (March 30, 1966).

56 The finished crown left East Berlin on November 27, 1965. It was so heavy that it needed to be transported in two parts and assembled on-site. Documents allow us to trace the saga. For example, Kühn needed special permission to travel to Hannover on July 27, 1964. (In accordance with the Rule of July 19, 1965, dealing with works of art (*Deutschen Innen-und Aussenhandel*) the work had to be transported by a GDR company to Konstanz, and from there by a West German company to Italy). The gates to the crypt, as well as the entrance gate, the railing for the court of honor, and two candleholders (the total weighing 3,365 kilograms) were also transpored. The West German company reminded the East German company that the manifest must be listed as wrought iron art works and that the words "soldiers'cemetery" needed to be avoided, else it could not cross the border. Fritz Kühn, a successful East-Berlin artist who received public commissions, was allowed to travel to the Futa Pass on September 7,

into parts and reassembled on-site.[57] Germany's division into two adversarial governments could not, in the end, impede the flow of art.

Fig. 64. Crypt interior with Kühn's "Crown of Thorns."

In the beginning the VDK had wanted groups of crosses at 100-meter intervals.[58] They would jut upwards from the rim of the wall and form silhouettes noticeable from afar. Early sketches show these groups of crosses of differing size, reminiscent of the VDK emblem of five crosses (and also reminiscent of Tischler's "symbolic crosses").[59] Very soon this concept

1965 and viewed a wood model for the crown in place. The AdK Archive, the Dieter-Oesterlen Archive, Büro Ordner-AR-Nr. 66.

57 Letter, Ernst Cramer to Kühn. Sculpture must be divided into two pieces (500 kg each) because it cannot be transported as one to the construction site. On-site, it was reassembled and transported as a whole to the crypt. The AdK Archive, the Dieter-Oesterlen Archive, Büro Ordner-AR-Nr. 76, (July 3, 1965).

58 These groups of crosses were discussed during a special session with Oesterlen. The VDK Archive, A.100–885, (February 17, 1960).

59 Design of sets of three crosses that rose above the wall at ca.100-meter intervals. All together there were twenty sets. The AdK Archive, the Dieter-Oesterlen Archive, Büro Ordner–AR-Nr. 80, plans, (February 11, 1960).

was abandoned and the cross motif evolved into a simple Latin cross in *Pietra Serena*, at regular intervals (67 altogether) affixed to, and forming an intrinsic part of the wall. The elegant simplicity of their final form is the result of an intensive and collaborative design effort.[60] Oesterlen consulted with Lander to perfect its form.[61] Similarly, benches were envisioned from the very beginning. These also evolved in form, from detached benches to niches with benches to single stone slabs jutting from the wall at regular intervals serving as seats, suitable for a single person to sit and ponder amidst the eternal winds.

To express his vision, Oesterlen's primary vehicle was architecture, not planting.[62] He wrote: "The structure's emphatic power will be affected by the wall, and not by the planting."[63] Within the angular, ever-shifting aesthetics of his design, sixteen modestly large and perfectly round pools of still water lie on the terraces amidst the graves, reflecting the sky. Born of necessity, to collect rain and snow melt run-off for irrigation, they are also a brilliant design feature. Their roundness, the cemetery's sole non-linear feature, works in counterpoint to angularity. Still waters, horizontal amidst canted fields and towering verticality, symbolize life.

60 Protocol of the trip to the Futa Pass site. Lander was present and to judge from the style of three different forms of crosses in thick pencil drawings, these drawings seem to be by his hand (the AdK Archive, the Dieter-Oesterlen Archive, Büro-Ordner AR–Nr. 80). The VDK Archive, A.100–885, (July 22, 1962). Plans by Oesterlen from April 29, 1962 and from June 19, 1962. Both are at the AdK Archive, the Dieter Oesterlen Archive, Büro Ordner –AR-Nr. 80.

61 One cross would have been inserted into the wall. It was rejected. Further, a design with three crosses was rejected by the VDK because of its proportions and its "primitive and fractal-like composition", *("primitive and gestückelte Zusammensetzung.")* A cross, attached to the wall, with the horizontal arm above the vertical was carried out. AdK Archive, the Dieter Oesterlen Archive, Büro Ordner –AR-Nr. 80, (February 15, 1960).

62 For full quote see fn. 10. *("...Die eigenwillige Form, die Herr Prof. Oesterlen für das zentrale Mal gewählt hat, findet volle Zustimmung, weil sie sich so ganz aus dem Grundgedanken entwickelt.")*

63 Oesterlen explained his plans to the VDK in a statement that makes clear that architecture would rule supreme: *("Heraushebung des Bauwerks als Markierungspunkt. Haupteindruckskraft wird von den Mauern und nicht vom Bewuchs ausgehen.")* The VDK Archive, A.100–885, (February 17,1960).

From Windswept Wilderness to the Mountain of the Dead

"… like a patterned gown splayed upon earth's jagged landscape."
Eberhard Schulz[1]

Fig. 65. The cemetery embedded in the Appenine Landscape

For the VDK, the cemetery at the Futa Pass was to be their final in Italy. There would be no "follow-up" projects, and thus it was of primary importance that it be able to accommodate a huge number of remains in accordance with their standards of decent respect. On February 17, 1960

1 Eberhard Schulz, *("…ein Kleid aus unregelmäßigen Feldmustern hängt so über der Geometrie der Erde, die auch nicht regelmäßig ist."* "Der Totenberg am Futa Pass.") *Frankfurter Allgemeine Zeitung,* (7 Juni 1969) Nr. 129.

Oesterlen presented his one-and-only model. Inspired by the landscape, his design treated the mountain as a coherent landscape sculpture.[2]

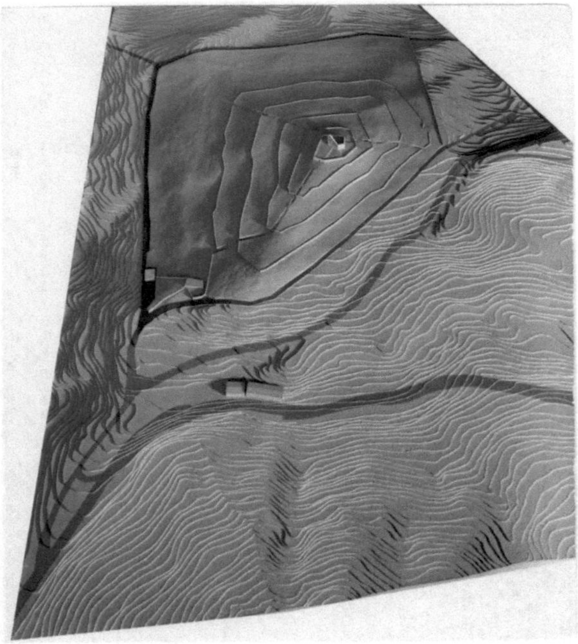

Fig. 66. Oesterlen's Model showing an inseparable relationship between land and architectural form. Note the fundamental abstract conception.

Natural topography would be smoothed and terraced for grave fields as "big steps in the landscape."[3] Landscape and architecture would be inseparable. Oesterlen's architect Loebnitz (1959) began by dividing the land into six blocks of grave fields.[4] Architect Schumann's (March 30, 1960) plan would accommodate 33,760 burials. Grave fields would have a co-parallel orientation (southeast to northwest), not following the spiral wall's course, with one square grave field touching upon another, each

2 Letter reiterating his document of November 2, 1960. The AdK Archive, the Dieter-Oesterlen Archive, Oest-Ord.-300, (October 1, 1982).

3 Ibid.

4 See *Vorentwürfe* (first drafts) plan from Oesterlen's office. The AdK Archive, the Dieter-Oesterlen Archive, Vorentwürfe, (February 10, 1960).

framed distinctly with stone rims.[5] There would be almost no space left for planting. Their plan could only be characterized as "claustrophobic," and it became obvious that a very experienced landscape architect would be needed. (Landscape designs for Tischler's cemeteries had been created within his office, and Tischler himself was also a landscape architect.) In the end, Oesterlen's collaboration with landscape architect Walter Rossow resulted in taking possession of the mountaintop in a way that harmonizes with the surrounding Apennines.[6]

Landscape Design: Landscape Architect Walter Rossow and Horticulturist Helmut Bournot

Rossow was well known and successful, having created the landscaping for the prestigious *Hansa-Viertel*, Berlin's modernist complex designed by internationally prominent architects.[7] He also had experience with soldiers' cemeteries in Germany and at that time was designing two cemeteries in France.[8] Oesterlen sent him material,[9] a study of the vegetation zones and native plants, produced by a professional botanist,[10] as well as his design plans (that would not change). Rossow accepted the offer and

---

5 The architect was Werner Schumann, who worked in Oesterlen's office from 1956 to 1959. Oesterlen, *Bauten*, p. 264, gives a complete list of Oesterlen's office staff.) For the plan see The AdK Archive, the Dieter-Oesterlen Archive, Vorentwürfe (March 30, 1960).

6 Rossow sent a photo of the finished cemetery to Oesterlen. The AdK Archive, the Dieter Oesterlen Archive, Pläne, Futa Pass, (April 18, 1968). See Fig. 65.

7 Walther Rossow: "Koordinierung und Gesamtplanung der Grünflachen im Hansaviertel Berlin." *Garten + Landschaft*, 67, (1957).

8 Plougerneau/Lesneven Bretagne and Mont de Huisnes, Normandy, inaugurated September 24, 1963. See <*www.volksbund.de/kriegsgraeberstaetten*> under the entry of cemeteries.

9 Oesterlen sent Rossow plans and sections, a total plan titled "design of crosses and tombstones" (5238–5239) dated February 7, 1960. Plan indicates his basic layout superimposed onto the contour map. The AdK Archive, the Dieter-Oesterlen Archive, Büro Ordner-AR.- Nr. 77, (February 26, 1969).

10 Ibid. A plant report on the area by K. Walther dated 1959 (more specific date not given) titled *Pflanzengutachten der Bundesanstalt für Vegetationskartierung*. It was commissioned by the VDK and given to Oesterlen.

retained Helmut Bournot of Berlin (1925–1980), a planting specialist. (Rossow and Bournot, referred to herein as R/B, worked for Oesterlen as a team and would collaborate with him on two subsequent projects.)[11] On March 31, 1960 Oesterlen and Rossow visited the site together.

R/B's scope of work was specified in detail and a cost estimate came to DM 1,124,080, and was then revised upward to DM 1,184,000.[12] These figures are mentioned here because they amount to an astounding ca. twenty-five percent of the total project cost. Rossow was to undertake the design of the grave fields, the pathways, the planting, and the irrigation system. (The latter task turned out to be so complicated that he outsourced it to a separate specialist.) Archives contain copious correspondence between Oesterlen and R/B. All indications affirm that they shared an aesthetic vision, and would not let natural challenges stand in the way of realizing it. In addition, from their honoraria, Rossow and Oesterlen would pay for the services of Swiss landscape architect Ernst Cramer to serve as construction manager.[13] These events attest to the importance all the designers ascribed to landscaping. A short history of their overcoming challenges as the project proceeded follows.

It is striking how the cemetery was built into the landscape, and how the landscape penetrates into the cemetery. This seamless interaction was one of the features that earned Oesterlen the commission: "It is particuraly impressing, in which natural way the structures fit into the site."[14] Rossow understood, and went on to develop his design for the site's topography to compliment the architecture's inherent simplicity. Slightly canted grass-covered grave fields' rhythm and wooded vegetation would give structure and order, and create focal points. The result would be an aesthetic unit, with the

11 The Andreanum gymnasiums in Hildesheim, 1960–62 Oesterlen, *Bauten,* pp. 168–174 and private residence Bergmann in Lage/Lippe 1960–62. See Koch, *Dieter Oesterlen,* pp. 128–131.

12 Document specifies tasks to be undertaken by Rossow and Bournot, including choosing vegetation and materials, and honoraria: Rossow would receive DM 33,000, Oesterlen DM 53,000. The AdK Archive, the Dieter-Oesterlen Archive, Büro Ordner-AR.-Nr. 74 (March 3, 1960).

13 Letter from Oesterlen to Rossow. The AdK Archive, Walter-Rossow Archive, box 28, (November 27, 1962).

14 *("Dabei ist es besonders eindrucksvoll, in welch natürlicher Weise sich die einzelnen Bauwerke in das Gelände einfügen.")* Directorate's session report. The VDK Archive, A.100–885, (February 26, 1960). The landscape is quasi an equal partner with the architecture of the cemetery.

wild Apennines visible from every point. All elements (materials, plants, grading) would work toward the goal that man-made and natural elements would "grow together" in time.[15] Both men thought in terms of "the fullness of time." (Their US counterparts made no provision for the effects of time passing. Maintenance would need to be relentless in order to preserve each cemetery's integrity.)

The size of grave fields and the arrangement of the graves needed to accommodate a "frightful number of dead" (*ungeheure Zahl der Toten*).[16] The VDK convened a special meeting to finalize its grave field guidelines, stressing the importance of the individual and of "full" grave markers.[17] Oesterlen saw the design of grave fields as a bearer of the message: "Graves, organized in manageable groups, express comradeship in the suffering they shared."[18] The sheer number of these grave fields (72) covering the whole expanse of the mountain would be awe-inspiring, while the court of honor with its triangular wall on the mountain's apex would hold every element of the site together.

As mentioned, accommodating a huge number of graves was the VDK's prime directive, but for the designers, this was a matter of both art and professional pride.[19] There would be several grave field plans (*Aufteilungspläne*) over three years as well as many trips to the site by R/B. Between the first "location drawing" (*Gesamtansicht*) in the spring of 1960[20] and the final plan (*Ausführungsplan*) in the summer of 1964 lay a complicated

15 Correspondence between Rossow and Oesterlen. Rossow agrees with Oesterlen's inspiration and sees his design in support of it. Date not given, but from context it was likely in early spring 1960. "*Anmerkung zur Landschaftsplanung.*" The AdK Archive, Walter-Rossow Archive, Box Oest-01-301.

16 The AdK Archive, the Dieter Oesterlen Archive, Oest-Ord-300, (October 1, 1982).

17 Minutes (*Niederschrift*), special session of VDK's Committee of Experts of the VDK. The AA Archive, (March 25, 1960). Befund 92, 602–85.20/271.

18 *("… in Gruppen zu unterteilen, überschaubar, auffindbar, auch als Ausdruck des in Kameradschaft erlittenen Geschickes.")* The AdK Archive, the Dieter-Oesterlen Archive, Oest-Ord-300, (October 1, 1982).

19 Letter from the VDK's grave service, from Müller to Oesterlen literally begging him for assurance that graves would be orderly and no planting on them would be planned. *("Wir bitten um ausdrückliche Bestätigung, dass diese Gräber in fortlaufender Reihenfolge … werden können und dass auf keinem dieser Gräber eine Anpflanzung vorgesehen ist.")* The AdK Archive, the Dieter-Oesterlen Archive, Büro Ordner-AR-Nr. 66, (October 4, 1961).

20 Grading map of the whole site with Oesterlen's design. The AdK Archive, The Water-Rossow Archive, Rossow /Bournot, Pläne # 514, (February 7, 1960).

"balancing act," where aesthetic vision confronted and overcame nature's stubborn hindrances.[21]

As at Caira, reburial had to be postponed for more than a year from the VDK's initial estimated date.[22] Bodies already exhumed from their temporary graves were being kept in storage at a considerable cost. A copious correspondence on the subject of plans and schedules deals with issues surrounding exhumation and reburial.[23] The VDK staff members, who managed these sensitive matters, are the "unsung heroes" of the Futa Pass project.

R/B's first two tentative plans were based on Oesterlen's model, and drafted prior to Rossow's first site visit (with Oesterlen on March 31, 1960).[24] These make clear the magnitude of the challenge. Rectangular fields (blocks) completely covering irregular topography appear like a viscous maelstrom extending to the property line, with edgings set somewhat askew in response to the natural topography, and paths zigzagging through the terraces.[25] The plans provided little space for planting, accommodating only singular trees at the site's edges. The second plan shows even more densely arranged fields than the first, but also reveals Rossow's attempt to harmonize with Oesterlen's aesthetics, and to avoid monotony.

Only after Rossow's personal experience at the Futa Pass site were R/B able to draft a new plan (April 5, 1960) that, in a stroke of genius, harmonized grave fields and landscape. Somehow, they created more free land for planting while increasing the number of graves to 35,820. Their plan involved grave field blocks of five different sizes, arranged amidst larger, more irregular planted areas, especially along the site's

21 Ibid. Plans from July 24, 1964 and August 20, 1964.

22 Stiebeling's (VDK) intent to begin reburials on July 31, 1960 proved unrealistic. The AdK Archive, the Dieter-Oesterlen Archive, Büro Ordner-AR-Nr. 66 (March 31, 1960).

23 Burial scheduled to begin in July of 1961 was postponed to September 19, 1961. The strategy was to begin filling graves that were ready (north end block 31 and 32), but new planning was needed there, blocks 30 and 33 could be used. After munitions were found, burial shifted to the south side, blocks 5 and 8. The pressure increased. By the end of 1963, 18,000 were to be buried. The AdK Archive, the Dieter Oesterlen Archive, R/B Belegungspläne.

24 The plan indicates an area of 96,725 square meters of total area with five percent slope. It already shows cisterns. The AdK Archive, the Walter-Rossow Archive, WRA-R+B-Pläne, Plan n-514-Lageplan-1, (March 26, 1960).

25 The plan shows pencil corrections. The AdK Archive, Walter Rossow Archive, WRA-R+B-Pläne, Plan 515 Variante, (March 27, 1960).

periphery. It became the standard from which all following modifications would be derived (under Oesterlen's ever-present supervision).[26] Modifications, however, would be constant as subterranean surprises were discovered, as well as an ever-increasing number of remains that needed to be accommodated.[27] Through all this, Rossow's vision of the surrounding untamed landscape penetrating into the designed landscape held fast: 'The grave fields' "outward facing borders were to be eliminated and thus will permit the environment to merge into the cemetery."[28] (The plan also shows that R/B envisioned circular surface cisterns. Construction was set to begin in the spring of 1961, but this estimate proved to be overly optimistic.)

On the north side, a fence of so-called Tessino plates (resembling standing grave markers) was erected in order to fulfill the contract with the former owner.[29] An autumn of 1961 modification shows grave fields in biased arrangement in this area.[30] Oesterlen was not pleased: "Your overview plan for arrangement of grave blocks seems to me a bit crazy…and about those grave fields, one might imagine it possible to find a somewhat more harmonious solution."[31] R/B immediately traveled to

26 This plan was officially accepted by the VDK in the spring of 1961. The AdK Archive, Walter Rossow Archive, WRA-R+B-Pläne, Plan 526 (and 527) R/B Arch. Mappe 2; preceeding plan: The AdK Archive, the Walter Rossow Archive, WRA-R+B-Pläne, Mappe 1, (April 5, 1960).

27 Plan 527 shows a large area to the west hatched in red that would need to be filled. 70 cm are the minimum for planting anything. Accordingly, the supporting walls would have to be higher and explosions might be necessary. Accompanying text evidences that Rossow recognized the difficulties of the terrain: rocks and uneven soil covers. The AdK Archive, the Dieter-Oesterlen Archive, Büro Ordner-AR-Nr. 74, (May 8, 1960).

28 Rossow to Oesterlen. *("…die scharfen Abgrenzungen nach außen sind aufgehoben …. günstige Verbindung der Umgebung mit dem Friedhofsgelände (kommt) zustande…")* The AdK Archive, Passo della Futa Archive, Mappe I, Rossow I and II, (May 10, 1960).

29 To erect a fence there by 1961 was part of the contract. Letter, v. Münchhausen to the VDK to remind them of the contractual duty to erect a fence there. The AdK Archive, the Dieter-Oesterlen Archive, Büro Ordner-AR-Nr. 66, (December 21, 1960).

30 The AdK Archive, the Dieter-Oesterlen Archive, Büro Ordner-AR-Nr. 74, (September 22, 1961).

31 Letter from Oesterlen to Bournot calling the plan confused. *("Auf Ihrem Übersichtsplan der Blockeinteilung fällt mir die etwas wirre Anordnung der Gräberfelder 35–37 sowie 44 bis 45 auf… ob sich nicht auch hier eine Lage der übrigen Gräberfelder entsprechende harmonische Lösung finden lässt.")* The AdK Archive, the Dieter-Oesterlen Archive, Büro Ordner-AR-Nr. 74, (January 1, 1962).

the site (January 18, 1962).[32] A different arrangement of smaller blocks followed.[33]

Again, along the cemetery's eastern property line,[34] soil conditions necessitated small changes in the course of the spiral wall, changing the available space for burial.[35] When Rossow and Oesterlen visited the site together (August 23, 1962) they realized that a grave field to the southeast, closest to the entrance, had to be abandoned despite several attempts to save it. Also on the north side, the design and orientation of grave field blocks became a point of contention. Oesterlen disapproved of the aesthetics, summarily stating: "Cannot stay this way" (*Können nicht so bleiben*).[36] Purchasing more land was not an option, so Oesterlen and R/B together developed a creative solution using smaller blocks, seven in total.[37] The ultimate, final plan still awaited an exact number of graves.

---

32 See note regarding that trip. The AdK Archive, the Dieter Oesterlen Archive, Büro Ordner-AR-Nr. 80, (February 23, 1962).

33 Blocks appear arranged seemingly without order. The AdK Archive, the Dieter-Oesterlen Archive, Büro Ordner-AR-Nr. 74; also see the VDK, Futa Pass_Altplaene, 00014594, (January 26, 1962).

34 The Eastern side has a local road as a boundary, the south and west have areas planted with native shrubs and trees that transition to the existing landscape. See fn. 24 for plan.

35 See correspondence between Oesterlen and Bournot in the spring of 1962. The AdK Archive, the Dieter-Oesterlen Archive, Büro Ordner-AR-Nr. 74 R/B, (January 2, 1962).

36 Letter, Oesterlen to Rossow. He also disliked the arrangement of access paths to the grave fields and to the spiral main path. *("Anschlüsse an Spiralweg nicht gut, wie an dem Tessiner Zaun wie auch untereinander, dicht aufeinander folgende Stufen bis 2 Meter.")* The AdK Archive, the Dieter -Oesterlen Archive, Büro Ordner-AR-Nr. 80, (October 10, 1962).

37 The final plan shows that blocks 44 and 45 remained biased, 44 more so because it was situated at a place where the wall turned, and their rims became oriented toward the east rather than toward the wall. Blocks 38 and 39 had to given up altogether. Blocks 35, 36, and 37 were built and oriented toward the wall but had no rims, probably for aesthetic reasons.

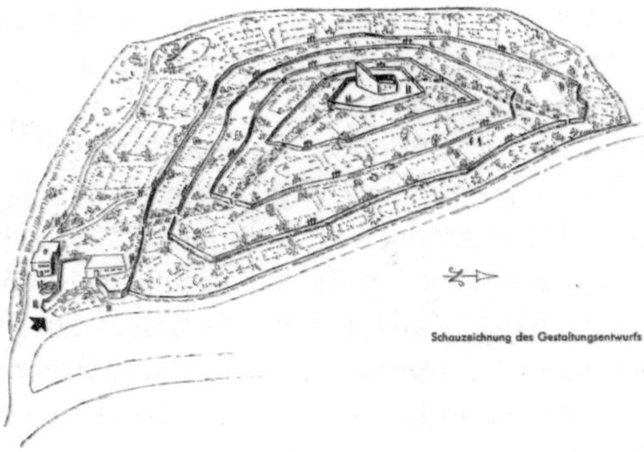

Fig. 67. Final Lay-out plan of the grave fields. Note grave-field one has been eliminated due to landslides and note the concept of groups of crosses on the wall.

For the spiral wall itself, Rossow had submitted a design sketch showing round irregular boulders, as used by Italian farmers.[38] Oesterlen would have none of it and insisted on angularity with horizontally layered stones. He would tolerate nothing curved (except after some resistance, the circular surface cisterns). He demanded that the course of the main pathway closely accompany the wall.[39] Rossow consequently adjusted his plan, but not without a good-natured protest: "Your objections with regard to the pathway were excessive. I love nothing that you detest. Round is not just pleasant."[40] Compromising, Oesterlen allowed a rounded-off space for planting where an access path to graves branched off the main path.[41]

38 A type used by Italian farmers where irregular natural boulders are stacked loosely to render a wall with an irregular top-line. The AdK Archive, the Walter Rossow Archive, WRA-R+B-Pläne, Plan-508-Stützmauer, (March 2, 1960); The VDK Archive, Futa Pass_Altplaene, 00014609, (March 22,1967).

39 Landscape design plans from 1963–1964 show many corrections in red and green pencil. Boundaries of grave fields were moved, steps on the spiral pathway eliminated, the rim of a grave field omitted, green areas for planning adjusted in green pencil. The pathways straightened in order to follow the course of the wall. Some round lines remained where an entrance path to the grave fields met the spiral path. The AdK Archive, the Walter Rossow Archive, WRA-28-9-Plan 62521, (December 2, 1962); the AdK Archive, Futa Pass II. The AdK Archive, the Walter Rossow Archive, WRA-28–9, Plan 62521 (December 2, 1962); The AdK Archive, Futa Pass II, (December 2, 1962).

40 (*"Ihre Argumentationen zum Weg waren stark überhöht. Nichts von dem was Sie verabscheuen liebe ich. Rund ist nicht nur gemütlich"*) The AdK Archive, the Dieter-Oesterlen Archive, Büro Ordner-AR-Nr. 74, (May 30, 1963).

41 The VDK, Futa Pass_Altplaene, 00014334, (May 18, 1963).

Rossow insisted on his own ideas where he felt they were important. For example, he insisted that a belt of vegetation at the west side have priority over further expansion of grave fields.[42]

Such "back-and-forth" attention to detail continued into the fall of 1963.[43] It is testimony to the great dedication on the parts of Oesterlen, Rossow, and Bournot.[44] They even allotted part of their own honoraria for the on-site supervision services of Ernst Cramer, who would later say that the Futa Pass was the work of his lifetime.[45] (US artists showed such dedication as well. Evidently, designing a soldiers' monumental cemetery was something very personal and special for all of these artists.)

Contractually, R/B were in charge of drainage. A plan shows that they envisioned a large pond at the southwest corner to collect and store water.[46] They soon realized that they would need the services of an irrigation specialist.[47] They engaged Professor M. Seyberth, who designed a comprehensive drainage system to include the site's extensive, subterranean buttressing against landslides.[48] Perhaps Rossow's most brilliant contribution was the circular surface cisterns (two meters deep) situated on the

---

42 Letter from Oesterlen to the VDK's Stiebeling telling him that Rossow did not want to have new fields on the west side. It would be a problem to find space for the 411 bodies from the emptied cemetery in Cervia since the VDK insisted that the 411 remains should be buried together. The AdK Archive, the Dieter-Oesterlen Archive, Büro Ordner-AR-Nr. 66, (September 9, 1963).

43 The AdK Archive, the Walter Rossow Archive, WRA-R+B-Pläne, Mappe 5, R/B Plan, (November 1, 1963).

44 The AdK Archive, the Walter Rossow Archive, WRA- R/B -Arch.-Plans: 8324, 5281,5278,5280,5225, show changes as to numbers of fields, orientation and location of rims.

45 See Udo Weilacher, Zwei skulpturale Gedenkstätten der 60er Jahre. Mutige Gestaltung Schweizer Architekten", in *Garten +Landschaft,* Nr. 1, 2002, P. 22. (Discusses the Futa-Pass cemetery).

46 The AdK Archive, the Walter Rossow Archive, WRA-B/R-Pläne, Plan 8262, (July 22, 1961).

47 Letter from Bournot to engineer Seyberth specifying his task. ("*Die Verteilung der Zisternen und Stränge überlassen wir Ihnen. [30 Stück] Bitte Leitungen in die grünen Zwischenräume zwischen Grabfeldern legen.")* The archive allows one to trace this subterranean building activity. The AdK Archive, the Dieter Oesterlen Archive, Büro Ordner Arch.-Nr. 74, (September 9, 1964).

48 One of several detailed drainage plans reveals the extent of the underground work needed. The AdK Archive, the Walter Rossow Archive, B/R Arch. Mappe 10, 8261, (November 23, 1963).

terraces.[49] There are also subterranean cisterns (six meters deep), but the fourteen surface cisterns are open and are an aesthetic feature.[50] Oesterlen, keeping watch over design alterations, seemed to begin to appreciate them only after a trip to the Futa Pass with Bournot (August 31 1961).[51] Later Oesterlen stood fully behind the idea, describing them as being "like eyes in the landscape."[52] The cemetery recalls ancient ruins, with a citadel at its summit. Openness to the surrounding Apennines invites nature's presence.

Fig. 68. Futa Pass aspect showing all design elements, note one of Rossow's water basins in foreground. Note also the rim of grave-field toward the side of the monument as a design feature.

49 Oesterlen at first rejected the round basins/cisterns because he did not want anything round, but came around to endorse them fully. The AdK Archive, the Dieter Oesterlen Archive, Oest-Or-300.

50 Letter from Rossow to Oesterlen. The AdK Archive, the Dieter-Oesterlen Archive, Büro Ordner-AR-Nr, 77, (May 8, 1960).

51 Note about Oesterlen's trip to the Futa Pass in order to inspect grave field designs. The AdK Archive, the Dieter-Oesterlen Archive, Büro Ordner-AR-Nr. 66, (August 31, 1961).

52 See fn. 49.

Mass Death Visualized

"For soldiers' cemeteries, precisely thought-through ordering of their graves has the implicit effect on future generations of blocking out war's utter senselessness and death's anonymity."
Werner Fuchs-Heinritz[1]

Fig. 69. Mass death visualized, with monument appearing needle-like.

1 Werner Fuchs-Heinritz, *Todesbilder*, ("… *die Anlage von Soldatenfriedhöfen ließe sich ebenso aus der Aussicht … erklären, die Sinnlosigkeit und Anonymität des Kriegstodes nachträglich aufzuheben durch die genau durchdachte Gliederung der Kriegsgräber.*") The author furthermore points out that mass death cannot be grasped with conventional responses to death. p. 146. Ergo, Rossow departs from this convention.

Fields of Graves

Grave fields fitted into the mountainous terrain with geometric harmony are landscape art not intended to minimize pain, but rather, they are intended to focus awareness on the senseless and unnecessary death that most of those buried at the Futa Pass suffered during the already-lost war's final weeks. R/B had no model to follow. The mountain landscape presented unknown challenges. Determined to leave nothing to chance, R/B constructed two grave fields or "blocks" for testing. Without the lessons gained from these try-out blocks, costs would have gone into "the immeasurable."[2] One lesson was that the stone edges along their sides could be eliminated. However, stone rims oriented toward the monument gave visual structure.[3] R/B insisted on their aesthetic authority, agreeing with elimination edgings on the more gently sloping north side,[4] but protested against eliminating it altogether: "Doing away with stone edging on the uphill edges will not be discussed unless the whole project is to be fundamentally altered."[5] Once it became evident that these needed to be

2 Letter, Bournot to Oesterlen telling that the try-out blocks have proven their worth. (*Musterblöcke haben sich bewährt*). The AdK Archive, the Walter Rossow Archive, WRA-R+B-Pläne, Mappe 7, (July 21, 1965). *("Dadurch die restliche Blöcke einwandfrei...Rutschungen, drainage hinter den Steinkanten, notwendig an den Böschungskronen erhöhter Auftrag...verwendete Platten stabil...ohne diese Proben wären die Schäden an sämtlichen Steinkanten in Unermeßliche gegangen. Musterblöcke kosteten DM 18, 409.")*

3 For the design of each grave field with regard to rims see the VDK Archive, Futa Pass_Altplaene. On some fields, the reinforced rim was executed only toward the spiral wall. The VDK Archive, Futa Pass_ Altplaene, 00014593, (January 26, 1962). VDK Archive, Futa Pass_ Altplaene, 00014582, (January 26, 1962). On some blocks, the rim had to be placed perpendicular rather than parallel to the spiral wall. See final R/B plan, the VDK Archive, Futa Pass_ Altplaene, 00014329. A plan from September 1963 shows simplifications. The VDK Archive, Futa Pass_ Altplaene, 00014294 and 00014568, (May 9, 1963). Adjusting to topographical conditions, rims were placed at the outside end of the grave field rather than toward the spiral wall or on the small side, as in Block 21. The VDK Archive, Futa Pass_ Altplaene, 00014568 (May 5, 1963).

4 Letter from Rossow to Oesterlen. Some stone edges in the north, however, were omitted later for aesthetic reasons. *(... Steinkante in den neuen Feldern weggelassen, weil sie dort störend wirken würden.)* The fields were 35, 36, and 37. The AdK Archive, the Dieter-Oesterlen Archive, Büro-Ordner-AR-Nr. 74, (August 9, 1963).

5 *("Das Wegfallen der Steinkante an der Oberkante der Gräberfelder ist gar nicht zu diskutieren, ohne damit das gesamte Projekt grundsätzlich anders einrichten zu*

strengthened, their re-design included extending their reach to beyond the breadth of the grave field. Stone edging remained, and detailed instructions were given to the Italian firm executing them in order to ensure that they be robust enough.[6] As such, they provide an aesthetically pleasing accent in the total pattern. Oesterlen approved of this design feature.[7] Perfection would rule.

R/B, proceeding with their plans, adjusted or omitted various pathways to make space.[8] Specific plans specifying dimension, orientation, number, and exact location of each individual grave were in place by mid-1962.[9] For the interested reader, the following paragraph outlines the design details:

> Stone rims (30 cm above grade) on the grave fields' uphill-facing edges define the grave field's width, and act as low retaining walls. Within each grave field, individual graves are arranged in perpendicular rows and columns. Shared marker stones defined walkways for foot access. Soldiers are buried one meter beneath the surface in individual sarcophagi. A grave's standard dimension was 140 cm x 70 cm. A single marker (70 x 40 x 10 cm) served (initially) two adjoining and was placed at the foot of both graves, centered on the line the two shared. This square (140 cm x 140 cm) unit of two graves containing four sarcophagi and one grave marker was joined to another, head-to-head, forming a unit of four graves (280 cm x 140 cm). Thus, the two marker stones were at the opposite ends, 280 cm apart. These rectangular units, with their marker stones at each end, are placed beside one another to form rows without interruption. A mall that was edged by grave markers was (initially) formed on the

müssen.") Letter from Rossow to Oesterlen. The AdK Archive, Reg.No. 74. (May 30, 1963). In support of this view Oesterlen literally quotes Rossow in his letter to the VDK's Ehrke, (July 5, 1963), the AdK Archive, the Dieter-Oesterlen Archive, Büro-Ordner-AR-Nr. 66.

6 Letter from Rossow to Crescon, (construction supervisor under Cramer). *(Firma Scapellini sollte die Kanten machen, weil "die Steine sollen aus demselben Material sein wie die Steine der Mauer, ohne sichtbare Schlagstellen, mit dem Hammer so glatt wie möglich außen behauen, Stoßfugen so gerade geschlagen...Die Stoßfugen können 1–1,5 cm breit sein. Ohne Zementmörtel lose verlegt.")* The AdK Archive, the Dieter-Oesterlen Archive, Büro-Ordner-AR-Nr. 74, (June 20.1963).

7 For a detailed commentary to R/B's grave design concept see the AdK Archive, the Walter Rossow Archive, WRA- R+B Pläne, Plan 514, (July 10, 1960).

8 For example, Block 10 had neither pathway nor edge because the upper spiral wall changes direction sharply at this area and framed it on two sides (leaving room for only a narrow strip of grass or planting).The AdK Archive, the Walter Rossow Archive, WRA-R+B-Pläne, Plan from June 23, 1962.

9 This can be deduced from dates on these plans. VDK Archive, Futa Pass_Altplaene.

> surface. These markers appear lined up in a row, and since each marker stone reaches only halfway into each grave's 70 cm width, they alternate with intervals of grass surface of equal lengths. Treading upon graves was (initially) to be avoided. Grass pathways 40 cm wide were inserted as corridors between the rows of marker stones. These marker stones were replaced in 1988 with single markers, most bearing four names, and pathways between marker stones were eliminated.[10] Walking over graves, by that time, had apparently become acceptable.

A design with different grasses became a financial matter. "One cannot apply the style of the Americans and English. Even if costs should not be put in the foreground, it is nevertheless an urgent matter in the background."[11] Originally, the grave rows were to be planted with finer grass than the pathways between them.[12] Walking was to be limited to the areas of rougher grass. Later, for financial reasons, this was replaced with uniform grass cover. Still later, native wild species of grasses invaded. Grave fields themselves, with rhythmic striped patterns, are not monotonous and paved pathways at their periphery frame a sequence of green stripes. Single rows of grave markers at the edges, and double rows of these in between, form bright lines.[13] The original rhythmic layout with different textures of grass was lost, but the original solidity of ratios and patterning still resonates, despite today's more uniform look. The expanse of multiple grave fields is punctuated by corridors with marker stones, giving numbers for the location of graves. Of luminous *pietra serena*, they contrast with the grasses.

---

10 Two tombstones were unified into one. A so-called "*Kissenstein*" had slanted sides on each of which the two names were inscribed. Important: The pathways between the markers were thus eliminated. The visitor from then on had to walk on the graves proper in order to walk to the respective marker. (The VDK's guidelines of 1958 stressed that it was to be avoided to walk over graves in order to access grave markers.) The AA Archive, VdK _richtlinien, p. 2, (March 28, 1958), Befund 92, 602–85.20/271.

11 Quote from Herr Stammer (Konsulatsekretär): *("Man könne nicht den Stil der Amerikaner und Engländer bei uns anwenden. Wenn die Kostenfrage auch nicht in den Vordergrund gestellt werde dürfe, so stehe sie doch sehr eindringlich im Hintergrund.")* Speical session, the VDK Archive, A.100–883, (March 5, 1960).

12 The grass mixture for the blocks was 40 percent merion blue, 50 percent festuca rubra genuina, 10 percent agrostis alba stolonifera. Today there is little noticeable from this variation of grasses.

13 This pattern characterizes the grave fields. See VDK Archive, Futa Pass_Altplaene 00014499. That plan is called "*Belegungsplan*" (occupation plan) Type 1, (September 18, 1961).

In 1988, the new grave markers were installed, each replacing two markers with one so-called pillow marker (*Kissenstein*). The corridor between the graves was thereby eliminated, doing away with deference to the traditional German abhorrence to treading upon graves. (By comparison, treading upon graves was not seen as something to be discouraged in the US cemeteries.) One might see in this a fading of a Germanic quasi-religious atavism that held burial places to be sacred.

Oesterlen designed one grave field himself. Additional remains continued to be discovered, usually of soldiers that had died together, after the crypt had been sealed, and hence another *Kameradengrab* was needed, this time in the open. The VDK 1965 guidelines stated that: "Special value shall be given to a dignified design of the comrades' grave"[14] (although in 1964, the VDK had renamed this more euphemistically as "group grave" [*Gruppentotengrab*]).[15] Oesterlen's intention was to honor, through the clear grouping of graves, the fates of soldiers who suffered in "comradeship."[16] He personally designed Block 53, the group grave that is a much smaller grave field, located near the zenith.[17] A paved pathway encloses it, and no stone rim. One hundred seventy-seven grave markers of five different sizes are arranged in rows, with and without intervals of grass patches,[18] and graves between the rows of markers have varying widths.[19] Here, one sees Oesterlen's rejection of standard patterning. His personal creation went toward "expressionism." (This design was not executed).

14 *("Auf eine würdige Gestaltung der Kameradengräber ist besonderer Wert zu legen.")* These guidelines were formulated for cemeteries in France, when it became clear that the VDK would be allowed to work in France (1966). The VDK Archive, A.10–183, (March 22, 1965).

15 See the VDK Archive, Futa Pass_Altplaene, 00014485/84, (November 25, 1964). Not all of the soldiers buried in comrades' grave are unknown. The important aspect is that they fought and died together.

16 Letter, VDK's Ehrke to Oesterlen. The question remains open whether the remains buried in this grave came from the Cervia cemetery. In any case the VDK's Ehrke reminds that special efforts were made. The AdK Archive, the Dieter-Oesterlen Archive, Büro-Ordner-AR-Nr. 66, (June 6, 1963).

17 Graves are arranged side-by-side, 65 x 40 x 13, next to doublewide graves 135 x 44 x 13. See VDK Archive, Futa Pass_Altplaene, 00014485/84, (November 25, 1964).

18 Three are 135 x 95 x 13, one 185 x 95 x 134, the largest is 255 x 95 x 13 cm. The VDK Archive, Futa Pass_Altplaene, 00014485/84, (November 25, 1964).

19 See the VDK Archive, Futa Pass_Altplaene, 00014485/84, (November 25, 1964).

The Planting Design

Oesterlen, firmly in control, expressed firm ideas about landscape design, planting, and nature's forces: "...Apennines in view from every point...the knoll, that bears the cemetery...rhythm of trees and shrubbery define and accent architecture...vegetation binds landscape and architecture...landscape part of the whole...nature and the whole shall grow together as one..."[20] Not immodestly, his written words imply nature not overcoming his creation, but in the end joining with it. Rossow embraced his concept. His landscape design, his designed rhythms of the grave fields, and his placement of trees and shrubs would mediate between nature and Oesterlen's architecture. Though the VDK stressed the need for simplification, Rossow envisioned extensive planting. The visitor should be able to walk on "walkable vegetation."[21] An enclosure with native growth was part of the conception from the beginning, as evident in R/B's first plan of 1960, drafted without firsthand knowledge of the site. Dense areas of shrubs and trees are indicated to create an unnoticeable transition that would fit the cemetery into the totality of the landscape. Native species were proposed and a botanist was retained to carefully catalog these.[22] Concern for pleasing transitions went so far as to suggest planting on the neighbor's land (after Count Pontello was assured that the planting would belong to him).[23] (This unsolicited benefit for the Count was followed by other such instances where improvements to the site benefited

20 *("Die bewegte, großräumige Landschaft des Appeningebirges ist von jeder Stelle des Hügels, der den Friedhof trägt, zu erleben, er ist ein Teil davon geblieben. Er hebt sich ab durch den Rhythmus der Gehölzpflanzungen, welche abgrenzen und gliedern und Schwerpunkte bilden. Die Vegetation bindet das Gesamtbauwerk in die Landschaft ein. Das gärtnerische Konzept ist Teil des Ganzen. Bau und Landschaft sind im Laufe der Zeit miteinander verwachsen.")* The AdK Archive, the Dieter- Oesterlen Archive, Oest.-Ord.-300.

21 The demand for walkable vegetation ("*begehbarer Pflanzenwuchs*") is spelled out in guidelines from August 4, 1959. The AA Archive, Befund 92, 602–85.20/271.

22 A thorough study of native vegetation, six pages long, includes photos. He had travelled to the site. Plants needed to be able to survive in high altitude and climate. He included blooming varieties. K. Walther Stolzenau/Weser 1959. The AdK Archive, the Dieter-Oesterlen Archive, Büro-Ordner- AR- Nr. 77, (1959). This study was given to Oesterlen when he received the commission.

23 Letter, v. Münchhausen to the VDK saying that Count Pontello was impressed with the design of the cemetery. The AdK Archive, the Dieter-Oesterlen Archive, Büro-Ordner- AR- Nr. 66, (September 23, 1964).

Italian neighbors. For example, a water main (*Wasserleitung*) was connected to a nearby restaurant. Its proprietor, Signor Cetica, who could expect an easement in the form of a new path for his carts (*Karrenweg*), considered his contractual rights more than fulfilled when he became connected to said water system.[24])

Bournot's work culminated in a detailed planting plan in May 1963 that consisted of 24 sheets that gave detailed lists of plants.[25] He envisioned every possible pocket of space decked with a variety of groundcovers and blooming varieties, including roses.[26] The list of plants for the entrance area alone names 15 varieties of roses and five different kinds of bulbs. A novelty in this plan was to "splurge" on a four-meter-tall solitaire tree (*Carpinus Betulus*) for the entrance court.[27] (This tree remains as of this writing, but little remains from the original plan.) Within the VDK, Bournot's plan met with strong criticism. So many kinds of plants would not be practical for a cemetery in the free landscape, as they would require too much care. "Every cemetery (erected) in open landscape should be planted such that even if left in an untended condition, it would remain a memorial."[28] The plan seemed excessive. "I see it as illogical to imagine planting such a diverse assortment of horticulture on a site that, in the long run, will need to get along with minimal maintenance."[29]

24 See letter, v. Münchhausen to the VDK saying that building a new cartway will be replaced by a water line for Mr. Cetica. The AdK Archive, the Dieter-Oesterlen Archive, Büro-Ordner- AR- Nr. 67, (after 1965), (January 28, 1966).

25 The VDK Archive, Futa Pass_Altplaene, 00014331–00014355. The latter dated May 18, 1963, slightly corrected on November 10, 1963 gives list of plants.

26 Roses were considered "Un-German" for World War I cemeteries under the NS regime. See Brands, "From World War I." p. 239.

27 This tree was already marked in a planting plan that dates from January 1963 (day illegible). The AdK Archive, the Walter Rossow Archive, R+B Pläne, Map 11 6336, (January 1963).

28 Misgivings came from Herrn Lendholt (Hannover), Institut für Grünplanung und Gartengestaltung der T.H. Hannover, *("… jeder Friedhof in der freien Landschaft sollte m. E. so bepflanzt werden, dass er auch in einem pflegelosen Zustand ein Denkmal bleibt.")* The AdK Archive, the Dieter Oesterlen Archive, Büro-Ordner- AR- Nr. 66, (June 18, 1963).

29 Letter from Ahlers to VDK's Ehrke. *("Ferner halte ich es für abwegig, bei einer Anlage, die sich auf weite Sicht ohne eine gewisse Pflege halten muss, ein so vielseitiges Sortiment von gärtnerischen Pflanzen vorzusehen.")* Also soil improvements and feeding would be excessive. The AdK Archive, the Dieter-Oesterlen Archive, Büro-Ordner- AR- Nr. 66, (June 26, 1963).

The VDK's intention (that pre-dated the second World War) remained to design for a level of maintenance that would assure permanence while at the same time allow for a managed reversion to nature. In response to very real concerns about future maintenance, plant-loving Bournot responded that it was not the "variety," but rather the "quantity" of plants that should be reduced.[30] (US cemeteries need continuous costly maintenance and during the US government shutdown in the fall of 2013 all US foreign cemeteries were closed to the public.) Bournot traveled again to the site.[31] Subsequent planting plans show alterations.[32] His plan of August 20, 1964 shows an almost radical simplification, and this became binding.[33] The spiral pathway's undulations, which provided for pockets for planting (shown in the May 1963 plan), were mostly eliminated. Nevertheless R/B, in this last simplified plan, were able to salvage some of Bournot's ambitious variety. Planting in areas near the entrance was retained, including the fourteen varieties of roses and of five types of bulbs, various ground covers and jasmine.[34] Dense planting remained at the cemetery's south and east edges, allowing the cemetery to blend into the vegetation growing naturally nearby. Varieties making up this transition zone needed to be carefully selected. When distribution of the grave fields had become fixed and a planting plan accepted, a contract was let to an Italian company for building the grave fields as well as for the landscaping.[35] Bournot was responsible for meticulous supervision. Some plant varieties were not available as promised, so they were delivered from Germany.[36]

30 Letter from Bournot to the VDK. The AdK Archive, the Dieter-Oesterlen Archive, Büro-Ordner- AR- Nr. 74, (June 21, 1963); (June 6, 1963).

31 Letter, Bournot to the VDK informs that low-growing trees will no longer be planted on terraces, but only at the cemetery's periphery. The AdK Archive, the Walter Rossow Archive, R+B Pläne, Map 4, (July 17, 1963).

32 See VDK Archiv, Futa Pass_Altplaene, 00014294, section of plan issued by R/B and dated September 29, 1963, a few months later than the elaborate May plan.

33 For this plan see the AdK Archive, Walter Rossow Archive, R+B Pläne, Map B/R 8335, 8245 to 8257; and see the VDK Archive, Futa Pass_Altplaene, 00014326 to 29, (August 20, 1964).

34 The VDK Archive, Futa Pass_ Altplaene, 00014355.

35 The size of the land to be landscaped and planted (excluding the paths) was ca. 33, 144 square meters. The AdK Archive, the Walter Rossow Archive, Mappe 28–3, from budget plan, (March 26, 1960.).

36 R/B travelled to the Futa Pass site on May 4, 1964 in order to meet Sgaravatti, an Italian firm that carried out the landscaping, stoneworks for rims and pathways for DM 310,381.79. Letter of Bournot to Sgaravatti, telling him that since the ordered

By 1963 wages in Italy had risen, and cost estimates followed suit. Actual planting could not begin before August 1965. Grasses were first sown, and Rossow recommended waiting with further planting until May 1966.[37] The Italian contractor threatened to cease work unless the VDK agreed to modify the contract, revising fees upward,[38] with the result that by 1965, financial strains necessitated further simplification of the planting plan.[39]

Oesterlen monitored the planting plans and their various revisions and seldom interfered, but following a visit to the site (August 23, 1962) he observed that the view on the site from the highway was becoming obstructed.[40] The edge planting was therefore changed. In another instance, Oesterlen summarily and tersely informed the VDK or his vision, as if citing a design principle: "Groups of trees only on singular places, all grave fields in grass, between them low growing brush."[41] Trees, originally planned to set accents between grave fields, were quickly replaced by shrubs. Despite his appreciation of landscape and planting design, Oesterlen held architecture to be the principal carrier of the message. (Similarly, US architects Gugler and Smith left no doubt that architectural design was primary for their cemeteries at Anzio/Nettuno and Impruneta.)

As of this writing, a half-century on, the site has a bleak look. The visitor sees grave fields becoming part of a stubborn wilderness of mountain grasses and shrubs. R/B's wonderful (though simplified) plant variety is not in evidence. Some trees next to the spiral wall have resisted the rough

and promised plants were not available, he ordered them from Germany. (… *Pflanzen nicht da, obwohl seit September vorher bestellt. Werden nun aus Deutschland geliefert.*) The AdK Archive, the Dieter-Oesterlen Archive, Büro-Ordner- AR- Nr. 74, (February 25, 1965).

37 Ibid.

38 Letter from Bournot to the VDK saying that Sgaravatti demanded enormous wage increase and would travel to VDK Office in Kassel to negotiate. Twenty-seven percent was accepted. This can be deduced from Bournot's visit to the site on October 20, 1964. The AdK Archive, the Dieter-Oesterlen Archive, Büro-Ordner- AR- Nr. 74, (September 26, 1964).

39 Report by Rossow reduces planting. The AdK Archive, the Walter Rossow Archive, R+B Pläne, Map 7, (August 26 to 28, 1965).

40 Letter from Oesterlen to Rossow. The AdK Archive, the Dieter-Oesterlen Archive, Büro-Ordner- AR- Nr. 74, (November 1, 1962).

41 Letter, Oesterlen to the VDK reporting his planting decisions. Trees only for emphasis. He adds a short list of plants. Crataegus, Sarothamnus, Salix, Alnus, Ligustrum, Rosa, Corylus, Prunus, Acer, Evonymus. The AdK Archive, the Dieter-Oesterlen Archive, Büro-Ordner- AR- Nr. 66, (November 27, 1962).

climate, with eight months of frequent precipitation, which includes winter snow. Roses refuse to bloom. Constant cold winds have reduced the blooming plan to its basic elements. The mixture of grasses sown on the grave fields is irrigated and cut regularly such that it is discernable from the natural vegetation, and at the peripheries, a mix of planted and existing vegetation now forms a green zone. Lichens grow on Oesterlen's spiral wall. Wild grasses and flowers have invaded and seem to push upward against it. Rossow, accepting this "fading" of his garden vision, continued to recognize and defend the singularity of Oesterlen's design. In 1987, when gravestones needed replacement, standing crosses were (once again) proposed and Rossow made a personal appeal to the VDK not to alter Oesterlen's vision.[42] "The basic element of his concept of planar grave fields is to be maintained, like 'cut-outs' from the mountain vegetation stacked against the great wall that, as the only vertical element, marks the knoll."[43]

42 These can be found in many VDK cemeteries such as at Pomezia. Grave stones were replaced on the Futa Pass in 1988 but remained flush with the ground.

43 Letter from Rossow to a Mr. Barth. The AdK Archive, the Dieter-Oesterlen-Archive, Büro-Ordner AR-Nr. 4, persönliche Briefe, (February 2, 1987).

XI. The Futa Pass Cemetery: An Exoneration of Weimar Republic's "Aberrant Art?"

"Beaux Arts design principles, even when appearing in Bauhaus designs, cannot adequately express our times."
Martin Wagner[1]

Creation of an Architectural Style. The Modern from Tradition

The Futa Pass's design came at time when German culture was in upheaval, but how radical was it? Stylistic analysis locates it in an art-historical context and demonstrates that it stands on the shoulders of precedents, mainly those that had been damned during Germany's Nazi period. In Oesterlen's work two main sources of influence can be recognized: expressionism and modernism (both dominant during the Weimar Republic era). Oesterlen integrated both in his larger body of work, creating architecture that expressed synchronicity with its time, and that was recognized as such at its time.[2]

As a teacher (professor at the Technical University of Braunschweig), Oesterlen taught architecture that was functional and that was characterized by perpendicular lines, cubic structures, and clear geometry.[3] Glass,

---

1 *("Das Prinzip der École des Beaux Arts—auch wenn es im Bauhaus—erscheinen sollte!—ist nicht genug, um unsere Zeit zu gestalten.")* Letter from Martin Wagner to architect Richard Doecker (December 8, 1947), (the former, at that time was professor of city planning at Harvard University, and had been the city planner responsible for the *Hufeisensiedlung* by Bruno Taut). Werner Durth, *Deutsche Architekten,* p. 349.

2 See Schmedding, "Die Moderne," in *Architecture.* Taking Oesterlen as an example, Schmedding identifies modernism and expressionism as constituents of a new architectural tradition for the post-WW II period.

3 Roland Böttcher, Kristiana Hartmann, and Monika Lemke-Kokkelink, "die Architekturlehrer der TU Braunschweig", in *Hildesheimer Jahrbuch für Stadt and Stift Hildesheim*, pp, 249–250. (Braunschweig, 1995).

concrete, and brick were favored construction materials. In this so-called *Braunschweiger Schule* the *Neues Bauen* dictum and *Bauhaus* paternity were evident. Oesterlen and others found rich inspiration in such modernist traditions of pre-Nazi time. Modernism became the tradition from which to proceed.[4] Prestigious architects Mies von der Rohe and Walter Gropius (among others) both had emigrated (to the US) and later returned to build again in Germany. This lent added prestige to German modernism. Such revisiting (in the 1950s and 1960s) of expressionism and "degenerate art" was later examined in light of suppression of the NS memory (*Verdrängung*) and rehabilitation (*Wiedergutmachung*).[5] Oesterlen defended German post-WW II artistic reorientation toward early twentieth century styles as an act of "liberation" from the hypocritical "blood-and-soil" (*Blut und Boden*) sanctimony and NS classicism, which he demeaned as "… the State presenting itself in the form of the 34th re-make of threadbare classicism."[6] Sculpture's move toward abstraction was, in his view, also liberation from figural sculpture's "ordered, indistinct realism."[7]

Modernism and expressionism cannot be neatly separated. Oesterlen's teacher, Hans Poelzig (1869–1936) blended both movements in various ways.[8] Although Poelzig's name and work are typically associated with expressionism, he also used the angular severity and clear shapes that are characteristic of modernism. Oesterlen's angular stylistic regulative at his Futa Pass might well be rooted in Poelzig's strict angularity, as seen in the latter's

4 See fn 2.

5 Conversation with Christian Fuhrmeister about post-WW II German art and architecture as an act of rehabilitation (*Wiedergutmachung*), (November 2013). Judgements of NS art lack consensus, and hence its refutation is selective.

6 See Koch, *Dieter Oesterlen*, p. 250, from a speech by Oesterlen at the Berlin Academy of Art in response to Günther Grass on May 8, 1985. *("…Befreiung von der realistischen Blut- und Bodentümelei bzw. von dem staatsrepräsentierenden 34sten Aufguß eines fadenscheinigen Klassizismus.")*

7 An obvious revival of pre-NS art was at work. Oesterlen viewed his works as heir to abstract cubism.

8 Julius Posener ed., *Hans Poelzig, Gesammelte Schriften und Werke,* (Berlin: Mann, 1970); Wolfgang Pendt and Matthias Schirren (eds.), *Hans Poelzig. Architekt, Lehrer, Künstler.* (DVA, München, 2007); and Wolfgang Pendt, *Expressionist Architecture*, (New York: Preager, 1973).

"zigzagging" motif at his sulfuric acid plant at Luban (Poland, 1911–12). Angular zigzagging, not just for ornamentation but as the principle of arranging blocks, was a central element in *Neues Bauen* as seen in Ernst May's work in Frankfurt.[9] Expressionism in this form was evident in Oesterlen's remarkable (protestant) churches (twenty) characterized by sharply angled choirs, roofs, and walls folded in zigzag manner and bold "crystalline" interior spaces (e.g., *Christuskirche*, Bochum 1957–59).[10]

The Futa Pass design reveals expressionistic inspiration in several ways.[11] Its monument appears to dominate its canted platform, and due to the slanted walls, the eye cannot find rest on calm horizontals.[12] This recalls Poelzig's characteristic "crookedness," the shifting of direction and angles, with walls and ceilings neither horizontal nor vertical, manifested in, for example, his famous sets for the film, *Golem.*[13] Innovations of other architects as well may be recognized at Oesterlen's Futa Pass.[14] Of great influence was "the crystalline," a central spatial design principle of the 1920s.[15]

---

9 Susan R. Henderson, *Building Culture: Ernst May and the New Frankfurt Initiative, 1926–1931*, Peter Lang, Publishing (New York: 2013). For an example of a "zigzagging" of the roofline, see Ernst May's entrance hall of the IG Farben administration building in Frankfurt a.M. (1928–31).

10 For illustration of the Christuskirche, Bochum see Schmedding, "Die Moderne," in *Architecture,* p. 518.

11 Julius Posener, "Max Taut and Otto Bartning, Expressionistische Architektur," in Vorlesungen zur Geschichte der Neuen Architektur. *ARCH+journal, Arch+Verlag*, Aachen, vol. 53, (September 1980), pp. 69–76.

12 Giacomo Calandra di Roccolino, "Paessaggio per la memoria. Il cimitero tedesco del Passo della Futa di Dieter Oesterlen", *La Rivista di Engramma*, 95, (December 2011), p. 2, relates the aesthetics of changing viewpoints to "the poetry of Poelzig."

13 *Der dramatische Raum. Hans Poelzig*. Malerei, Theater, Film. Catalogue of exhibition in the Museum Haus Lange and the Museum Haus Esters. (Krefeld: 1986). (No author or publisher given, published probably by the museums.)

14 One can see elements of major architectural geniuses in Oesterlen's work, such as Hans Scharoun designs (non-executed), Fritz Höger's Chile House (1923, Hamburg), which was executed. This is not to suggest an imitation of these, but rather Oesterlen's capability to integrate others' inspirations, while creating something new and inspired in its own right.

15 Schmedding, *Dieter Oesterlen,* chapter Der Kristalline Raum, pp. 138, 140. Author gives a summary of the influence of the concept of the "crystalline" and its influence on modernist German architecture, such as the Bauhaus and German expressionism.

Fig. 70. Water color design for a skyscraper by Hans-Scharoun.

A watercolor by Hans Scharoun takes this principle to logical extreme, showing a dense conglomeration of sharp-edged cubes jutting upward, and triangular-shaped stacked wall panels standing on their tips, as if being "sucked" into the picture plane.[16] The concepts of the "organic" in energy and movement, represented by Scharoun's example, could not have been foreign to Oesterlen since it was an expression of the zeitgeist. Sharouns's rehabilitation in post-war Germany might have given additional impetus to Oesterlen's concept of his spiral wall as organic.

Expressionism's characteristics include mass set at an unusual angle, an abrupt alteration of planar orientation, and a severe line forming anti-Vitruvian collages, seemingly borne from imagination and spontaneity. Hugo Häring's farm buildings at Gut Gurkau (near Lübeck, 1923–24) pre-figured such collages. A section of this complex had a triangular roof in steep diagonal. In Häring's complex, the same style was used for both utilitarian and public buildings. (In the US, classicism, full or stripped-down, was generally considered the only appropriate style for public buildings.) Oesterlen's bold abrupt turns, as an aesthetic principle, had impressive predecessors.

One of the most salient features of the Futa Pass cemetery is the concept of architecture as sculpture, which was a salient feature of the zeitgeist of the 1920's.

This is seen impressively in the Wasili Luckhardt's model for a peoples' theater (never executed).[17] The upward-sweeping building is radical

---

16 Julius Posener ed. *Hans Poelzig,* p. 67.

17 Dennis Sharp, *Modern Architecture and Expressionism*, (New York: George Braziller, 1966). Illustration, p. 74.

and daring, and once seen, unforgettable. Poelzig encouraged his pupil (Oesterlen) toward sculptural architecture and his Futa Pass belongs to this conception. (Architecture's expressing a sculptural totality also characterized Mendelsohn's postage-stamp-sized architectural sketches, which he did during the first World War.[18] Research and stylistic analysis do not single Mendelsohn out as a source of inspiration for Oesterlen.)

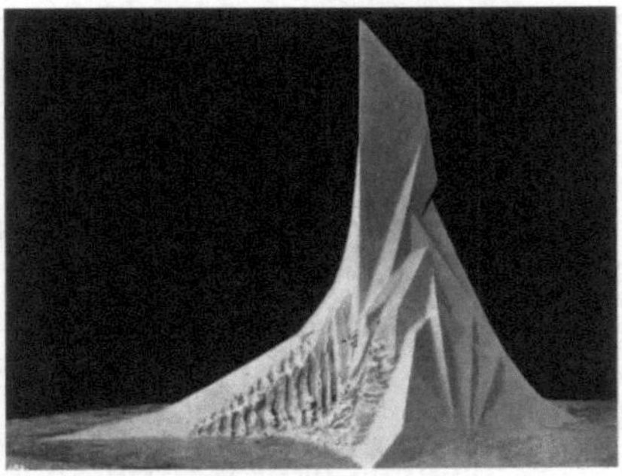

Fig. 71. Design for "Theatre for the People" by Wasily Luckhart.

Oesterlen was in a formative stage of youth during the Weimar Republic period, a time that witnessed bold, socially relevant architectural experiments reflecting utopian visions, which might have made an impression on Oesterlen. The prominent architect Bruno Taut's founding the "*Gläserne Kette*," a union of various artists (mostly architects) (1919–1920) to which Walter Gropious and Hans Scharoun (among others) belonged. Taut presented a utopian vision of a town where glass buildings are imbedded in and intimately intertwined with nature.[19] Taut's *Stadtkrone* concept (1919), a monumental structure on a prominent place dominating the surrounding city and resting on walled substructures, attracted much attention and is considered one of the most extraordinary visions in architectural history of this period.[20] It revived the medieval preference for locating politically

18 Kathleen James, *Erich Mendelsohn and the Architecture of German Modernism*, (Cambridge: Cambridge University Press, 1997). Mendelsohn's sculptural architecture (Einstein Tower, Potsdam), expresses a zeitgeist.

19 Posener, "Hans Poelzig, Bruno Taut und der Expressionismus", pp. 57–64; "Max Taut II –Visionen, Gläserne Kette" pp. 64–68. "Max Taut, Otto Bartning, expressionistische Architektur", pp. 69–76.

20 Richard Ingersoll and Spiro Kostof, *World Architecture: A Cross-Cultural History,* (New York: Oxford University Press, 2013).

and spiritually important buildings such as churches and castles on hills. Even though such architecture's medieval mission had disappeared, Taut's inventive influence remained. Strong interest in this vision was made manifest (in the 1920s) through a competition held by the city of Halle in 1927 to design a kind of *Stadtkrone*. (Oesterlen was sixteen at the time.) Many grandees in architecture participated, including Poelzig, Oesterlen's teacher, and Gropius.[21] The authority of the *Stadtkrone* might safely be cited as having influenced Oesterlen's crowning the hill with a monument and indeed, one of Taut's sketches from 1921 shows an abstract edifice consisting of triangular shapes standing on composite terraces. The sketch shows nothing perpendicular and everything in motion.[22]

In the Weimar Republic period's architecture, "place" took on an important role.[23] Oesterlen was immediately and powerfully inspired upon his first viewing of the Futa Pass site. Its topography seemed like a huge sculpture and it immediately brought to his mind "The Tower of Babel" by Pieter Brueghel the Elder (1563, on display at the *Kunsthistorisches Museum* in Vienna). The painting depicts a spiraling edifice under construction.[24] (The spiral invites theories about its symbolism, which remain intriguing conjecture.[25]) Oesterlen explained his having drawn inspiration for Futa Pass's spiral wall from it, and credited the painting with having animated his ideological approach, interpreting it as an allegory of human hubris leading to conflict.[26] As the model that Oesterlen prepared shows, his spiral path has echoes of that painting even though Bruegel's spiral is rounded. Oesterlen's spiral crosses the site's contour lines at a constant

21 Web page under "Halle 127 Wettbewerb" April 24, 1914. The site shows designs. Members of the *Neues Bauen* movement participated: Paul Bonatz (an important voice in the *Darmstädter Gespräche*), Walter Gropius, Hans Poelzig, Peter Behrens, Emil Fahrenkamp, and Wilhelm Kreis, who designed many Totenburgen (non-executed) during the NS regime.

22 For illustration see Posener, fn.11. "Max Taut, Otto Bartning, expressionistische Architektur", p. 71.

23 About the importance of the landscape as a source of inspiration see Posener, "Bruno Taut II", pp. 64–68.

24 Commentary by Oesterlen on his sources of inspiration for his Futa Pass design. The AdK Archive, the Dieter Oesterlen Archive, Oest-Or-300, (October 1, 1982).

25 Michel Duinker, "Der Totenberg. Un Cimetiere Miltaire Allemand Au Col de la Futa / Italie", (Diplomarbeit: Zürich, 1994), The author discusses the symbolism of the spiral in the context of death. p. 11.

26 See fn. 24.

grade, keeping them always on the right as the visitor ascends clockwise. (However, his starting point for architectural imagination that resulted in a spiraling retaining wall may well have been, at least in part, a simple, freestanding wall that was pre-existing at the site.[27])

Direct messaging through architecture distinguishes the Futa Pass from other German, and all US examples (where, as was discussed, an established Beaux-Arts aesthetics was seen as the carrier of an ideological message that supposedly could be understood by the common man). At Futa Pass, Oesterlen intentionally broke with *Totenburg* traditions, explicitly distancing himself from Castel del Monte and Tannenberg, both aesthetically and ideologically. Futa Pass's characteristically expressionist fragmentation and unpredictability expresses the opposite of *Totenburg's* settled and immutable eternity.

The conception and the realization of the monument as an abstract entity occurred during the fertile Weimar Republic years, as evidenced by Mies van der Rohe's memorial (1926) to the socialists Rosa Luxemburg and Karl Liebknecht, who were murdered, likely by thugs at the behest of their political enemies. The uneven restlessness of the stone surface counterpointed the severe straight-lined composition with projecting rectangular sections, rendering an aura of unpleasantness, incompletion, and challenge, and presaged Futa Pass's stonework and Lander's boulders jutting from the plane of his mighty mosaic in the court. The stone's intrinsic expressiveness as the principal aesthetic carrier is also made manifest in Otto Kohltz's design for a WWI memorial, a truly gigantic structure with an open court (1919).[28]

Walter Gropius's monument to the "victims of March" (*Märzgefallenen*, 1922) killed in the rightist *Kapp Putch* of 1920 offers the strongest model for Futa Pass's design.[29] This was located at Weimar's main cemetery (*Hauptfriedhof*) (and was destroyed by the Nazis). A wall surrounded the graves and from it, a triangular body of concrete arose with several breaks, reaching upward in diagonal thrusts. The monument had no precedents. Its

27 Discussion with Giacomo Calandra di Roccolino in Berlin, (March 1, 2013).

28 Posener, "Max Taut", pp. 69–76 and see fn. 11.

29 Schmedding, "Die Moderne," in *Architecture*, pp. 517–518, introduces Gropius's monument as a source of Oesterlen's inspiration. Gropius called his work "Blitzstrahl," (Flash of lightning); In Schmedding, *Dieter Oesterlen*, p. 117. She references Prange, *Das Kristalline.* In Gropius's monument, Prange sees the crystal as having become autonomous and abstracted form. Schmedding argues that Oesterlen's impulse toward abstraction, which resulted in sculptural architecture, preceded his explanation of the sudden end of a soldier's life through the Futa Pass's monument's sudden drop off.

daringness left an indelible impression on Lander, who resided in Weimar. Gropius's referred to his *Märzgefallenen Denkmal* as a "lightning bolt" (*Blitzstrahl*)[30] and Lander's calling his Futa Pass mosaic "lightning bolts" (*Blitzstrahle*). Still another link between Gropius and the Futa Pass cemetery may be recognized in the use of Bauhaus-derived lettering.

Fig. 72. Gropius's monument to fallen freedom-fighters, named "Märzgefallenen", Weimar (destroyed 1936).

The influence of Gustav August Munzer's marine monument (at Laboe, begun in 1927, re-started in 1933 and finished in 1936) is to be noted. This monument's astonishing sculptural quality, representing a wind-filled sail, presents abstracted shape as a bearer of the message. Whereas Tischler's post-WW II cemeteries in Italy all have a figural sculpture playing a central role (as do US examples), the Futa Pass has only a "crown of thorns," a non-figurative abstraction as Oesterlen wished. Oesterlen relied on architecture to be the primary vehicle of message. As well as conforming to the VDK's mandate to emphasize Christian symbols and their power to

30 Klaus–Jürgen Winkler and Hermann van Bergeijk, *Das Märzgefallenen Denkmal in Weimar*, p. 30, (Weimar, 2004). The monument was destroyed in 1936, and was rebuilt in slightly changed form.

console, taken together with a quote from Jesus's "Sermon on the Mount," the two communicate that suffering is the central message, and not glory.

The search for newness of "meaning" played an important role in post-WW II Germany (e.g., the *Darmstädter Gespräche*). For Oesterlen, good architecture linked form and content through "spiritual demands."[31] This conception superseded the "form follows function" dictate of the Bauhaus and it was understood internationally in the 1950s that spiritual values were inherent to good architecture.[32] In 1956, Oesterlen articulated architecture's spiritual components using four categories: "expressiveness, atmosphere, character, and symbolism" (*Ausdruck, Atmosphäre, Charakter, Sinnbild*).[33] For him, spirit was intrinsic in modernism's rational character. He referred to Mies van der Rohe's work as "*zeitgeist* expressed in spatial form" (*raumgefaßter Zeitwille*).[34] Abjuring symbols, Oesterlen's obvious intention was to imbue the Futa Pass with a subtle and ineffable "spirit." This shows itself also in the Futa Pass's relationship to nature, a revival of the traditional German respect for it. Lacking was the mythical veneration for nature that stood for "Germanness" in WWI war cemeteries, where landscape, often expressely left untouched, served as the main bearer of national identity and symbol of eternity.[35] This presents an interesting contrast to the US examples, where no archival evidence could be found to indicate that such was discussed, even as a notion.

Germany's moral reorientation is seen in commemorative forms, traditions, and styles being, as Oesterlen put it, "liberated" from their previously politicized connotations. The Futa Pass gives form to said reorientation, so succinctly articulated in the first sentence of Germany's constitution (1949):

---

31 "*Geistige Ansprüche*" from Oesterlen's dedication speech of the Technische Hochschule, Braunschweig in 1956.) See Koch, *Dieter Osterlen*, p. 212.

32 Ibid. Oesterlen quoted the Scheveningen charter of the *Union Internationale des Architectes* (The UIA) in the above speech of 1955: *("Der Architekt, das ist der Mensch, der die Kunst des Bauens meistert und so die Stätten an denen die Menschen ruhen oder sich regen, aufs beste gestaltet und beseelt.")* "The architect is the man who gives his best to master the art of building and endowing it with spirit, and thus designs the places where men may rest or act."

33 See fn. 31.

34 Ibid. "*Zeitwille*" literally translated means "will of the time." This has a Hegelian touch, as if the time had its own will. Oesterlen saw himself as the agent of change. Therefore, I approximate the term to something like "*Zeitgeist*."

35 Brands, Gunnar, From World War I, in *Places of Commemoration*. The author discusses the central role of nature for WWI German cemeteries. pp. 217–240. See Chapter II, fn. 5.

“The dignity of the individual is unassailable” (*Die Würde des Einzelnen ist unantastbar*). For all of Robert Tischler’s prodigious creativity, he no longer fit the VDK’s (and Germany’s) commemorative paradigm, having evolved to eschew all meaning except tragedy. Oesterlen found a way to honor mass death in a “liberated” fashion devoid of troublesome associations. His Futa Pass design created new from old, in a single stroke relegating soldierly funerary traditions, which had proved to be so tenacious, to history.

And yet, history itself proves to be tenacious. Futa Pass’s restlessness expresses an ambiguity. Might this unique work have pre-figured the passionate emotional and intellectual uproar, the questioning of every assumption and presumption that would very soon come to the fore, beginning in 1968 (the year after Alexander Mitscherlich’s generational work *Die Unfähigkeit zu Trauern* appeared)?[36]

Fig. 73. Architecture as sculpture. End of an age, or new beginning?

36 Alexander und Margarete Mitscherlich, *Die Unfähigkeit zu Trauern. Grundlagen kollektiven Verhaltens,* (Piper: München, 1967).

Postlude: The Reception of the Futa Pass Cemetery by the Press

When dedicated, nearly twenty-five years after the war's end (June 28, 1969), and likely due to its extraordinary architecture, the Futa Pass cemetery attracted press coverage. In contrast to the US press' derogation of the art and architecture presented in the US examples, the German press's reception was generally positive. The following is a short sample that gives insight into Germany's cultural climate of the day.

The influential Frankfurter Allgemeine lauded the VDK's efforts, the cemetery's aesthetics, and its message. Eberhard Schulz captured the work's essence, which he described as "pure expression in architecture" (*reine Ausdrucksarchitektur*).[1] Schulz described its wall's shooting skyward like an arrow as architectural ecstasy without triumphalism. Oesterlen, in stark (*sachlicher*) form, had presented the "monumental" without "pathos." Schulz also addressed the point that the cemetery's destination was mass burial and that this was successfully achieved. He noted that the "final" and the "terrible" were not evaded, and that the individual was not subsumed in the mass.[2]

Faustmann, writing for "*Die Fackel*," expressed that with the Futa Pass cemetery the VDK had made a thought-provoking leap forward.[3] Faustmann especially noted the monument's court, its provocative emptiness and the way it stimulated his own musings on the interrelatedness of "life and death, death and life, justice and injustice, fate and coincidence, war and peace, freedom and slavery. Here man is made accountable."[4] He reminded his

1 Eberhard Schulz, "Der Totenberg," the AdK Archive, the Oesterlen Archive, Büro-Ordner-Nr. 301 (June 7, 1969).

2 Ibid. "… how the majesty of finality and horror was not avoided at this place." *(„… wie man doch der Majestät des Entgültigen und Furchtbaren an dieser Stelle nicht ausgewichen ist.")*

3 *Die Fackel* Nr. 8, 1969, Verband der Kriegsbeschädigten, Kriegshinterbliebenen, (Association of War Maimed and War Survivors).

4 Ibid. *("Die Frage nach der von Leben und Tod, von Tod und Leben, Recht und Unrecht, Fügung und Zufall, die Frage um Krieg und Frieden, Freiheit und Sklaverei. Hier muss der Mensch Rechenschaft ablegen.")*

readers that 56,293,500 human lives were lost as a direct result of WW II and that another 35 million were maimed. Oesterlen and Lander's work clearly spoke to him. He concluded with the admonition that the present bears responsibilities for the future, for humanity.

Others voiced different views. Although favorably impressed, Bernd Manfred Woyt writing for a church-related press agency, in an article titled "… the Law of Active Peace"[5] questioned whether the beauty of the landscape setting did not suppress the awareness of war's cruelty. His reaction further reflected issues surrounding the VDK that had not yet come to rest. He questions whether the VDK had distanced itself sufficiently from its previous NS notions, and asks what it imparted to the young about war. Should the VDK not be more politically active in contributing to Germany's re-democratization process, beyond promoting its motto "Work for Peace?" His voice stands for ongoing criticism of the VDK, often from left leaning circles, particularly by the young generation.

This sample of the German press's thoughtful and generally positive reaction makes an interesting counterpoint to the US press's very dismissive reactions to their cemeteries (in Italy) as documented in the chapter Sic Transit Gloria. While the US gives examples of style and message, in a temporal sense, they are static and oriented toward preserving the past; the Futa Pass cemetery breaks with even its German antecedents and is able to project its message into the future.

---

5 "… das Gesetz des aktiven Friedens," *Ausgabe für kirchliche Presse* No. 26, (July 25, 1969). The VDK Folder A.100–865.

Epilogue

What the US cemeteries say …

The post-war ABMC was military-oriented and well connected to political power. After the conclusion of WW II and as soon as re-authorized by congress, it formed a coterie of east coast upper-class, conservative, well educated design elite whose relationships had been formed well before the outbreak of the war. There is no evidence that the war brought major harm to any of these individuals. They socialized together, shared tastes, and refused even to entertain the possibility that any style other than classicist styles they had refined at the exclusive American Academy in Rome was appropriate for public art at the national level. There is no record of any discussion of moral issues to parallel the German VDK's lively discourses. The coterie considered architectural style and artistic message as if pre-ordained, as if their pre-war world and their own relevance in it would continue unchanged.

The ABMC believed itself to be the guardian of American values. That an elite could and should interpret for the common person was a strong cultural strain of the day. The US population at large accepted such "top-down" thinking. There was consensus that classicist public architecture echoing Washington guaranteed continuity, expressing authority and nationhood. There was also consensus that naturalist figural sculpture, enhanced through classical and Art Nouveau elegance, was understandable by the common man and that through it, comforting patriotic notions could be channeled.

US cemeteries are highly idealized "little Americas", couched in beautiful lands, insular and closed unto themselves through superb landscaping. Each has a cozy reception room representative of an idealized home of the day, as if quoting a Shirley Temple or a Judy Garland film, with sofas, a generous fireplace, and art on the wall over the mantle. Their tasteful classicism, even today, bespeaks comfort, beauty, and plenty, but also a 1940's vision of authority, nationhood, and patriotism. Figural sculpture expresses, at each cemetery's memorial, values that Americans agreed on. Today's visitor might well ask why there are no graves for those whose remains could not be separated. Nothing is to be found in archives to address such inconvenient questions. The experience of battle, the agony of dying young and perhaps slowly, and of killing one's fellow human being are effectively anaesthetized and sanitized. Battlefields are "hallowed" by the two "little Americas" in Italy, two artifacts of a nostalgic vision of US America standing as one nation, indivisible and unambiguous. The art and architecture is effective, the messaging distinctly readable and clearly understandable by older US American visitors whose memories go back far enough. As to the future, it seems the design coterie hoped their vision would remain relevant. Archived correspondence reveals no discussion other than great pride in having footholds in Europe through such emblems of victory, but also a resigned acceptance that few from the US would visit.

As of this writing, nearly seventy years after completion, these two cemeteries lie remote from major travel routes and are somewhat difficult to locate, although both can be reached via public transportation. Anzio/Nettuno is an isolated island among shopping centers and the suburban maze that sprang up around Rome after the war. For the occasional motorist traveling the old road between Florence and Sienna, Impruneta still jumps into view with verdant, bright foreignness that attracts interest, but the way to it is poorly marked. Most motorists take the *Autostrada* from which it is not visible. Entries in guest books by non-American visitors praise their park-like beauty with irrigated green lawns, and often thank the US for its sacrifices, but the prominent battle maps and didactic verbiage is seldom noted, likely because it seems patronizing to European visitors who have enjoyed their own post-war peace and prosperity.

Do these two cemeteries truly represent US America's self-view of the late 1940's? The US public facilitated allied victory through organization and teamwork. Everyone had played his or her part in what would be the greatest public experience of everyone's lifetime, a truly nationalizing experience. Post-war America satiated itself on this experience. Its cemeteries clearly, and with gratitude, extol soldiers' sacrifices in service of America's path toward global hegemony. Patriotic notions prevailed through the 1950's as mass culture formed, and its associated movement toward less individualism and more convention. The US became more outwardly religious and the words "under God" were added to the Pledge of Allegiance (1954). The meanings of duty, honor, and country were clearly understood by all. War movies proliferated and films in general became highly formulaic. Fault lines however began to appear in the form of, for example, Alan Ginsberg's "Howl" (1955) and the "beat generation." A decade later, as the Vietnam situation ensued, the values enshrined in WW II cemeteries seemed more ambiguous. This schism was generational to a great extent, evidenced by the expression "generation gap" which became common. Older people were often shocked to find that they had not passed these cemeteries' enshrined values on to their children.

What the German cemeteries say …

In the early post-war years, the German people struggled to survive, to adapt, and to rebuild shattered lives. People needed shelter, work, and to eat. Konrad Adenauer, the Federal Republic's first chancellor and head of its ruling parties, the conservative union of the CDU/CSU, provided steady, capable leadership. He came to symbolize Germany's rise from ashes, its astonishing economic recovery, and its re-entry into the community of nations, and he was much revered. The public at large avoided discussing the past and its crimes, some were too busy, some actively suppressing the painful and shameful subject. People preferred to dream of a world intact and free from stress, as presented in superficial feel-good films called "*Heimatfilme.*" The political mood was conservative, the society conventional, if not stuffy. The VKD's first WW II cemeteries reflect post-war Germany of the mid 1950's. It was not until 1959 that the famous film "The Bridge" portraying the madness of sending adolescent males to their sure death was released. Prior to this, such madness was too familiar and too painful to remember. The VKD represented the conservative, Christian side of the society. Throughout the land, former NS party members held many positions requiring professional competence. The VDK had such persons. The many archived letters, concepts, memos, and protocols arguing positions on decisions to be made often contain terminology that recalls Nazi times. One therefore needs to place VDK's re-orientation and accomplishments within a conservative societal context to understand, for example, its loyalty to Robert Tischler, part of post-war Germany's "functionary elite." But it must be noted that the VDK leadership constantly attempted to restrain

Tischler's urges for monumentality and his outdated notions of soldierdom. German society's difficult progression toward moral renewal is reflected in VDK's gradual imposition of decisively democratic tenets, even though basic re-orientations, such as treating the soldier as an individual rather than as a member of the national collective, can be found in the immediate post war years. The VDK's acceptance of Dieter Oesterlen's radical and extraordinary design for the Futa-Pass cemetery is evidence of its eventual siding with progressive elements of German society. This, rather than its hesitating steps toward freeing itself from its past NS-tinged ideology, is the gauge of the VDK's post-war achievements.

Contrary to the US situation there was no official clear and binding standard of soldiers' cemetery design in Germany. Political turbulence during the twentieth century's first half had resulted in various concepts. Nevertheless, Robert Tischler, VDK's chief architect since 1926, had established conventions in his many designs for WWI cemeteries. The post-war VDK inherited his idiosyncratic mixture of neo-medieval style and modernist cubistic features. With this came his ideologizing linking of nature with eternity, his mass graves, and his grandiose vision of "German-ness" expressed in nature and architecture. Tischler's continuing hegemony provided the VDK with an aesthetic anchor as it sought new meanings. His Italian post-war cemeteries make striking contrast to the US cemeteries, which display bright travertine and marble. They witness a new spirit where mysticism surrounding fallen soldiers has all but disappeared and nature no longer bears muddled notions of eternity. They evidence Germany's growing abhorrence of quasi-religious irrationality. Germany at large was becoming less irrational and less outwardly religious. Narrow, cramped entrances became wider and more open. The VDK shied away from showy monuments. Death in battle had been de-sanctified. Neither could it be portrayed as pertinent to any aspect of national greatness. Propagandistic glorification of NS times was banished. Youthful death, and in such huge numbers, was tragedy plain and simple. Following an aborted design at Cassino, and hefty opposition from the Catholic Church, the practice of crypt burial was abandoned.

Tischler's symbolic group crosses prompted a great deal of discussion. While he argued that singular crosses in great numbers would present a dreadful sight, others recalled that group crosses had been glorified by NS ideologues as symbolizing the military and absolute obedience. Tischler prevailed, but his group crosses were only used at Pomezia (later replaced) and at Costermano. With emphasis on the individual, each soldier had his own grave, and up to six shared a marker stone. Mass graves, once symbols of the collective, were used only for those whose remains could not be separated. The VDK however did retain the "Comrades' Grave," comradeship being a cherished traditional notion. Christian symbols were used to convey suffering, and unlike US cemeteries where the dead have a special place in heaven, there is no resurrectionist promise or message. However, beginning in the late 1960's a simple tall metallic cross was added.

Death is present, not hidden. Beauty is not used to soften the visitor's experience. Planting tends to be somber, flowers rare. The visitor soon realizes that these soldiers died for nothing, their lives wasted for a wrong cause. Individuating the dead and articulating a family's loss through sculpture in generalizing, contemporary, at times abstracted style facilitates the individual's processing of tragedy. The experience is one of quiet mourning and search for consolation. "Blessed are they that suffer, for they will be comforted", from Jesus's Sermon of the Mount is inscribed at the Futa Pass cemetery. Soldiers' death in battle is thus politicized as an admonition to make peace and to reconcile conflict over the graves of those who perished in conflict, no matter which side they were on. German culture at large, especially the post-war youth, became passionately anti-war.

German tradition holds nature to be a sacred force, quasi God-like. This tradition had been co-opted and abused by NS ideologues. Still, the VDK preferred mountain sites where the open cemetery would dialogue with its surrounding landscape and the sky, establishing a spiritual connection, such as at Caira, but especially so at the Futa Pass in the wild and rugged, virtually uninhabited Apennine Mountains. Today, irrigation is limited and indigenous grasses turn brown in summer. Nature re-asserts itself, promising eventually to become one with the cemetery, thus conferring a natural, non-resurrection eternity.

German post-war cemeteries in Italy were not designed and built at once, as were the American examples, but were sequenced over a period of time, and so reflect changes within German society. One reads in them Germany's return to world-class technical skills after near total destruction. One sees in them the results of growing prosperity. Through their aesthetics and their messaging, they witness Germany's progress toward democratization and the hard, serious work that went into moral rehabilitation. One realizes that it was the trauma of destruction and total moral disgrace that propelled Germany toward aesthetic invention and new ideology.

The VDK's final cemetery (in Italy) at the Futa-Pass demonstrates that Germany had finally found its long-sought means to express cultural intangibles, such as the dialogs between human kind and nature, between landscape and architecture, between mass death and the individual. Radically new as it may seem, its aesthetics are clearly traceable to art and architecture that pre-dated the NS period, mainly Weimar Republic expressionism. Today, the cemetery hosts performances of Greek classical drama.

Both US cemeteries and their German counterparts make political statements. US successes, and especially its cold war mindset resulted in officialdom's two decade-long rejection of international modernism, introduced into the US by foreigners and hence tainted by association with communism. The result was US official public art's becoming hostage to US hegemony. It is notable that ABMC's artists and architects of the early post-war years, locked in stasis, succeeded in being so creative and individual. This stylistic stasis persisted until a new monument, the Vietnam Memorial, was built in Washington. Initiated privately, its brilliant abstract design was selected in a blind competition won by a young woman architectural student of Chinese parentage.

Destruction set German designers free and propelled them toward stylistic innovation. The Futa Pass cemetery's mighty rising wall seems to shout defiance, evoking a strong emotional shock. Its central monument suggests a book closing and/or opening. Its architecture expresses restlessness and reminds one of moral ambiguities persisting in post-war Germany, prefiguring the passionate emotional and intellectual uproar that would come to the fore as student revolts erupted in 1968.

Selected Bibliography

Archives

AdK Archiv, Akademie der Künste, Archiv für Bauwesen, Berlin, referred to as AdK Archive.

Archivio di Stato, Rome, Italy, referred to as State Archive, Rome.

Archivio die Stato, Florence, Italy, referred to as State Archive, Florence.

Archivio Storico del Ministero degli Affari Esteri, Rome, referred to as Archive of the Ministry of Foreign Affairs.

Avery Architecture and Fine Arts Library, Columbia University, New York City.

College of Environmental Design Archives, University of California, Berkeley, CA.

Columbia University Archives, New York City.

NARA National Archives and Record Administration, College Park, Maryland.

New York Historical Society Library Manuscript Collection, New York, referred to as NYHS.

Politisches Archiv des Auswärtigen Amtes, referred to as AA Archive.

Smithsonian Institute American Art, Archives, Washington D.C.

University of Pittsburgh, Archives, Pittsburg, Pennsylvania.

Volksbund Archiv, Kassel, referred to as VDK Archive

Books and Journals

Adams, Michael C.C. *The Best War Ever: America and World War II*. Baltimore and London: The John Hopkins University Press, 1994.

Ahrend, Max. *Die Bauten des Volksbundes in ihrer geschichtlichen und kulturellen Bedeutung*. Berlin W 15, Kurfürstendamm 165/166. Volksbund Dt. Kriegsgräberfürsorge, 1935.

Bernstock, Judith E. *Under the Spell of Orpheus: The Persistence of a Myth in Twentieth Century Art.* Southern Illinois University Press, 1991.

Birnbaum, Charles A., and Robin Karson, eds. *Pioneers of American Landscape Design.* New York: McGraw Hill, 2000.

Bodnar, John. *Remaking America: Public Memory, Commemoration, and Patriotism in the Twentieth Century.* Princeton University Press, 1992.

Böttcher, Roland, Kristiana Hartmann, and Monika Lemke-Kokkelink. *Die Architekturlehrer der TU Braunschweig.* Braunschweig, 1995.

Brands, Gunnar. "From World War I Cemeteries to the Nazi Fortresses of the Dead: Architecture, Heroic Landscape, and the Quest for National Identity in Germany." Wolschke-Bulmahn, Joachim; ed. *Places of Commemoration: Search for Identity and Landscape Design.* Dunbarton Oaks: 2001.

Calandra di Roccolino, Giacomo. "Paessaggio per la memoria. Il cimitero tedesco del Passo della Futa di Dieter Oesterlen." *La Rivista di Engramma*. (December 2011).

Campanella, Thomas J. "American Curves: The Public Works Legacy of Gilmore D. Clarke, and Michael Rapuano." *Harvard Design Magazine.* Summer 1997.

Canaday, John. "Our National Pride: War Memorials." *Architectural Forum.* 91, (December 1949).

Cazzato, Vincenzo et al. *Ville e Giardini Italiani. Instituto Poligrafico e Zecca Dello Stato*. Rome: Libreria Dello Stato, 2004.

Colotti, Francesco. *Il paesaggui dei Caduti: Dieter Oesterlen, cimitero monumentale germanico*. Casabella: 2013.

Costanzo, Denise Rae. "The Lessons of Rome: Architects at the American Academy 1947–1966." PhD diss., Pennsylvania State University, August 2009.

Craven, Wayne. *Sculpture in America*. New York: Thomas Y. Crowell Co., 1968.

Dogliani, Patrizia. "Il cimitero tedesco della Futa." La montagna dopo la guerra. Continuità e rotture nell'Apennino bolognese tra Idice e Setta-Reno: 1945–2000. Aspasia, 2009, Bologna: Isrebo, 2009.

Duinker, Michel. „Der Totenberg. Un Cimetière Miltaire Allemand Au Col de la Futa, Italie." Diplomarbeit, Zürich: 1994.

Dülffer, Jost. „Im Schatten des zweiten Weltkrieges. Der Soldatentod in deutscher und internationer Erinnerungskultur." *Loccumer Protokolle* 25, Corinna Hauswedell, Rehburg ed. (Loccum, 2008).

Durth, Werner. *Deutsche Architekten: Biographische Verflechtungen 1900–1970*. Braunschweig: Vieweg, 1986.
Etlin, Richard A., ed. *Nationalism in the Visual Arts: Studies in the History of Art.* Center for Advanced Study in the Visual Arts, Symposium Papers XIII, 29. National Gallery of Art, Washington D.C. Hanover and London: University Press of New England, 1991.
Exhibition Catalogue. *The Thirties: the Arts in Italy Beyond Fascism*. Florence: Tuscany Palazzo Strozzi, September 2012.
Faulkner, Barry. *Sketches from an Artist's Life.* Dublin: William L. Bauhan, 1973.
Feist, Peter H. *Figur und Objekt, Plastik im 20. Jahrhundert.* Leipzig: Seemann, 1996.
Fischer, Hannelore, Ulrich Grober. *Kollwitz, Käthe. Die trauernden Eltern; ein Mahnmal für den Frieden*. Köln: Dumont, 1999.
Fischer, Norbert. *Geschichte des Todes in der Neuzeit.* Erfurt: Sutton, 2001.
Fuchs-Heinritz, Werner. *Todesbilder der modernen Gesellschaft.* Frankfurt: Suhrkamp, 1969.
Fuhrmeister, Christian. „Die ‚unsterbliche Landschaft', der Raum des Reiches und die Toten der Nation. Die Totenburgen Bitoli (1936) und Quero (1929) als strategische Memorialarchitektur." *Kritische Berichte, Zeitschrift für Kunst-und Kulturwissenschaften,* Heft 2, Jahrgang 29, 2001. Marburg: Jonas Verlag, (2001).
——— "Klatchmohn und Ochsenblut, Zur Ikonographie der Kriegsgräberstätten des Volksbundes Deutsche Kriegsgräberfürsorge." in *Gartenkultur und Nationale Identität.* Gröning, Gert, and Uwe Schneider eds. Worms: Grüne Reihe, (2001).
——— "Robert Tischler. Chefarchitekt. 1926–1959. Ein Desiderat." *RIHA JOURNAL*, special issue, *Bauaufgabe Soldatenfriedhof/Kriegsgräberstätten.*
——— Kappel, Kai and Christian Fuhristere (2017). "Sketching a Potentially Global Field of Research" *RIHA JOURNAL*, special issue, *Bauaufgabe Soldatenfriedhof/Kriegsgräberstätten*. Fuhrmeister, Christian, and Kai Kappel, eds. (2017).
General Order 125, War Department.
Geneva Convention No. 1, Art. 17 (§ 330).
Geneva Convention No. 2, Art. 20 (§ 330).
Geneva Convention No. 3, Art. 120 (§ 330).

Geneva Convention No. 4, Art. 130 (§ 330).
Gillis, J.R. *Commemorations: The Politics of National Identity.* Princeton, NJ: Princeton University Press, 1994.
Glaser, Hermann. *Kleine deutsche Kulturgeschichte von 1945 bis heute.* Frankfurt am Main: S. Fischer, 2007.
——— *Deutsche Kultur, 1945–2000.* München: Ullstein, 2000.
——— *Aus den Trümmern zur Post-Moderne: Zur Kulturgeschichte der Bundesrepublik Deutschland*. München: Max Hueber Verlag, 1986.
Gobel, David and Daves Rossell, eds. *Commemoration in America: Essays on Monuments, Memorialization, and Memory*. Charlottesville and London: University of Virginia Press, 2013.
Grossman, Elizabeth G. "Architecture for a Public Client: The Monuments and Chapels of the American Battle Monuments Commission." *Journal of the Society of Architectural Historians* 43, (May 1984).
Grzimek, Waldemar. *Deutsche Bildhauer des zwanzigsten Jahrhundert: Leben, Schulen, Wirkungen.* München: Moos, 1969.
Gstettner, Hans. *Deutsche Soldatenmale, Erbaut vom Volksbund Deutsche Kriegsgräberfürsorge.* Berlin: 1940.
Harbeson, John. "Our Memorials Abroad." *National Sculpture Review*. American Battle Monuments Issue IV, no. 4 (Winter 1955).
——— "A Collaborative Undertaking." *AIA Journal* 36, (August 1961).
Hettling, Manfred, and Jörg Echternkamp, eds. *Gefallenengedenken im globalen Vergleich, Nationale Tradition, politische Legitimation und Individualisierung der Erinnerung*. München: Oldenbourg, (2013).
Hitchcock, Henry Russell. *Architecture: Nineteenth and Twentieth Centuries*. Baltimore: Penguin Books, 1958.
Hofmann, Werner. *Die Plastik des 20. Jahrhunderts*. Frankfurt a.M.: 1958.
Hütt, Michael et al.; *Unglücklich das Land, das Helden nötig hat: Leiden und Sterben in den Kriegsdenkmälern des Ersten und Zweiten Weltkrieges. Anmerkungen zur Geschichte und Ideologie des Volksbundes Deutsche Kriegsgräberfürsorge nach 1945.* Studien zur Kunst-und Kulturgeschichte BD 8 Marburg: Jonas, 1990.
Huxtable, Ada Louise. "$24,000,000 Shrine is Proposed for Capital." *New York Times*, June 3, 1960.
——— "History of Nation to be Sculptured in Shrine on a Mountain in Georgia." *New York Times.* August 7, 1953.
Jackson, John Brinckerhoff. *American Space: The Centennial Years 1865–1876.* New York: Norton, 1972.

James, Kathleen. *Erich Mendelsohn and the Architecture of German Modernism.* Cambridge: Cambridge University Press, 1997.
Jewiss, Virginia. "Picarda's Peace in a German War Cemetery." *MLN.* (Published by the John Hopkins University Press.), 127, no. 1 (January 2012).
Kähler, Gert. "Architektenporträt Dieter Oesterlen." *Der Architekt.* (August 7, 1987).
Kaiser, Walter. "A Well Known Terrifier." *New York Review of Books.* July 11, 2013.
Kammerer, Herbert. "Paul Manship." *National Sculpture Review* 15, (Fall 1966).
Kammerer, Willi. *Volksbund Deutsche Kriegsgräberfürsorge Erinnern und Gedenken: Kunst auf Kriegsgräberstätten.* Kassel: VDK, 2002.
King, Alex. *Memorials of the Great War in Britain: The Symbolism and Politics of Remembrance.* Oxford: Berg, 1998.
Koch, Alexander; ed. *Dieter Oesterlen, Bauten und Planungen 1946–63,* vol. 1. Stuttgart: 1964.
Kohler, Sue A. *The Commission of Fine Arts: A Brief History 1910–1976.* Washington, D.C.: 1977.
Koselleck, Reinhart. *Kriegerdenkmale als Identitätsstiftung der Überlebenden.* München: 1979. In Identität, Odo Marquard, and Karl-Heinz Stierle eds. München: Fink, 1994.
Koselleck, Reinhart, and Jeismann, Michael, eds. "*Der politische Totenkult. Kriegerdenkmale der Moderne.*" München, Fink, 1994.
Kostof, Spiro. *The Third Rome, 1870–1950.* Berkeley: University of California Press, 1973.
Kruse, Kevin M. *One Nation under God.* New York: Basic Books, 2015.
Kuberek, Monika. „Die Kriegsgräberstättem des Volksbundes Deutsche Kriegsgräberfürsorge." In *Unglücklich das Land, das Helden nötig hat: Leiden und Sterben in den Kriegsdenkmälern des Ersten und Zweiten Weltkrieges. Anmerkungen zur Geschichte und Ideologie des Volksbundes Deutsche Kriegsgräberfürsorge nach 1945.* Michael Hütt, ed. Marburg: Jonas, 1990.
Küppers, Barbara. *Fritz Schmoll genannt Eisenwerth.* Weimar: Geisteswissenschaften, 2003.
Kurth, Willy: Käthe Kollwitz. Geleitwort zum Katalog der Ausstellung in der Deutschen Akademie der Künste, 1959.
Law 389 of the 66th Congress.

Lehman, Milton. "We Visit the Ghosts of Anzio." *Saturday Evening Post* 222, (September 24, 1949).

Lemay, Kate Clarke. *Triumph of the Dead: American World War II Cemeteries, Monuments, and Diplomacy in France.* Tuscaloosa, AL: University of Alabama Press, 2018.

Lichtenstern, Christa. *Plastik als Kraft, Beobachtungen zur Lehmbruck-Rezeption in der Bildhauerei im 20. Jahrh.* Berlin: Georg Kolbe Museum, 2002.

——— „Voraussetzungen und Entwicklungen naturästhetischer Perspektiven in der Skulptur und Plastik nach 1945" *Marburger Jahrbuch für Kunstwissenschaft.* (Marburg: 1993).

Lindner, Werner. *Gestaltung von Kriegsgräberstätten. Hinweise und Richtlinien.* VDK Kassel: 1955.

Linenthal, Edward Tabor. *Sacred Ground: Americans and Their Battlefields.* Urbana and Chicago: University of Illinois Press, 1991.

Littman, William. "A Century of Teaching Architecture at the University of California, Berkeley, 1903–2003." *Design on the Edge.* Waverly Lowell, Elizabeth Byrne and Betsy Frederick-Rothwell, eds. Berkeley: University of California Press, (2009).

Loeffler, Jane. *The Architecture of Diplomatic Building: America's Embassies.* New Jersey: Princeton University Press, 2011.

Loucheim, Aline. "Memorials to Our War Dead Abroad." *New York Times.* January 15, 1960.

Lovell, Margaretta, M. "Death, War and Public Memory: Iwo Jima, Vietnam, and the Wild West." *Journal of Modern Art History Department, Faculty of Philosophy.*, 8. University of Belgrade, (2012).

Lurz, Meinhold. "…ein Stück Heimat in Fremder Erde. Die Heldenhaine und Totenburgen des Volksbunds Deutsche Kriegsgräberfürsorge." *Archplus_ausgabe.* 71 Aachen, Arch+ (1983).

Manship, John. *Paul Manship.* New York: Abbeville Press, 1989.

Marshall, George C. "Our War Memorials Abroad: A Faith Kept." *National Geographic.* (June 1957).

Mauch, Christof, ed. *Nature in German History*, New York: Oxford, 2004.

Mayo, James M. *War Memorials as Political Landscape.* New York: Praeger, 1988.

Michalski, Sergiusz. *Public Monuments: Art in Political Bondage 1870–1997.* London: Reaktion Books, 1998.

Ministerio Della Difesa. *Sacrari Militari della Seconda Guerra Mondiale Rome*. Rome: 1974.

Mitscherlich, Alexander and Margarete. *Die Unfähigkeit zu trauern: Grundlagen kollektiven Verhaltens*. München: Piper, 1967.

Mosse, George L. *Fallen Soldiers: Reshaping the Memory of the World Wars*. Oxford: Oxford University Press, 1990.

Mumford, Louis. "In Search of a New Monumentality." *Architectural Review* 104 (September, 1948).

——— "Monumentalism, Symbolism, and Style." *Architectural Review* 105 (April, 1949).

Murtha, Edwin. *Paul Manship*. New York: Macmillan, 1957.

North, Thomas. "Our Memorials Abroad." *National Sculpture Review*. American Battle Monuments Issue. IV, no. 4 (Winter 1955).

Oesterlen, Dieter. *Bauten und Texte 1946–1991*. Tübingen and Berlin: Ernst Wasmuth Verlag, 1992.

——— "Der Totenberg am Futa Pass." *Bauwelt* (November 1982).

Paver, Chloe, and Bill Niven, eds. *Memorialization in Germany Since 1945*. London: Palgrave Macmillan, 2010.

Pendt, Wolfgang and Matthias Schirren, eds. *Hans Poelzig. Architekt, Lehrer, Künstler.* München: DVA, 2007.

Pershing, General John, "Forever America." *American Legion* 2, no. 5 (May 1927).

Posener, Julius, ed. *Hans Poelzig, Gesammelte Schriften und Werke.* Berlin: Mann Verlag, 1970.

——— „Vorlesungen zur Geschichte der Neuen Architektur. Die Architektur der Reform (1900–1924)." *ARCH+journal* 53, Arch+Verlag, Aachen, (September 1980).

Prange, Regine. *Das Kristalline als Kunstsymbol: Bruno Taut und Paul Klee. zur Reflexion des Abstrakten in Kunst und Kunsttheorie der Moderne*. Hildesheim: Olms, 1991.

Price, Renée. *New Worlds: German and Austrian Art 1890–1940*. New York: Neue Galerie, 2001.

Rand, Harry. *Paul Manship*. Washington, D.C.: Smithsonian Institution Press, 1989.

Reynolds, Donald Martin. *Masters of American Sculpture: The Figurative Tradition from the American Renaissance to the Millennium*. New York: Abbeville Press, 1993.

Robin, Ron. "A Foothold in Europe: The Aesthetics and Politics of American War Cemeteries in Western Europe." *Journal of American Studies* 29, no. 1 (April 1995).

——— *Enclaves of America: The Rhetoric of American Political Architecture Abroad, 1900–1965*. New Jersey: Princeton University Press, 1992.

Roosevelt, Eleanor. "My Day." *New York Times*. August 15, 1953.

Saure, Gabriele und Gisela Schirmer, eds. *Käthe Kollwitz. Die Eltern (1926–1932) in Kunst gegen Krieg und Faschismus*. Weimar: 1999.

Savage, Kirk. *Monument Wars: Washington D.C., the National Mall and the Transformation of the Memorial Landscape*. University of California Press, 2005.

Anne Schmedding. „Die Moderne als Tradition. Architektur in der Nachkriegszeit der BDR am Beispiel Dieter Oesterlens (1911–1994)," in Architecture in the Age of Empire. Weimar: Bauhaus-Kolloquium Bauhaus-Universtiät, 2011.

Sforza, Carlo. "Italy, the Marshall Plan and the 'Third Force.'" *Foreign Affairs* 26, no. 3 (April 1948).

Scholz, Kristina. *The Greatest Story Ever Remembered. Erinnerung an den Zweiten Weltkrieg als sinnstiftendes Element in den USA*. Frankfurt A.M.: Peter Lang, 2008.

Shanken, Andrew M. "Planning Memory: Living Memorials in the United States during World War II." *Art Bulletin* 84. New York: College Art Association, (2002).

Sharp, Dennis. *Modern Architecture and Expressionism.* New York: George Braziller, 1966.

Shenton, Edward. "They Will Never Be Forgotten." *Saturday Evening Post* 227 (August 14, 1954).

Simonides. *The Greek Dead at Thermopylae.* K.E. Diehl ed. Rome: Ateneo, 1970.

Sloane, W. David. *The Last Great Necessity: Cemeteries in America*. Baltimore, 1991.

Stamp, Gavin. „Almost Unbearable." November 2007, in *Apollo*, London: (February 2012).

Steere, Edward. *Shrine of the Honored Dead: A Study of the National Cemetery System.* Department of the Army, Historical Section, Office of the Quartermaster General. Washington, D.C.: 1956.

Tietz, Jürgen. *Das Tannenberg-Nationaldenkamal: Architektur, Geschichte, Kontext.* Verlag Bauwesen, 1999.

Treaty of Versailles, Article 225.

Ulio, J.A. *Graves Registration.* War Department, January 15, 1945.

Valentine, Lucia and Alan. *The American Academy in Rome, 1894–1969.* Charlottesville: University of Virginia Press, 1973.

van der Zande, Johan. "Orpheus in Berlin: A Reappraisal of Johann Georg Sulzer's Theory of the Polite Arts in Central European History." *Central European History* 28 (1995).

Vitry, Paul. *Paul Manship, Sculpteur American.* Paris: Editions de la Gazette des Beaux-Arts, 1927.

Weilacher, Udo. *Visionäre Gärten. Die modernen Landschaften von Ernst Cramer. An architectonic landsculpture. Military cemetery at the Passo la Futa*, Italy, pp. 126–131. Basel: Birkhäuser, 2001.

Willmann, Georg. *Kriegsgräber in Europa: Ein Gedenkbuch.* München: Bertelsmann, 1980.

Winkler, Klaus-Jürgen und Hermann van Bergeijk. *Das Märzgefallenen Denkmal in Weimar.* Weimar: Univ. Verl., 2004.

Winter, Jay M. *Sites of Memory, Sites of Mourning: The Great War in European Cultural History.* Cambridge and New York: Cambridge University Press, 1995.

Wolf, Fridolin. *Kunstschmiedepraxis.* Lübeck: Charles Colemann, 1980.

Wright, David H. "The First World War Memorial in the Cortile of the American Academy in Rome 2000." Dickinson, Sheila K. and Hallett, Judith P.; eds. *Rome and Her Monuments: Essays on the City and Literature of Rome in Honor of Katherine A. Geffcken.* Wauconda, IL: Bolchazy-Carducci, (2000).

Yalom, Marilyn. *The American Resting Place* (photographs by Reid Yalom). Boston: Houghton Miffin Harcourt, 2008.

Yegül, Fikret K. *Gentlemen of Instinct and Breeding: Architecture at the American Academy in Rome, 1894–1940.* New York: Oxford University Press, 1991.

Zelinsky, Wilbur. *Nation Into State.* Chapel Hill: The University of North Carolina Press, 1988.

Transatlantic Aesthetics and Culture

General Editors

Boris Vejdovsky, Agnieszka Soltysik & Jьrg Schwyter

Vol. 1 Beverly Maeder, Jьrg Schwyter, Ilona Sigrist & Boris Vejdovsky (eds.)
The Seeming and the Seen. Essays in Modern Visual and Literary Culture.
2006. 373 pages.
ISBN 3-03910-997-9 / US-ISBN 0-8204-8373-7

Vol. 2 Lucy Kay, Zoл Kinsley, Terry Phillips & Alan Roughley (eds.)
Mapping Liminalities. Thresholds in Cultural and Literary Texts.
2007. 232 pages.
ISBN 978-3-03911-455-9

Vol. 3 Irene Ramalho Santos & Antynio Sousa Ribeiro (eds.)
Translocal Modernisms. International Perspectives. With a preface by Franзoise Meltzer.
2008. 276 pages.
ISBN 978-3-03911-690-4

Vol. 4 Christine Reynier (ed.)
Cross-Cultural Encounters between the Mediterranean and the English-Speaking Worlds.
2011. 216 pages.
ISBN 978-3-0343-0604-1

Vol. 5 Marina Camboni, Andrea Carosso, Sonia Di Loreto, and Marco Mariano (eds.)
Translating America. The Circulation of Narratives, Commodities, and Ideas between Italy, Europe, and the United States.
2011. 377 pages.
ISBN 978-3-0343-0395-8

Vol. 6 Isabel Caldeira, Maria Josй Canelo, Irene Ramalho Santos (eds.)
America Where? Transatlantic Views of the United States in the Twenty-first Century.
2012. 287 pages.
ISBN 978-3-0343-1239-4

Vol. 7 Francesca de Lucia
Italian American Cultural Fictions. From Diaspora to Globalization.
2017. 180 pages.
ISBN 978-3-0343-2565-3

Vol. 8 Birgit Urmson
German and United States Second World War Military Cemeteries in Italy. Cultural Perspectives.
2018. 398 pages.
ISBN 978-3-0343-3516-4